Applied Hydrogeology

C. W. FETTER, JR.
University of Wisconsin—Oshkosh

Charles E. Merrill Publishing Company
A Bell & Howell Company
Columbus   Toronto   London   Sydney

Published by Charles E. Merrill Publishing Co.
A Bell & Howell Company
Columbus, Ohio 43216

This book was set in Optima
Production Editor: Ann Mirels
Cover Design Coordination: Will Chenoweth
Cover Map: Hydrogeological features of south central Great
    Basin, Nevada. Courtesy U.S. Geological Survey. Hydrology
    by I. J. Winograd, 1965; geology by William Thordarson,
    1965.

Library of Congress Catalog Card Number: 79–92805
International Standard Book Number: 0–675–08126–2
Printed in the United States of America
        8   9   10—87

for Nancy

Preface

Applied Hydrogeology is intended to serve as a textbook for an undergraduate course in hydrogeology or geohydrology. The reader is expected to have a working knowledge of college algebra. Calculus would be helpful, but it is not a prerequisite for the reader's practical understanding. The application of mathematics to problem solving is stressed, rather than the derivation of theory. To this end, many example problems are included, with step-by-step solutions provided. Case studies are presented in a number of chapters to enhance the reader's understanding of the occurrence and movement of groundwater in a variety of geologic settings. Because of its applied emphasis, this book should be useful to practicing geologists, engineers, planners, geographers, foresters, ecologists, and others. To assist those who might use the text as a reference, a glossary has been included. Tables of various functions, as well as tables for unit conversions, make up the appendices.

One of the problems faced by anyone working in hydrogeology is the large number of units and systems of units in common usage. In *Applied Hydrogeology,* the primary units are according to the SI system (meter, kilogram, second), generally followed by English equivalents. Example problems are worked in a variety of units so that the reader may gain familiarity with them as they are likely to be encountered in real-world applications.

I am grateful to all those who helped me. Dr. Mary Anderson of the University of Wisconsin—Madison and Dr. Sam B. Upchurch of the University of South Florida, Tampa, carefully reviewed the draft manuscript and made many helpful suggestions. Dr. James Hoffman of the University of Wisconsin—Oshkosh reviewed Chapter 9 and Dr. David Stephenson of Woodward-Clyde Consultants, San Francisco, reviewed Chapter 11 and parts of other chapters. Mrs. Pam Spaulding cheerfully typed several drafts of the manuscript. Dr. Richard Parizek of Pennsylvania State University furnished the photographs used in Chapter 12.

PREFACE

My thanks to the American Geophysical Union for permitting me to use figures from a number of AGU publications, including *Water Resources Research*, *Journal of Geophysical Research*, and *Transactions (EOS)*. I am also grateful to Dr. Jay Lehr of the National Water Well Association for granting permission to include figures from the journal, *Ground Water*.

C. W. Fetter, Jr.

Contents

1 WATER

1.1 Water   2
1.2 Hydrology and Hydrogeology   3
1.3 The Hydrologic Cycle   4
1.4 Energy Transformations   6
1.5 The Hydrologic Equation   7
1.6 Hydrogeologists   7
1.7 Applied Hydrogeology   8
1.8 Sources of Hydrogeologic Information   9
References   10

2 EVAPORATION AND PRECIPITATION

2.1 Evaporation   14
2.2 Transpiration   17
2.3 Evapotranspiration   18
2.4 Condensation   21
2.5 Formation of Precipitation   21
2.6 Measurement of Precipitation   23
2.7 Snow Measurements   24
2.8 Effective Depth of Precipitation   25
References   30

3
RUNOFF AND STREAMFLOW

3.1   Events During Precipitation   34
3.2   Hydrograph Separation   37
      3.2.1   Baseflow Recessions   37
      3.2.2   Storm Hydrograph   40
      3.2.3   Gaining and Losing Streams   42
3.3   Stream Order   44
3.4   Rational Equation   44
3.5   Floodwaves   47
3.6   Duration Curves   48
3.7   Determining Groundwater Recharge from Baseflow   50
3.8   Measurement of Streamflow   52
      3.8.1   Stream Gauging   52
      3.8.2   Weirs   54
3.9   Manning Equation   56
      References   57

4
SOIL MOISTURE AND GROUNDWATER

4.1   Porosity of Earth Materials   60
      4.1.1   Definition of Porosity   60
      4.1.2   Porosity of Sediments   61
      4.1.3   Porosity of Sedimentary Rocks   64
      4.1.4   Porosity of Plutonic and Metamorphic Rocks   65
      4.1.5   Porosity of Volcanic Rocks   66
4.2   Specific Yield   67
4.3   Hydraulic Conductivity of Earth Materials   70
      4.3.1   Darcy's Experiment   70
      4.3.2   Hydraulic Conductivity   71
      4.3.3   Permeability of Sediments   74
      4.3.4   Permeability of Rocks   75
4.4   Forces Acting on Groundwater   76
4.5   Water Table   77
4.6   Capillarity and the Capillary Fringe   79
4.7   Infiltration   81
4.8   Soil Moisture   83
4.9   Theory of Unsaturated Flow   88
4.10   Water-Table Recharge   91
4.11   Aquifers   92
4.12   Aquifer Characteristics   94
4.13   Homogeneity and Isotropy   97
4.14   Hydrostratigraphic Units   100
      References   101

5
PRINCIPLES OF GROUNDWATER FLOW

5.1   Introduction   *106*
5.2   Mechanical Energy   *108*
5.3   Hydraulic Head   *110*
5.4   Force Potential and Hydraulic Head   *112*
5.5   Darcy's Law   *113*
    5.5.1   Darcy's Law in Terms of Force and Potential   *113*
    5.5.2   The Applicability of Darcy's Law   *114*
    5.5.3   Discharge and Seepage Velocities   *116*
5.6   Permeameters   *117*
5.7   Equations of Groundwater Flow   *120*
    5.7.1   Confined Aquifers   *120*
    5.7.2   Unconfined Aquifers   *124*
5.8   Solution of Flow Equations   *125*
5.9   Gradient of Hydraulic Head   *125*
5.10   Flow Nets   *127*
5.11   Refraction of Streamlines   *129*
5.12   Steady Flow in a Confined Aquifer   *130*
5.13   Steady Flow in an Unconfined Aquifer   *132*
5.14   Freshwater-Saline Water Relations   *139*
    5.14.1   Coastal Aquifers   *139*
    5.14.2   Oceanic Islands   *144*
5.15   Tidal Effects   *146*
    References   *147*

6
REGIONAL GROUNDWATER FLOW

6.1   Introduction   *152*
6.2   Steady Regional Groundwater Flow in Unconfined Aquifers   *152*
    6.2.1   Recharge and Discharge Areas   *152*
    6.2.2   Groundwater Flow Patterns in Homogeneous Aquifers   *153*
    6.2.3   Heterogeneous Aquifers   *160*
    6.2.4   Anisotropic Aquifers   *162*
6.3   Confined Aquifers   *162*
6.4   Transient Flow in Regional Groundwater Systems   *164*
6.5   Noncyclical Groundwater   *164*
6.6   Springs   *165*
6.7   Geology of Regional Flow Systems   *167*
6.8   Groundwater-Lake Interactions   *177*
    References   *182*

7
GEOLOGY OF GROUNDWATER OCCURRENCE

7.1   Introduction   186
7.2   Unconsolidated Aquifers   187
        7.2.1   Glaciated Terrain   188
        7.2.2   Alluvial Valleys   197
        7.2.3   Alluvium in Tectonic Valleys   198
7.3   Lithified Sedimentary Rocks   205
        7.3.1   Clastic Sedimentary Rocks   214
        7.3.2   Carbonate Rocks   218
        7.3.3   Coal and Lignite   226
7.4   Igneous and Metamorphic Rock   227
        7.4.1   Intrusive Igneous and Metamorphic Rock   227
        7.4.2   Volcanic Rocks   229
7.5   Groundwater in Permafrost Regions   236
7.6   Coastal Plain Aquifers   239
7.7   Groundwater in Desert Areas   244
7.8   Groundwater Regions of the United States   245
        1. Western Mountain Region   245
        2. Alluvial Basins   245
        3. Columbia Lava Plateau   247
        4. Colorado Plateau and Wyoming Basin   247
        5. High Plains   248
        6. Unglaciated Central Region   248
        7. Glaciated Central Region   249
        8. Unglaciated Appalachian Region   250
        9. Glaciated Appalachian Region   250
        10. Atlantic and Gulf Coastal Plain   250
            References   251

8
GROUNDWATER FLOW TO WELLS

8.1   Introduction   258
8.2   Unsteady Radial Flow   258
8.3   Well Hydraulics in a Completely Confined Areally Extensive
        Aquifer   260
        8.3.1   Theis Method   261
        8.3.2   Jacob Straight-Line Method   266
        8.3.3   Distance-Drawdown Methods   268
        8.3.4   Slug Tests   270

8.4   Flow in a Semiconfined Aquifer   *274*
    8.4.1   No Storage in the Leaky Confining Layer   *275*
    8.4.2   Storage in the Leaky Confining Layer   *276*
    8.4.3   Pumping Tests for a Leaky Artesian Aquifer with
            No Storage in the Confining Layer   *279*
    8.4.4   Pumping Test for a Leaky Artesian Aquifer with Storage
            in the Confining Layer   *284*
8.5   Effect of Partial Penetration of Wells   *285*
8.6   Water-Table Aquifer   *287*
8.7   Steady-State Radial Flow   *292*
8.8   Intersecting Pumping Cones and Well Interference   *293*
8.9   Effect of Hydrogeologic Boundaries   *294*
8.10   Pumping-Test Design   *296*
    8.10.1   Single-Well Pumping Tests   *297*
    8.10.2   Pumping Tests with Observation Wells   *299*
    References   *301*

⑨
WATER CHEMISTRY

9.1   Introduction   *306*
9.2   Units of Measurement   *306*
9.3   Types of Chemical Reactions in Water   *307*
9.4   Ideal Gas Law   *308*
9.5   Law of Mass Action   *309*
9.6   Common Ion Effect   *311*
9.7   Chemical Activities   *311*
9.8   Ionization Constant of Water and Weak Acids   *314*
9.9   Carbonate Equilibrium   *317*
    9.9.1   Carbonate Equilibrium in Water with Fixed Partial
            Pressure of CO_2 *318*
    9.9.2   Carbonate Equilibrium with External pH Control   *320*
9.10   Free Energy   *322*
9.11   Oxidation Potential   *322*
9.12   Surface Phenomena   *326*
    9.12.1   Adsorption   *327*
    9.12.2   Ion Exchange   *329*
9.13   Collection of Water Samples   *331*
9.14   Designing Water-Sampling Programs   *332*
    9.14.1   Groundwater   *332*
    9.14.2   Surface Water   *334*
9.15   Age Dating of Groundwater   *335*
9.16   Presentation of Results of Chemical Analyses   *336*
9.17   Solute Movement in Groundwater   *338*
    References   *344*

10
QUALITY OF WATER

10.1   Introduction   *350*
10.2   Water Quality Criteria for Metals   *352*
     10.2.1   Arsenic   *352*
     10.2.2   Barium   *353*
     10.2.3   Beryllium   *353*
     10.2.4   Boron   *353*
     10.2.5   Cadmium   *353*
     10.2.6   Chromium   *354*
     10.2.7   Copper   *355*
     10.2.8   Iron   *355*
     10.2.9   Lead   *355*
     10.2.10 Manganese   *356*
     10.2.11 Mercury   *356*
     10.2.12 Nickel   *357*
     10.2.13 Silver   *357*
     10.2.14 Zinc   *358*
10.3   Water Quality Criteria for Inorganic Nonmetals
     10.3.1   Chlorine   *358*
     10.3.2   Fluoride   *359*
     10.3.3   Nitrogen, Inorganic   *359*
     10.3.4   Phosphorus   *360*
     10.3.5   Selenium   *361*
     10.3.6   Sulfate and Sulfide   *361*
10.4   Water Quality Criteria for Organic Compounds
     10.4.1   Chlorinated Hydrocarbons   *362*
     10.4.2   Chlorophenoxys   *363*
     10.4.3   Halomethanes   *363*
10.5   Additional Water Quality Critera   *363*
10.6   Pathogenic Organisms   *365*
10.7   Groundwater Contamination   *365*
     10.7.1   Septic Tanks and Cesspools   *367*
     10.7.2   Landfills   *368*
     10.7.3   Mining   *370*
     References   *370*

11
GROUNDWATER DEVELOPMENT AND MANAGEMENT

11.1   Introduction   *378*
11.2   Dynamic Equilibrium in Natural Aquifers   *378*
11.3   Groundwater Budgets   *381*

11.4 Management Potential of Aquifers   *383*
11.5 Paradox of Safe Yield   *385*
11.6 Legal Constraints to Groundwater Development   *387*
    11.6.1 Federal Aspects of Water Law   *387*
    11.6.2 State Surface-Water Law   *388*
    11.6.3 State Groundwater Law   *388*
11.7 Artificial Recharge   *390*
11.8 Protection of Water Quality in Aquifers   *392*
11.9 Groundwater Mining and Cyclic Storage   *394*
11.10 Conjunctive Use of Ground and Surface Water   *397*
11.11 Trends in Water Resources Management   *399*
    References   *400*

12

FIELD AND COMPUTER METHODOLOGY

12.1 Introduction   *406*
12.2 Fracture-Trace Analysis   *406*
12.3 Surficial Methods of Geophysical Investigations   *412*
    12.3.1 Direct-Current Electrical Resistivity   *413*
    12.3.2 Seismic Methods   *417*
    12.3.3 Gravity and Magnetic Methods   *426*
12.4 Geophysical Well Logging   *427*
    12.4.1 Caliper Logs   *430*
    12.4.2 Temperature Logs   *430*
    12.4.3 Single-Point Resistance   *430*
    12.4.4 Resistivity   *432*
    12.4.5 Spontaneous Potential   *432*
    12.4.6 Nuclear Logging   *434*
12.5 Models of Groundwater Flow   *438*
    12.5.1 Basic Model Types   *438*
    12.5.2 Finite-Difference Methods   *440*
    12.5.3 Illinois State Water Survey Model for Nonleaky Confined Aquifers   *443*
    12.5.4 Illinois State Water Survey Model for a Water-Table Aquifer   *451*
12.6 Hydrogeologic Site Evaluations   *453*
    References   *456*

APPENDICES   *459*

GLOSSARY   *469*

INDEX   *481*

CASE STUDIES:

Regional Flow Systems in the Great Basin   *167*
Regional Flow Systems in the Coastal Zone of the Southeastern
   United States   *172*
Hydrogeology of a Buried Valley Aquifer at Dayton,
   Ohio   *190*
Tectonic Valleys—San Bernardino Area   *200*
Sandstone Aquifer of Northeastern Illinois-Southeastern
   Wisconsin   *207*
Volcanic Plateaus—Columbia River Basalts   *229*
Volcanic Domes—Hawaiian Islands   *230*
Alluvial Aquifers—Fairbanks, Alaska   *238*
Chemical Geohydrology of the Floridan Aquifer System   *341*
Deep Sandstone Aquifer of Northeastern Illinois   *380*

Water

chapter 1

1.1 WATER

Water is the elixir of life; without it life is not possible. Although many environmental factors determine the density and distribution of vegetation, one of the most important is the amount of precipitation. Agriculture can flourish in some deserts, but only with water either pumped from the ground or imported from other areas. Civilizations have flourished with the development of reliable water supplies—and then collapsed as the water supply failed. This is a book about the occurrence of water, both at the surface and in the ground.

A person requires about 3 liters of potable water per day to maintain the essential fluids of the body. Primitive people in arid lands exist with this amount as their total consumption. A single cycle of a flush toilet may use 23 liters of water. In New York City the per capita water usage exceeds 1000 liters daily. Much of this amount is for industrial, municipal, and commercial purposes; for personal purposes, the typical American uses 200 to 300 liters per day. Even greater quantities of water are required for energy and food production. Total per capita usage in the United States in 1970 was almost 7000 liters (1850 gallons) per day (1).

A common goal of all countries is to increase economic production. It is generally thought this will result in a better lifestyle for the citizens. While the validity of this assumption has been questioned by some individuals in the heavily industrialized countries, the goal of governments at all levels still remains to promote economic expansion. This will naturally increase per capita water usage, although water conservation may help. Inasmuch as the population of most countries is growing, it is likely the total use of water will increase, even with conservation measures.

The United States has a history of increasing water usage. Figure 1.1 illustrates the withdrawal of water in the United States for all uses except hydroelectric power generation. From 1955 to 1970 total usage increased by 54 percent, while per capita usage also increased (Figure 1.2).

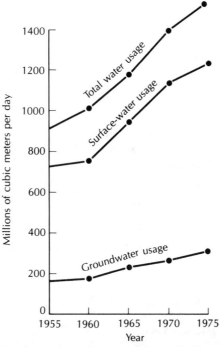

FIGURE 1.1.   Withdrawal of water in the United States for all uses except hydroelectric power generation. SOURCE: U.S. Geological Survey (References 1, 7, 8, and 9).

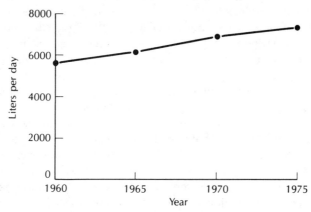

FIGURE 1.2.   Per capita water usage in the United States for all uses except hydroelectric power generation. SOURCE: U.S. Geological Survey (References 1, 7, 8, and 9).

HYDROLOGY AND HYDROGEOLOGY   1.2

As viewed from a spacecraft, the earth appears to have a blue-green cast due to the vast quantities of water covering the globe. The oceans may be obscured by billowing swirls of clouds. These vast quantities of water distinguish Earth from the other planets in the solar system.

Hydrology is the study of water. In the broadest sense, hydrology addresses the occurrence, distribution, movement, and chemistry of all waters of the earth. **Hydrogeology** encompasses the interrelationships of geologic materials and processes with water. (A similar term, **geohydrology,** is sometimes used as a synonym for hydrogeology.)

The physiography, surficial geology, and topography of a drainage basin, together with the vegetation, influence the relationship between precipitation over the basin and water draining from it. The creation and distribution of precipitation is heavily influenced by the presence of mountain ranges and other topographic features. Running water and groundwater are geologic agents that help shape the land. The movement and chemistry of groundwater is heavily dependent upon geology.

Hydrogeology is both a descriptive and an analytic science. The development and management of water resources are an important part of hydrogeology as well.

1.3 THE HYDROLOGIC CYCLE

An account of the water supply of the world would reveal that saline water in the oceans accounts for 97.2 percent of the total. Land areas hold 2.8 percent of the total. Ice caps and glaciers hold 2.14 percent; groundwater to a depth of 4000 meters accounts for 0.61 percent of the total; soil moisture 0.005 percent; freshwater lakes 0.009 percent; rivers 0.0001 percent; and saline lakes 0.008 percent (5). Over 75 percent of the water in land areas is locked in glacial ice or is saline.

Only a small percentage of the world's total water supply is available to humans as fresh water. More than 98 percent of the available fresh water is groundwater, which far exceeds the volume of surface water. At any given time, only 0.001 percent of the total supply of water is in the atmosphere. However, atmospheric water circulates very rapidly, so that each year enough water falls to cover the conterminous United States to a depth of 75 centimeters (2). Of this amount, 55 centimeters are returned to the atmosphere through evaporation and transpiration by growing plants, while 20 centimeters flow into the oceans as rivers (2). Although the previous sentence implies that the **hydrologic cycle** begins with water from the oceans, it actually has no beginning and no end. As most of the water is in the oceans, it is convenient to describe the hydrologic cycle as starting with the oceans.

Water evaporates from the surface of the oceans. The amount of evaporated water varies, being greatest near the equator where solar radiation is more intense. Evaporated water is pure, since when it is carried into the atmosphere the salts of the sea are left behind. Water vapor moves through the atmosphere as an integral part of the phenomena we term "the weather." When atmospheric conditions are suitable, water vapor condenses and forms droplets. These drops may fall to the sea, or on land, or may revaporize while still aloft.

Precipitation that falls on the land surface enters into a number of different pathways of the hydrologic cycle. Some of the water will drain across the land into a stream channel. This is termed **overland flow**. If the surface soil is porous, some water will seep into the ground by a process termed **infiltration**. The water clings to soil particles, and this soil moisture may be drawn into the rootlets of growing plants. After the plant uses the water, it is transpired as vapor into the atmosphere.

Excess soil moisture is pulled downward by gravity. At some depth, the soil or rock is saturated with water. The top of the saturated zone is the **water table,** and below the water table is **groundwater.** Groundwater flows through the rock and soil layers of the earth until it discharges as a spring or as seepage into a stream, lake, or ocean (Figure 1.3).

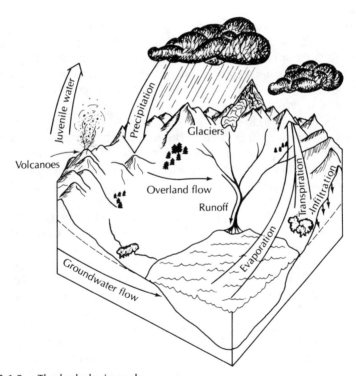

FIGURE 1.3.   The hydrologic cycle.

Water flowing in a stream can come from overland flow or from groundwater that has seeped into the stream bed. The groundwater contribution to a stream is termed **baseflow,** while the total flow in a stream is **runoff.**

Evaporation is not restricted to open water bodies, such as the ocean, lakes, streams, and reservoirs. Precipitation intercepted by leaves and other vegetative surfaces can also evaporate, as can water detained in land-surface depressions or soil moisture in the upper layers of the soil. Direct evaporation of groundwater can take place when the saturated zone is at or near the land surface.

1.4 ENERGY TRANSFORMATIONS

The hydrologic cycle is an open system in which solar radiation serves as a source of constant energy. This is most evident in the evaporation and atmospheric circulation of water. The energy of a flowing river is due to the work done by solar energy evaporating water from the ocean surface and lifting it to higher elevations where it falls to earth.

When water changes from one state to another (liquid, vapor, or solid), there is an accompanying change in the heat energy of the water. The **heat energy** is the amount of thermal energy contained by a substance. A **calorie** of heat is defined as the energy necessary to raise the temperature of one gram of pure water from 14.5° C to 15.5° C. The evaporation of 1 gram of water at 15° C requires an input of 590 calories of energy (the **latent heat of vaporization**). The water vapor retains this added heat energy. When water vapor condenses to the liquid form, 590 calories per gram are released (the **latent heat of condensation**).

In order to melt one gram of ice at 0° C, 80 calories of heat, the **latent heat of fusion**, must be added. The resultant temperature of the water is still 0° C, although a gram of water at 0° C has more heat energy than an equivalent weight of ice. Water can pass from the solid state to the vapor state (sublimation) with the addition of 700 calories per gram. Formation of frost and freezing of water result in the release of heat energy.

The transportation of water through the hydrologic cycle and the accompanying heat transfers are vital to the heat balance of the earth. At the equator, the amount of solar radiation is fairly constant through the year, while at the poles it varies from zero during the polar winter to amounts greater than those at the equator during the polar summer. During polar winters the land is in shadow, as the sun does not strike the ground. On the other hand, during the summers, the sun shines continuously. Over the year, the northern hemisphere northward of 38° latitude has a net heat loss, as the outgoing terrestrial radiation to space exceeds the incoming solar radiation that is absorbed. Between the equator and 38° N, there is more solar radiation absorbed than terrestrial radiation lost to space. In order to balance these anomalies, heat is transferred by currents in the oceans and through the atmosphere as movement of air masses and water vapor. This creates climatic conditions and changing weather patterns that profoundly affect the hydrologic cycle.

THE HYDROLOGIC EQUATION 1.5

The hydrologic cycle is a useful concept, but is quantitatively rather vague. The **hydrologic equation** provides a quantitative means of evaluating the hydrologic cycle. This fundamental equation is a simple statement of the **law of mass conservation**. It may be expressed as

Inflow = Outflow ± Changes in Storage

If we consider any hydrologic system—for instance, a lake—it has a certain volume of water at a given time. There are a number of inflows that add water: precipitation that falls on the lake surface, streams that flow into the lake, groundwater that seeps into the lake, and overland flow from nearby land surfaces. Water also leaves the lake through evaporation, transpiration by emergent aquatic vegetation, outlet streams, and groundwater seepage from the lake bottom. If, over a given period of time, the total inflows are greater than the total outflows, the lake level will rise as more water accumulates. If the outflows exceed the inflows over a time period, the volume of water in the lake will decrease. Any differences between rates of inflow and outflow in a hydrologic system will result in a change in the volume of water stored in the system.

The hydrologic equation can be applied to systems of any size. It is as useful for a small reservoir as it is for an entire continent. The equation is time-dependent. The elements of inflow must be measured over the same time periods as the outflows.

HYDROGEOLOGISTS 1.6

The professional hydrogeologist has a wide variety of occupations from which to choose. Employment may be found with federal agencies, United Nations groups, state agencies, and local government. Energy and mining companies may call upon the services of hydrogeologists to help provide water where it is needed, or perhaps remove it where it is unwanted. Private consulting organizations also employ many individuals trained in hydrogeology. Water resource management districts and planning agencies often include hydrogeologists on their staffs.

Hydrogeology is an interdisciplinary field. The hydrogeologist usually has training in geology, hydrology, chemistry, mathematics, and physics. Hydrogeologists are also being trained in such areas of engineering as fluid mechanics and flow through porous media, as well as in computer science. Such training is necessary, as hydrogeologists must be able to communicate effectively with engineers, planners, ecologists, resource managers, and other professionals. By the same token, an understanding of the basic principles of hydrogeology is useful to soil scientists, engineers, planners, foresters, and others in similar fields. Modeling of hydrologic systems is another area requiring knowledge of a number of disciplines.

1.7 APPLIED HYDROGEOLOGY

The argument has been made that "hydrogeology is more than a classical science" (10). Traditional studies in hydrogeology have focused on either the mathematical treatment of flow through porous media or on a general geologic description of the distribution of rock formations in which groundwater occurs. One occasionally even finds a paper describing the theoretical flow of fluids through an idealized porous medium that probably does not occur in nature. Likewise, many reports on the groundwater geology of an area make no attempt to evaluate how much water is available for use. Neither type of study has much practical value in and of itself.

Hydrogeologists are being employed as problem solvers and decision makers (10). They need to identify a problem, define the data needs, design a field program for collection of data, propose alternative solutions to the problem, and implement the preferred solution.

Hydrogeology is also being recognized by both hydrogeologists (11) and planners (12) as an important part of environmental planning. As population and economic growth expand, human use of the natural environment is becoming more intense. Incidents of environmental degradation are less likely to occur when the planner and the hydrogeologist cooperate.

The aid of the hydrogeologist is most often sought in evaluation of water resources. Groundwater is often available at a cost only one-tenth of that of a surface-water supply (11). Hydrogeologists can locate groundwater sources, evaluate the amount of water that can be developed, and help devise a management plan to utilize the resource. Land-use planning may be necessary to protect the recharge areas of the groundwater supply. Extensive paving of the land surface or the introduction of contaminants could reduce the availability of a groundwater source.

Waste disposal is another area of application for the hydrogeologist. Countless instances of groundwater pollution have been traced to land disposal of solid or liquid wastes. Groundwater has been contaminated with bacteria, viruses, chromium, nitrate, arsenic, chloride, chromate, cadmium, pesticides, and other toxic organic chemicals (13). Proper hydrogeologic siting of solid waste disposal areas can reduce the likelihood of groundwater contamination (14). Hydrogeological analysis is also important for preventing contamination from individual septic tanks, sewage lagoons, deep-well disposal of liquid waste, mining, and long-term storage and disposal of radioactive wastes.

Hydrogeological studies are an important part of the overall analysis of sites proposed for major construction projects. Coal and nuclear power plants, dams, tunnels, pumped storage reservoirs, and land-treatment systems for wastewater are typical examples of such projects. Solution mining of water-soluble uranium ore is another application of hydrogeology. Hydrogeology is also fundamental to environmental impact analysis.

The need for competent hydrogeologists has never been greater, especially in the applied areas of hydrogeology. A glance toward the future suggests there will be fewer natural resources, more intensive use of the land, and a greater desire to promote environmental conservation along with economic growth. The challenges to the hydrogeologist will grow, as will the opportunities.

SOURCES OF HYDROGEOLOGIC INFORMATION 1.8

Hydrogeologic information is available from a wide range of sources. In terms of sheer volume, the Water Resources Division of the U.S. Geological Survey is the leading source in the United States. This agency collects basic data on streamflow, surface-water quality, groundwater levels, and groundwater quality. The USGS also conducts water resources investigations and basic research. USGS publications are available in libraries that are designated depositories of federal documents; these publications are also available from the U.S. Government Printing Office.

The National Atmospheric and Oceanic Administration is the parent organization of the Weather Bureau. *The Climatic Record of the United States* is published for each state and contains precipitation, temperature, evaporation, and other climatic data. Other U.S. federal agencies that may conduct studies related to hydrogeology include the Corps of Engineers, Water and Power Resources Service, Soil Conservation Service, Environmental Protection Agency, Nuclear Regulatory Agency, and Department of Energy.

In most states, there are one or more agencies responsible for water-oriented research and other activities. The functions, responsibilities, and organizational format of state agencies in water resources activities vary from state to state. Typical agency designations include State Department of Water Resources or Water Survey, State Geological Survey, Department of Conservation or Natural Resources, and State Department or Board of Health. In many states, various responsibilities are allocated among several agencies. In addition, Congress has established provisions for a water resources research center or institute in each state and Puerto Rico. These are associated with a major university in each state.

Reports of current research and recent developments in hydrogeology and groundwater are included in the following journals:

☐ *Bulletin, International Association of Scientific Hydrology;*

9

☐ *Ground Water;*

☐ *Journal American Water Works Association;*

☐ *Journal of Hydrology;*

☐ *Transactions, American Society of Civil Engineers;*

☐ *Water Resources Bulletin;* and

☐ *Water Resources Research.*

A number of professional organizations sponsor symposia and meetings where technical sessions on hydrogeology or groundwater are held. These include the following:

☐ American Geophysical Union;

☐ American Society of Civil Engineers;

☐ American Water Resources Association;

☐ Geological Society of America;

☐ Geological Society of Canada;

☐ International Association of Scientific Hydrology;

☐ International Water Resources Association; and

☐ National Water Well Association.

References

1. MURRAY, C. R. and E. B. REEVES. *Estimated Use of Water in the United States, 1970.* U.S. Geological Survey Circular 676, 1972, 37 pp.

2. Federal Council for Science and Technology. *Scientific Hydrology.* Washington, D.C., U.S. Government Printing Office, 1962.

3. MEINZER, O. E. *Hydrology.* New York: McGraw-Hill Book Company, 1942, 712 pp.

4. DE WEIST, R. J. M. *Geohydrology.* New York: John Wiley & Sons, 1965, 366 pp.

5. FETH, J. F. *Water Facts and Figures for Planners and Managers.* U.S. Geological Survey Circular 601-I, 1973, 30 pp.

6. MAC KICHAN, K. A. *Estimated Use of Water in the United States, 1955.* U.S. Geological Survey Circular 398, 1957, 18 pp.

7. MAC KICHAN, K. A. and J. C. KAMMERER. *Estimated Use of Water in the United States, 1960.* U.S. Geological Survey Circular 456, 1961, 26 pp.

8. MURRAY, C. R. *Estimated Use of Water in the United States, 1965.* U.S. Geological Survey Circular 556, 1968, 53 pp.

9. MURRAY, C. R. and E. B. REEVES. *Estimated Use of Water in the United States, 1975.* U.S. Geological Survey Circular 765, 1977, 39 pp.

10. STEPHENSON, D. "Hydrogeology Is More than a Classical Science." *Ground Water*, 12, no. 3 (1974):148–51.

11. SOMMERS, D. A. "Put Hydrogeology into Planning." *Ground Water*, 8, no. 6 (1970):2–7.

12. MC HARG, I. L. *Design with Nature*. Garden City, N.Y.: The Natural History Press, 1969, 197 pp.

13. WALKER, W. H. "Where Have All the Toxic Chemicals Gone?" *Ground Water*, 11, no. 2 (1973):11–20.

14. ZANONI, A. E. "Ground-Water Pollution and Sanitary Landfills — A Critical Review." *Ground Water*, 10, no. 1 (1972):3–16.

Evaporation and Precipitation

chapter

2.1 EVAPORATION

Water molecules are continually being exchanged between a liquid and atmospheric water vapor. If the number passing to the vapor state exceeds the number joining the liquid, the result is **evaporation**. When water passes from the liquid to the vapor state, it will absorb 590 calories of heat from the evaporative surface for every gram of water evaporated. The vapor pressure of the liquid is directly proportional to the temperature. Evaporation will proceed until the air becomes saturated with moisture. The **absolute humidity** of a given air mass is the number of grams of water per cubic meter of air.

At any given temperature, air can hold a maximum amount of moisture: the **saturation humidity**. This is directly proportional to the temperature of the air. Table 2.1 gives the saturation humidity for several environmental temperatures. The **relative humidity** for an air mass is the percent ratio of the absolute humidity to the saturation humidity for the temperature of the air mass. As the relative humidity approaches 100 percent, evaporation ceases.

TABLE 2.1. Saturation humidity of air (grams per cubic meter)

Temperature °C	Humidity
−25	0.705
−20	1.074
−15	1.605
−10	2.358
− 5	3.407
0	4.874
5	6.797
10	9.399
15	12.83
20	17.30
25	23.05
30	30.38

SOURCE: *Handbook of Chemistry and Physics* (Cleveland, Ohio: CRC Publishing Company, 1976).

Condensation occurs when the air mass can no longer hold all of its humidity. This happens when an air mass is cooled and the saturation humidity value drops. If the absolute humidity remains constant, the relative humidity will rise. When it reaches 100 percent, any further cooling will result in condensation. The **dew point** for an air mass is the temperature at which condensation will begin. As condensation is the reverse of evaporation, the process of condensation releases 590 calories of heat to the surroundings per gram of water: the latent heat of condensation.

Evaporation of water takes place from free-water surfaces—lakes, reservoirs, puddles, dew droplets, etc. The rate is dependent upon factors such as the water temperature and the temperature and absolute humidity of the layer of air just above the free-water surface. Solar radiation is the driving energy force behind evaporation, as it warms both the water and the air. The rate of evaporation is also related to the wind—especially over land. The wind carries vapor away from the free-water surface and keeps absolute humidity low. By disturbing the water surface, the wind may also increase the rate of molecular diffusion from it.

Evaporation from lakes and reservoirs is an important consideration in water-budget studies. It can be computed for a lake or reservoir if all of the inflows (precipitation over the surface, surface-water inflow, and groundwater inflow) and the outflows (groundwater out-seepage, spillway discharge, and pumpage) and change in storage are known. The hydrologic equation (inflow = outflow ± changes in storage) is used. All of these factors, with the exception of the groundwater flux, can be measured with an error of perhaps ± 10 percent. In a carefully prepared water-budget study for Lake Hefner, Oklahoma, daily evaporation was computed to an accuracy of 5 to 10 percent (1). For many reservoirs, monthly or annual evaporation can be computed fairly easily. The most difficult factor to measure is the groundwater flux.

Free-water evaporation is measured quite simply by using shallow pans. The most commonly used is the **land pan**. The U.S. Weather Bureau maintains about 450 evaporation stations using Class-A land pans. Similar pans are used in Canada. They are 4 feet (122 centimeters) in diameter and 10 inches (25.4 centimeters) deep, made of unpainted galvanized metal. Land pans are placed on supports so that air can circulate all around. Water depths from 7 to 8 inches (17 to 20 centimeters) are maintained. Records are kept of the daily depth of water, the volume of water added to replace evaporated water, and the daily precipitation into the pan. Using the hydrologic budget, the daily evaporation can be computed. Errors may result from splash caused by heavy rainfall and drinking by birds. The wind movement is also measured, and expressed in units of miles per day. (A steady wind blowing at a velocity of 10 miles per hour would have a 24-hour wind movement of 240 miles per day.)

The water in a Class-A land pan will be warmed much more readily by solar radiation than the surface waters of a lake or reservoir. The chief reason is the difference between the water depth in the pan and the depth of the surface layer of reservoir water. The pan may also gain or lose heat through the sides and bottom, a process that does not occur in reservoirs. For

15

these reasons, observed pan evaporation is multiplied by a factor with a value less than 1.0, the pan coefficient, to estimate reservoir evaporation during the period of observation. Detailed studies in the United States Midwest have yielded monthly pan coefficients ranging from 0.58 in December to 0.78 in May, with an annual value of 0.75 (See Table 2.2).

TABLE 2.2. Class-*A* land pan coefficients for midwestern United States

January	0.62	July	0.76
February	0.72	August	0.75
March	0.77	September	0.73
April	0.77	October	0.69
May	0.78	November	0.63
June	0.77	December	0.58
	Annual 0.75		

SOURCE: W. J. Roberts and J. B. Stall, Illinois State Water Survey Report of Investigation 57, 1967.

The United States Weather Bureau has developed a lake evaporation nomograph (3). From this diagram, daily lake evaporation can be determined using mean daily temperature, solar radiation in langleys* per day, mean daily dew point temperature, and wind movement in miles per day. The graph in Figure 2.1 is entered from the left side at the mean daily air temperature. As an example, this is 75° F. A horizontal line is drawn across the chart along the 75° F axis. Perpendicular lines are dropped at the intersection of the values of solar radiation and mean daily dew point temperature. In the example, these are 500 langleys per day and 50° F. The right-hand perpendicular extends from the mean daily dew point temperature to the total daily wind movement. The example value is 200 miles per day. From this intersection, a horizontal line is drawn toward the left. This horizontal line and the left-hand perpendicular will intersect in a field indicating the mean daily lake evaporation. For the example in Figure 2.1, this is 0.25 inches per day.

In some instances, it may be necessary to estimate evaporation without the availability of evaporation pan data. Such estimates are possible via methods based on heat budgets (4). The energy budget for a reservoir may be used to find the amount of energy used for evaporation, which in turn can yield the amount of evaporation. Other methods use aerodynamic data and vapor pressures of the water and air in empirical formulas (5). Some investigators have been able to combine these two approaches (6).

---

*A langley is a measure of solar radiation equal to one calorie per square centimeter of surface. In the SI system (International System of Units based on the meter, kilogram, second, and ampere), the unit is the joule per square meter, which is equal to 4.184×10^4 langleys.

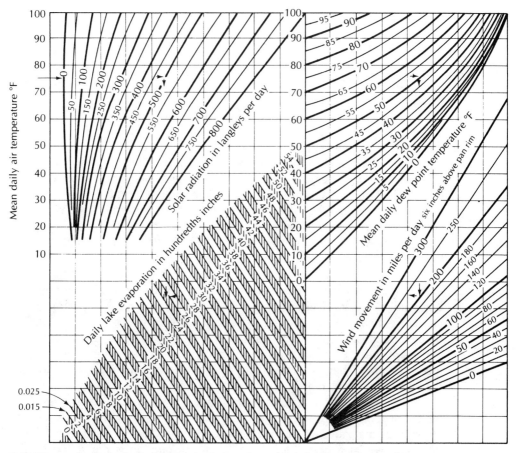

FIGURE 2.1. Nomograph used to determine the value of daily lake evaporation for shallow lakes if solar radiation, mean daily air temperature, mean daily dew point temperature, and wind movement are known. SOURCE: United States National Weather Service, Reference 3.

TRANSPIRATION 2.2

Free-water evaporation is only part of the mechanism for mass transfer of water to the atmosphere. Growing plants are continuously pumping water from the ground into the atmosphere through a process called **transpiration** (7). Water is drawn into a plant rootlet from the soil moisture due to osmotic pressure, whereupon it moves through the plant to the leaves. The turgidity of nonwoody

17

vascular plants is caused by the cellular pressures of the contained water. The water is passed as vapor through openings in the surface of the leaves known as **stomata**. Air also passes through these openings. A small portion (less than 1 percent) of the water is used to manufacture plant tissue, but most is transpired to the atmosphere. The process of transpiration accounts for most of the vapor losses from a land-dominated drainage basin.

The amount of transpiration is a function of the density and size of the vegetation. As an example, transpiration from a cornfield in May, when the plants are a few centimeters high, is much less than in August, when they may exceed two meters in height. Transpiration is obviously important only during the growing season; about 95 percent takes place during the daylight hours, when photosynthesis is occurring. Transpiration is also limited by available soil moisture. When the soil-moisture content becomes so low that the surface tension of the soil-water interface exceeds the osmotic pressure of the roots, water will no longer enter the roots. This is termed the **wilting point** of the soil.

When available water becomes limited, deep-rooted plants are more resistant to drought wilting than shallow-rooted plants, as the former can draw moisture from deeper layers. Also, some plants have fewer stomata and can close them through the use of special cells to reduce water loss during drought periods. Such drought-resistant species can transpire less water during periods of stress. **Phreatophytes** are plants with a tap root system extending to the water table. They can transpire at a high rate even in the desert, so long as the water table does not drop below the tap root.

Aquatic plants, or **hydrophytes**, are a special case. They exist with their root systems submerged, and the special cells some plants have to close the stomata are lacking. As long as adequate water is available, transpiration proceeds at a high rate. The rate of transpiration is controlled by the amount of solar energy and the heat content of the water. The water loss from a pond is about the same, whether or not emergent aquatic vegetation is present.

Measurement of transpiration can be performed under carefully controlled laboratory conditions. A **phytometer** is a sealed container partially filled with soil. Transpiration by plants rooted in the soil causes an increase in the humidity, which can be measured in the air space around the plant. However, such laboratory studies reveal little about the behavior of plants in natural or agricultural conditions.

2.3 EVAPOTRANSPIRATION

Under field conditions it is not possible to separate evaporation from transpiration totally. Indeed, we are generally concerned with the total water loss, or **evapotranspiration**, from a basin. Whether the loss is due to free-water evaporation, plant transpiration, or soil-moisture evaporation is of little importance.

The term **potential evapotranspiration** was introduced by Thornthwaite as equal to "the water loss which will occur if at no time there is a deficiency of water in the soil for the use of vegetation" (8). He made the assumption that potential evapotranspiration was dependent only upon meteorological conditions and ignored the effect of vegetative density and maturity. While this assumption is not correct, the method devised by Thornthwaite to compute potential evapotranspiration is still useful. The only necessary factors to input are mean monthly air temperature, latitude, and month (9,10). The latter two factors yield average monthly sunlight. The Thornthwaite method is reasonably accurate in determining annual values. As no factor for vegetative growth is included, values computed for spring and early summer are too high, as the crop is just emerging; mid-summer values may be too low.

Another method of estimating potential evapotranspiration was developed by Blaney and Criddle (11). This method introduces a crop factor, which varies as the growing season progresses. Thus, some of the objections to the Thornthwaite method are overcome, but the effects of wind and relative humidity on evapotranspiration still remain unaccounted for.

Evapotranspiration can be measured directly using a **lysimeter**— a large container holding soil and plants. The lysimeter is set outdoors, and the initial soil-moisture content is determined. Precipitation into the lysimeter, as well as any irrigation water added, are measured. Changes in soil-moisture storage reveal how much of the added water is lost to evapotranspiration. It is necessary to design the lysimeter so that any moisture in excess of that specifically retained by the soil is collected. The following equation can be used with the lysimeter:

$$E_T = S_i + P + I - S_f - D \qquad (2\text{-}1)$$

where

> E_T is the evapotranspiration for a period
> S_i is the volume of initial soil moisture
> S_f is the volume of final soil moisture
> P is the precipitation into the lysimeter
> I is the irrigation water added to the lysimeter
> D is the excess moisture drained from the soil

Lysimeters should be designed so that they accurately reproduce the soil type and profile, moisture content, and type and size of vegetation of the surrounding area. They should be buried so that the soil surface is at the same level inside and outside the container. Soil-moisture changes can be determined by sampling the soil, by means of moisture meters employing electrical current, or by weighing the entire mass of soil, water, and plants. Whatever method is employed, operation of a lysimeter is both time-consuming and expensive. If water is applied to the lysimeter at a rate sufficient to keep the soil at, or nearly at, field capacity (i.e., the maximum specific retention), the lysimeter will measure potential evapotranspiration.

19

When the soil moisture drops below the amount that it can hold against gravity by surface tension (the field capacity), available water may limit evapotranspiration to some value less than the potential evapotranspiration. The plants are required to draw upon soil moisture and, as this diminishes and less water is extracted, actual evapotranspiration falls below the potential. If the soil-moisture drops too low, the plants may wither and die. As previously mentioned, the soil-moisture content below which plants can no longer obtain moisture is the wilting point.

There is some uncertainty about the rate of evapotranspiration when the soil moisture is between the wilting point and the field capacity. Some have suggested that it proceeds at a rate equal to the potential evapotranspiration until the wilting point is reached (12), while others have suggested that the evapotranspiration rate is linearly proportional to the ratio of the remaining available soil moisture to the initial available soil moisture (9). Soil texture and unsaturated soil permeability play a major role in determining the rate of actual evapotranspiration (13).

Evapotranspiration is the major use of water in all but extremely humid, cool climates. If evapotranspiration were reduced, then runoff or groundwater infiltration or both could increase. This would increase the available water supply. Studies have shown that basin runoff from a forested watershed has increased following the timbering of the forest (14). The increase is greatest during the first year, when there is little reforestation. As the forest regrows, the runoff again decreases. Cutting of forests to increase runoff may also result in increased erosion from the uplands and concurrent sedimentation in the lowlands. Conversion of one plant cover to another can also affect the evapotranspiration rate. In arid Arizona, the conversion of a plot of land formerly covered with chaparral to grasses resulted in streamflow increases of several hundred percent. This was due in part to lower evapotranspiration, as the grass was not as deep-rooted as the chaparral (15). However, in Colorado, the conversion of sagebrush to bunchgrass had no appreciable effect on the amount of watershed runoff, although an increase in cattle forage did result (16).

In some areas of the humid eastern United States, which were originally wooded, marginal farms are being abandoned. The old fields are gradually reverting to a forest. There has been a concomitant decrease in streamflow from these watersheds. The replacement of deciduous forests with conifers results in an increase in evapotranspiration (17).

Experiments have shown that evaporation from small lakes and reservoirs can be reduced by applying a monolayer of a fatty alcohol to the water surface (18). This has not proven to be practical, however, due to the cost of a treatment and the rapid rate at which the fatty alcohol dissipates. Likewise, fatty alcohols have been used as antitranspirants in treating plants and soils. However, concentrations high enough to reduce transpiration also reduce crop growth (19). Chemical antievaporants and antitranspirants have not yielded the hoped for success.

20

CONDENSATION 2.4

When an air mass with a relative humidity lower than 100 percent is cooled without losing moisture, the relative humidity will approach 100 percent as the dew point temperature is approached. When the air mass is saturated, **condensation** may start to occur. Condensation generally requires a surface or nuclei on which to form. The morning dew or frost is the result of condensation taking place on plants or other surfaces. Rain or ice needs nuclei in the range of 0.1 to 10 microns. Particles serving as nuclei include clay minerals, salt, and combustion products.

In the absence of sufficient nuclei, the air mass may become supersaturated without the formation of raindrops or ice crystals. This is the theory behind artificial precipitation augmentation. "Cloud-seeding" procedures involve the addition of artificial nuclei to the atmosphere, including silver iodide and dry ice. Research has shown that even in severe draughts, atmospheric conditions conducive to successful seeding may sometimes occur during the summer (20).

Once droplets or ice crystals have formed, they initially grow by attraction (diffusion) of water vapor as well as additional condensation. Rising air masses or upward movements of clouds tend to keep newly formed fog and cloud elements aloft. These elements are in the size range of 10 to 50 microns. As cloud elements collide and coalesce, raindrops begin to form. When the raindrops start to fall, further collisions occur, so that some raindrops may grow as large as 6 millimeters in diameter. Rain that falls through an unsaturated air mass may evaporate before it reaches the ground. Falling ice crystals grow by diffusion and collision to form snowflakes. The largest snowflakes form when temperatures are close to freezing.

FORMATION OF PRECIPITATION 2.5

In order for precipitation to occur, several conditions must be met: (a) a humid air mass must be cooled to the dew point temperature, (b) condensation or freezing nuclei must be present, (c) droplets must coalesce to form raindrops, and (d) the raindrops must be of sufficient size when they leave the clouds to insure that they will not totally evaporate before they reach the ground.

Air masses are cooled by a process known as **adiabatic expansion**, which occurs when the air mass rises in the atmosphere. Since the atmosphere becomes less dense with altitude, a rising air mass must expand due to the lower pressure. If there is no exchange of heat between the air mass and its surroundings, the laws of thermodynamics dictate that the temperature will fall.

When the rising air mass is dry; that is, the relative humidity is lower than 100 percent, the rate of cooling is 1° C for every 100-meter rise in height. This is the **dry adiabatic lapse rate**. When the air mass reaches the dew point temperature, further lifting and cooling will cause condensation. The latent heat of vaporization is released; hence, the **wet adiabatic lapse rate** is lower than the dry rate. The exact value depends upon the amount of condensation occurring.

Under normal conditions, air temperature decreases with increasing altitude at a mean rate of 0.7° C for every 100 meters. Due to uneven or unsteady heating or cooling, the temperature gradient or lapse rate may be more or less than 0.7° C per 100 meters. **Temperature inversions**, or layers of warm air overlying cooler air, exist and are typically caused by warm air masses overriding cold fronts, or by conductive cooling of the earth's surface. Solar radiation during the day causes high temperature gradients.

Most rising air masses can be attributed to one of three causal factors: movement of weather fronts, convective processes, and orographic effects. **Frontal precipitation** is caused by the lifting of an air mass by a moving weather front. If a warm front is moving upward over a colder, more dense air mass, precipitation and cloudiness will extend for several hundred miles ahead of the surface front (Figure 2.2A). The slope of a front of this type is small and

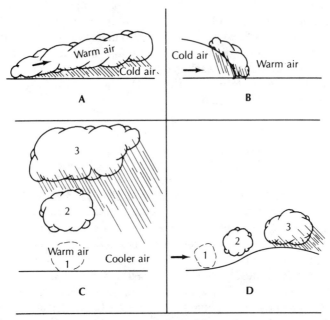

FIGURE 2.2. Precipitation caused by adiabatic lifting of an air mass may be the result of the following activity: **A.** A warm front pushing over a cold front; **B.** A cold front colliding with a warm front; **C.** Uneven heating near the surface causing a warm air mass to rise convectionally; **D.** Orographic lifting caused by prevailing winds blowing over a topographic high.

the rate of ascent of the warm air mass is slow; hence, precipitation is generally light. Should a cold front be moving, it will be typically faster than a warm front. The cold front is steeper, warm air is forced upward more rapidly, and heavier rain amounts may be recorded—especially near the surface front (Figure 2.2B).

Uneven heating of an air mass at the surface, or cooling at the top of an air mass, may cause it to be warmer than the surrounding air. The denser air will flow beneath it, causing the air mass to rise. It will continue to rise, and cool adiabatically, until the temperature is equal to its surroundings. This can cause **convectional** rising. Summer thunderstorms and associated cumulus clouds are a result of this process (Figure 2.2C).

If a moving air mass is forced upward over a mountain range, it must gain altitude. **Orographic** cooling and precipitation may result. The vegetation of the Black Hills of South Dakota is different from the surrounding grassland prairie. There is sufficient precipitation for forest to grow and, from a distance, clouds can be often seen hovering over the hills. This is orographically caused condensation, as air masses are forced upward as they move from west to east over the central Black Hills (Figure 2.2D).

MEASUREMENT OF PRECIPITATION 2.6

Any open container can be used to catch and measure rainfall. Experiments have shown that the size of the opening has little effect on the catch, except for very small (less than one-inch-diameter) gauges (21). The United States Standard Rain Gauge has an opening 8 inches (20.3 centimeters) in diameter, while the Canadian standard gauge is 9 centimeters (3.57 inches) in diameter. These are manually read gauges; the water is emptied and the gauges read once a day. The catch of precipitation gauges is affected by high winds. Such gauges generally catch less than the true amount of rainfall because of updrafts around the gauge opening. The location of the gauge is also critical. In one study, two identical 8-inch gauges were placed 3 meters (10 feet) apart on a ridge. One gauge consistently caught 50 percent more rainfall than the other (22). Gauges should be placed as close to the ground as possible in order to avoid wind. They should be in the open, away from trees and buildings. Low bushes and shrubs can provide a windbreak. Level ground is best, with the top of the gauge horizontal. On steep slopes, it may be desirable to have the orifice opening parallel to the slope.

The effect of wind is greatest for light rain or snow. Some rain gauges are equipped with a shield, or wind deflector, around the opening in order to overcome wind problems. This will improve the catch of snow, but it will still be less than 100 percent in substantial winds.

There are a number of different types of recording rain gauges available that can automatically measure or weigh the precipitation. The temporal distribution of precipitation through a day can thus be obtained. Such data are necessary for any studies of precipitation intensity. For remote areas, recording rain gauges can be used to record daily precipitation for long time periods. In such circumstances, manual gauges could only provide a total rainfall for the period between readings.

In the United States there are some 13,500 precipitation stations, for the most part operated by trained volunteers. Daily records from these weather stations are published monthly on a state-by-state basis in *Climatological Data*; data from recording stations are published in *Hourly Precipitation Data*. Both of these are publications of the U.S. Environmental Data Service. Canada has about 2000 precipitation stations, the data from which are published by the meteorological Canadian Atmospheric Environment Service in the *Monthly Record of Observations*.

As every viewer of local television news and weather programs knows, radar can be used to detect areas of precipitation. Rain droplets or snow reflects part of the directed radar beam back to the originating station. The amount of reflected energy is directly proportional to the intensity of the precipitation. The radar apparatus measures precipitation in the atmosphere. As the beam is at an oblique angle to the ground, the further the distance from the station, the greater the altitude of the precipitation being measured. Radar measurements of precipitation may not accurately indicate ground precipitation. Evaporation may occur between the point of measurement and the ground, or wind may cause the precipitation to drift so that it falls to earth at some place other than that indicated by the radar (23).

Some special radar equipment can convert the intensity of the radar reflection into precipitation rates. The rates are integrated over time to yield a depth of total precipitation over an area. This yields data about precipitation rates between ground stations. The use of radar in combination with conventional ground-station rain gauges can give improved areal measurement of precipitation (24).

2.7 SNOW MEASUREMENTS

The measurement of snowfall in standard rain gauges is subject to error due to turbulence around the gauge. The snow that is caught is melted, and the water equivalent reported. If only an approximation is required, a water content of 10 percent of the snow depth can be assumed. However, as anyone who regularly shovels snow knows, the density of newly fallen snow can vary considerably.

In northern and mountainous climates, the accumulation of snow on the ground is an important hydrologic parameter. In some areas, the runoff of melting snow in the spring is a predominant source of water for reservoirs used for water supply, irrigation, and power generation. A thick accumulation of snow can also mean a high flood potential when snowmelt occurs in the spring. Melting snow also recharges soil moisture and the water table.

Snow surveys are made periodically through the winter to measure the thickness and water content of accumulated snow. A thinwalled tube with a sharp leading edge is driven through the snow to the ground. The tube and the snow contained within are weighed, and the weight of the empty tube subtracted to determine the weight of the snow. A snow survey requires that someone make traverses, stopping at predetermined stations to make measurements. The snow courses should sample representative terrain, vegetative cover, and altitude of the catchment area.

The extent of snow cover can be mapped using satellite photography (25). The resulting data, combined with data from snow-course surveys can be used to determine the total volume of water in the snowpack. Melting of the snowpack can begin only when the temperature of the snow has risen to 0° C. Initial meltwater clings to snow granules by surface tension, so that at least 2 to 8 percent of the snowpack must melt before runoff begins. Energy-balance methods can be used to predict daily snowmelt (26).

EFFECTIVE DEPTH OF PRECIPITATION   2.8

In water-budget studies, it is necessary to know the average depth of precipitation over a drainage basin. This may be determined for time periods ranging from the duration of part of a single storm to that of a year. The data are generally measurements of precipitation and/or equivalent snowfall at a number of points throughout the drainage basin.

A problem is created if data are missing at one or more stations. This can occur due to equipment malfunction or operator absence. To solve the problem, three close precipitation stations with full records that are evenly spaced around the station with a missing record are used. The following equation yields an estimate of the missing data at Station Z. The mean annual precipitation (N) at Station Z and the three index stations, A, B, and C, as well as the actual precipitation at the index stations (P) for the time period over which data are missing are needed:

$$P_Z = 1/3\left[\frac{N_Z}{N_A}P_A + \frac{N_Z}{N_B}P_B + \frac{N_Z}{N_C}P_C\right] \qquad (2\text{-}2)$$

If the rain-gauge network is of uniform density, then a simple arithmetic average of the point-rainfall data for each station is sufficient to determine the **effective uniform depth (EUD)** of precipitation over the drainage basin (Figure 2.3).

If the rain-gauge network is not uniform, then some adjustment is necessary. The most accurate method, excluding use of radar data, is to draw a precipitation contour map with lines of equal rainfall (**isohyets**). In drawing the isohyets, such factors as known influence of topography on precipitation can

25

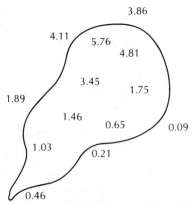

FIGURE 2.3. Precipitation-gauge network over a drainage basin. Precipitation amounts are given in inches. Station locations are at decimal points.

be taken into account. Simple linear interpolation between precipitation stations can also be used. The area bounded by adjacent isohyets is measured with a planimeter, and the average depth of precipitation over the area is the mean of the bounding isohyets. The effective uniform depth of precipitation is the weighted average based on the relative size of each isohyetal area (Figure 2.4). The drawback of the isohyetal method is that the isohyets must be redrawn and the areas remeasured for each analysis.

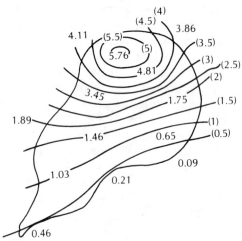

FIGURE 2.4. Isohyetal lines for the precipitation-gauge network of Figure 2.3. The isohyets show contours of equal rainfall depth with a contour interval of 0.5 inch. The contours are based on simple linear interpolation.

The **Theissen method** to adjust for nonuniform gauge distribution uses a weighing factor for each rain gauge. The factor is based on the size of the area within the drainage basin which is closest to a given rain gauge. These areas are irregular polygons. The method of constructing them can be described rather easily; however, it takes a bit of practice to master the technique. The rain-gauge network is drawn on a map of the drainage basin. Adjacent stations are connected by a network of lines (Figure 2.5A). Should there be doubt as to which stations to connect, lines should be between the closest stations. A perpendicular line is then drawn at the midpoint of each line connecting two stations (Figure 2.5B), and extensions of the perpendicular bisectors are used to draw polygons around each station (Figure 2.5C). It is best to start with a centrally located station and then expand the polygonal network outward. The area of each polygon is measured, and a weighted average for each station's precipitation is used to find the effective uniform depth (EUD).

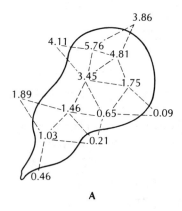

A

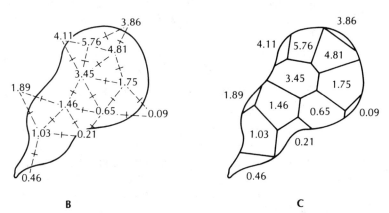

B

C

FIGURE 2.5.   Theissen polygons based on the rain-gauge network of Figure 2.3. **A.** The stations are connected with lines; **B.** The perpendicular bisector of each line is found; **C.** The bisectors are extended to form the polygons around each station.

27

EXAMPLE PROBLEM

Determine the effective uniform depth of precipitation using the arithmetic mean, isohyetal, and Theissen methods.

Arithmetic Mean Method

Figure 2.3 (page 26) shows a drainage basin with seven stations in its boundaries. An additional six stations are located outside the drainage divide. In the Arithmetic Mean Method, only the gauges inside the drainage basin boundary are considered.

Arithmetic Mean =
$$\frac{1.03 + 0.65 + 1.46 + 1.75 + 4.81 + 3.45 + 5.76}{7}$$

$$= 2.70 \text{ in.}$$

Isohyetal Method

The first step is to draw lines of equal precipitation (isohyets) on the drainage basin map. Isohyets are usually whole numbers or decimals (every 0.1 inch; every 0.5 inch; every 1 millimeter, etc.). The following rules apply:

1. Isohyets never cross.

2. Isohyets never split.

3. Isohyets never meet.

4. A station that does not fall on an isohyet will be between two isohyets. The isohyets will both be equal (either larger or smaller than the station value) or one will be larger and one smaller.

5. Adjacent isohyets must be equal or only one contour interval different in value.

6. Isohyets should be scaled between stations using linear interpolation.

Figure 2.4 (page 26) shows the isohyetal map of the problem area. The area between adjacent isohyets is determined by use of a planimeter. The eqivalent uniform depth of precipitation between isohyets is usually assumed to be equal to the median value of the two isohyets. For example, the EUD between a 1-inch isohyet and a 2-inch isohyet is 1.5 inches. For areas enclosed by a single isohyet, judgment should be used to estimate the equivalent uniform depth. The weighted average precipitation is based on the equivalent uniform depth of precipitation between adjacent isohyets and their areas.

A	B	C	D	E
		Net	Percent of	Weighted
Isohyet	Estimated	Area	Total	Precipitation (in.)
(in.)	EUD	(sq mi)	Area	(B × D)
5.5	5.6	1.1	0.7	0.043
5.0	5.25	7.6	5.4	0.281
4.5	4.75	10.6	7.5	0.355
4.0	4.25	9.5	6.7	0.285
3.5	3.75	8.6	6.1	0.227
3.0	3.25	8.3	5.9	0.190
2.5	2.75	10.7	7.5	0.208
2.0	2.25	12.3	8.7	0.195
1.5	1.75	15.1	10.6	0.186
1.0	1.25	23.8	16.8	0.210
0.5	0.75	31.2	22.0	0.165
> 0.5	0.3	4.0	2.8	0.008
TOTAL		141.8 sq mi		2.35 in. NET EUD

Theissen Method

This method provides for the nonuniform distribution of gauges by determining a weighting factor for each gauge. A weighted mean of the precipitation values can then be computed. Theissen polygons for the example problem are shown in Figure 2.5C (page 27). The area of each polygon is determined by a planimeter. A weighted mean of the EUD is found, based on the depth of precipitation and the area of the polygon within the basin boundary.

A	B	C	D
Station	Net	Percent of	Weighted
Precipitation	Area	Total	Precipitation (in.)
(in.)	(sq mi)	Area	(A × C)
5.76	16.9	11.9	0.686
4.81	16.1	11.4	0.546
4.11	3.4	2.4	0.098
3.86	1.6	1.1	0.044
3.45	19.3	13.6	0.470
1.89	2.5	1.8	0.033
1.75	12.0	8.5	0.148
1.46	19.8	14.0	0.204
1.03	18.0	12.7	0.131
0.65	17.0	12.0	0.078
0.46	6.0	4.2	0.019
0.21	7.2	5.1	0.011
0.09	2.0	1.4	0.001
TOTAL	141.8 sq mi		2.47 in. NET EUD

References

1. HARBECK, G. E. and F. W. KENNON. "The Water Budget Control." In *Water-Loss Investigations: Lake Hefner Studies Technical Report.* U.S. Geological Survey Professional Paper 269, 1954.

2. ROBERTS, W. J. and J. B. STALL. *Lake Evaporation in Illinois.* Illinois State Water Survey Report of Investigation 57, 1967, 44 pp.

3. KOHLER, M. A., T. J. NORDENSON, and W. E. FOX. *Evaporation from Ponds and Lakes.* U.S. Weather Bureau Research Paper 38, 1955.

4. PENMAN, H. L. "Natural Evaporation from Open Water, Bare Soil, and Grass." *Proceedings of the Royal Society* (London), ser. A, 193 (1948):120–45.

5. HARBECK, G. E. *A Practical Field Technique for Measuring Reservoir Evaporation Utilizing Mass-Transfer Theory.* U.S. Geological Survey Professional Paper 272-E, 1962, pp. 101–5.

6. KOHLER, M. A. and L. H. PARMELE. "Generalized Estimates of Free-Water Evaporation." *Water Resources Research*, 3 (1967):997–1005.

7. HENDRICKS, D. W. and V. E. HANSEN. "Mechanics of Transpiration." *American Society of Civil Engineers, Journal of Irrigation and Drainage Division*, 88 (June 1962):67–82.

8. THORNTHWAITE C. W. "Report of the Committee on Transpiration and Evaporation, 1943–1944." *Transactions, American Geophysical Union*, 25 (1944):687.

9. THORNTHWAITE, C. W. and J. R. MATHER. *The Water Balance, Publication 8.* Centerton, N.J.: Laboratory of Climatology, 1955, pp. 1–86.

10. THORNTHWAITE, C. W. and J. R. MATHER. *Instructions and Tables for Computing Potential Evapotranspiration and the Water Balance, Publication 10.* Centerton, N.J.: Laboratory of Climatology, 1957, pp. 185–311.

11. BLANEY, H. F. and W. D. CRIDDLE. *Determining Water Requirements in Irrigation Areas from Climatological and Irrigation Data.* U.S. Department of Agriculture, Soil Conservation Service Technical Paper 96, 1950.

12. VEIHMEYER, F. J. and A. H. HENDIRCKSON. "Does Transpiration Decrease as Soil Moisture Decreases?" *Transactions, American Geophysical Union*, 36 (1955):425–48.

13. MOLZ, F. J., I. REMSON, A. A. FUNGAROLI, and R. L. DRAKE. "Soil Moisture Availability for Transpiration." *Water Resources Research*, 4 (1968):1161–70.

14. HIBBERT, A. R. "Forest Treatment Affects on Water Yield." In *Forest Hydrology*, ed. W. E. Sopper and H.W. Lull. Oxford, England: Pergamon Press, 1967, pp. 527–43.

15. HIBBERT, A. R. "Increases in Streamflow after Converting Chaparral to Grass." *Water Resources Research*, 7 (1971):71–80.

16. SHOWN, L. M., G. C. LUSBY, and F. A. BRANSON. "Soil Moisture Effects of Conversion of Sagebrush Cover to Bunchgrass Cover." *Water Resources Bulletin*, 8 (1972):1265–72.

17. URIE, D. H. "Influences of Forest Cover on Groundwater Recharge Timing and Use." In *International Symposium on Forest Hydrology*, ed. W. E. Sopper and H. W. Hull. Oxford, England: Pergamon Press, 1967, pp. 313–24.

18. BARTHOLIC, J. F., J. R. RUNKELS, and E. B. STENMARK. "Effects of a Monolayer on Reservoir Temperature and Evaporation." *Water Resources Research*, 3 (1907):173–80.

19. GALE, J., E. B. ROBERTS, and R. M. HAGEN. "High Alcohols as Antitranspirants." *Water Resources Research*, 3 (1967):437–41.

20. HUFF, F. A. and R. G. SEMONIN. "Potential of Precipitation Modification in Severe Droughts." *Journal of Applied Meteorology*, 14 (1975):974–79.

21. HUFF, F. A. "Comparison between Standard and Small Orifice Rain Gauges." *Transactions, American Geophysical Union*, 30 (1955):689–94.

22. COURT, A. "Reliability of Hourly Precipitation Data." *Journal of Geophysical Research*, 65 (1960):4017–24.

23. STOUT, G. E. and E. A. MUELLER. "Survey of Relationships between Rainfall Rate and Radar Reflectivity in the Measurement of Precipitation." *Journal of Applied Meteorology*, 7 (1968):465–74.

24. WILSON, J. W. "Integration of Radar and Rain Gauge Data for Improved Rainfall Measurements." *Journal of Applied Meteorology*, 9 (1970):489–97.

25. BARNES, J. C. and C. J. BOWLEY. "Snow Cover Distribution as Mapped from Satellite Photography." *Water Resources Research*, 4 (1968):257–72.

26. PRICE, A. G. and DUNN, T. "Energy Balance Computations of Snowmelt in a Subarctic Area." *Water Resources Research*, 12 (1976):686–94.

Runoff and Streamflow

3.1 EVENTS DURING PRECIPITATION

During a precipitation event, some of the rainfall is intercepted by vegetation before it reaches the ground. This may later fall to the ground or evaporate. In a heavily forested area, most of the precipitation is caught by leaves and twigs. For a period at the start of a summer thunderstorm, no raindrops reach the forest floor, although drops can be heard striking the leaves overhead. When the storage capacity of the leaf surfaces is exhausted, water will run down tree trunks and drip downward (1, 2, 3). The amount of water intercepted by dense forests ranges from 8 to 35 percent of total annual precipitation (4). In a mixed hardwood forest in the northeastern United States, it averaged 20 percent in the summer and winter seasons (5). While evaporation of intercepted water reduces the net transpiration by the plants, in some cases most of the evaporated water is simply lost. One study concluded that only about 10 percent of the intercepted water actually reduced evapotranspiration (6).

The water reaching the ground can infiltrate into the soil, form puddles, or flow as a thin sheet of water across the land surface. Hydrologists refer to the water trapped in puddles as **depression storage**. It ultimately evaporates or infiltrates.

The overland flow process, sometimes called **Horton overland flow** after Robert Horton (7, 8), occurs only when the precipitation rate exceeds the infiltration capacity. In areas in which soils have a high infiltration capacity, this process may only occur during very intense storms, or when the soil is saturated or frozen. In order for overland flow to occur, the infiltration capacity of the soil must first be exceeded; then the depression storage must be filled (Figure 3.1).

If the unsaturated zone is uniformly permeable, most of the infiltrated water percolates vertically. Should layers of soil with a lower vertical hydraulic conductivity occur beneath the surface, then infiltrated water may move horizontally in the unsaturated zone. This **interflow** may be substantial in some drainage basins and contribute significantly to total streamflow. Thin permeable soil overlying fractured bedrock of low permeability could provide a geologic condition contributing to significant interflow (Figure 3.2).

Water will fall directly onto the surface of lakes and reservoirs during the period of precipitation. This amount might not be considerable for

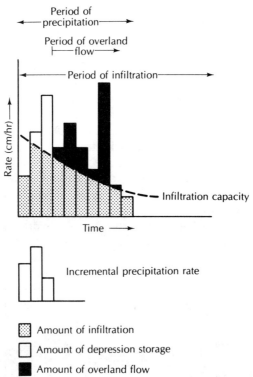

Incremental precipitation rate

Amount of infiltration

Amount of depression storage

Amount of overland flow

FIGURE 3.1. Incremental precipitation rate and its dissociation into amounts of infiltration, depression storage, and overland flow. Infiltration begins when the precipitation does. Overland flow does not begin until the depression storage is exhausted. Overland flow continues past the termination of precipitation. Infiltration will continue as long as there is any water in depression storage—usually past the period of overland flow.

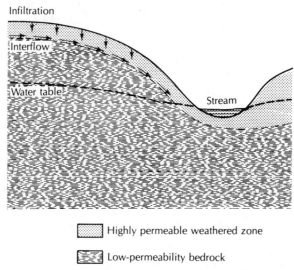

Highly permeable weathered zone

Low-permeability bedrock

FIGURE 3.2. Interflow developing where a highly permeable but thin layer of weathered rock overlies a bedrock unit of lower permeability.

35

streams, but for lakes and reservoirs it could be. Lake Michigan and its associated water bodies have a surface area of 22,300 square miles. The land area of the surrounding drainage basin is 45,000 square miles (9). Assuming equal distribution of precipitation over the entire Lake Michigan basin, about one-third falls as **direct precipitation** on a water body.

Infiltrated water that reaches the water table becomes stored in the groundwater reservoir. This is not static storage, as groundwater is in constant movement. While freshly infiltrated precipitation is entering the groundwater reservoir, other groundwater, known as **baseflow,** is discharging into the stream. If infiltration causes the water table to rise, groundwater discharge into nearby streams will also increase. For baseflow streams, the amount of groundwater discharge is directly proportional to the hydraulic gradient toward the stream (Figure 3.3).

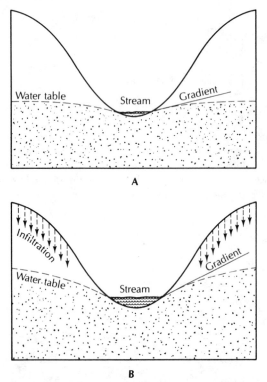

FIGURE 3.3. Influence of the water-table gradient on baseflow. The stream in Part A is being fed by groundwater with a low hydraulic gradient. A gentle rain does not produce overland flow, but infiltration raises the water table. The increased hydraulic gradient of Part B causes more baseflow to the stream, which is now deeper and has a greater discharge.

The runoff cycle in which so much emphasis is placed on Horton overland flow has been criticized on several fronts (10). Horton overland flow is rarely observed in the field, except after very heavy precipitation events. This is especially true if the ground is covered with vegetation or humus, such as leaf litter (11). Horton overland flow appears to be more common in arid regions or areas in which the soil has been compacted by vehicles, animals, etc. (12). Overland runoff can also occur when precipitation falls on soils that are saturated.

Water that infiltrates into the soil on a slope can move downslope as lateral unsaturated flow in the soil zone. This has been called **throughflow** (11). The difference between throughflow and interflow is that throughflow emerges as seepage at the foot of the slope rather than entering a stream, as does interflow. Thus, the throughflow appears as overland flow before entering a stream channel. This overland flow is called **return flow** (13) to distinguish it from Horton overland flow.

A more comprehensive concept of the hydrologic cycle has been proposed by Dunne (12). In arid to subhumid climates where there is thin vegetation, Horton overland flow is the main contributor to the storm peak and comprises most of the streamflow. In humid climates, Horton overland flow is not significant, but interflow, return flow, and direct precipitation on the channel are important. Where there are thin soils and gentle, concave slopes, direct precipitation and return flow are more important than interflow. On steep, straight slopes, interflow becomes much more important, although return flow and direct precipitation still cause the peaks.

HYDROGRAPH SEPARATION 3.2

A stream hydrograph shows the discharge of a river at a single location as a function of time. While the total streamflow shown on the hydrograph gives no indication of its origin, it is possible to break down the hydrograph into components such as overland flow, baseflow, interflow, and direct precipitation. The model presented in this section is based on the Horton runoff cycle; it would be most useful for arid-zone hydrology.

3.2.1   BASEFLOW RECESSIONS

The hydrograph of a stream during a period with no excess precipitation will decay following an exponential curve. The discharge is composed entirely of groundwater contributions. As the stream drains water from the groundwater reservoir, the water table falls, leaving less and less groundwater to feed the stream. If there were no replenishment of the groundwater reservoir, baseflow to the stream would become zero. Figure 3.4 shows a baseflow recession curve for a stream in a climate with a dry summer season.

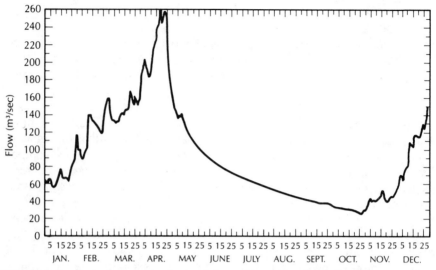

FIGURE 3.4. Typical annual hydrograph for a river with a long dry summer season: Lualaba River, Central Africa. SOURCE: C. O. Wisler and E. F. Brater, eds., *Hydrology*, 2nd ed. (New York: John Wiley & Sons, 1959). Used with permission.

The **baseflow recession** for a drainage basin is a hydromorphic characteristic. It is a function of the overall topography, drainage pattern, soils, and geology of the watershed. Figure 3.5 illustrates this by showing the annual summer recession of a river for six consecutive years. The start of the baseflow recession was considered to be the day when the annual discharge dropped below 3500 cubic feet per second. The recession is similar from year to year.

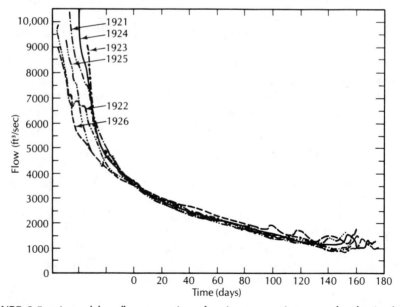

FIGURE 3.5. Annual baseflow recessions for six consecutive years for the Lualaba River, Central Africa. SOURCE: C. O. Wisler and E. F. Brater, eds., *Hydrology*, 2nd ed. (New York: John Wiley & Sons, 1959). Used with permission.

The baseflow recession equation is

$$Q = Q_0 e^{-at} \qquad \qquad \text{(3-1)}$$

where

Q is the flow at some time t after the recession started

Q_0 is the flow at the start of the recession

a is a recession constant for the basin

t is the time since the recession began

EXAMPLE PROBLEM

Part A: Find the recession constant for the basin of Figure 3.5.

If

$$Q = Q_0 e^{-at}$$

then

$$e^{-at} = Q/Q_0$$
$$-at = \ln Q/Q_0$$
$$a = -1/t \ln Q/Q_0$$

From Figure 3.5, $Q_0 = 3500$ cubic feet per second. After 100 days, $Q = 1500$ cubic feet per second.

$$a = -1/t \ln Q/Q_0 = -1/100 \ln 1500/3500$$
$$= -0.01 \ln 0.4286$$
$$= -0.01 \times (-0.847)$$

Therefore,

$$a = 8.47 \times 10^{-3}$$

if Q is in cubic feet and t is in days.

Part B: What would the baseflow be after 40 days of recession?

$$Q = Q_0 e^{-at}$$
$$= 3500 \exp(-8.47 \times 10^{-3} \times 40)$$
$$= 3500 \times 0.712$$

Therefore,

$$Q = 2490 \text{ ft}^3/\text{sec}$$

39

3.2.2 STORM HYDROGRAPH

While the baseflow component of a stream is somewhat constant, the total discharge of the stream may fluctuate greatly through the year. The difference is due to the episodic nature of precipitation events that contribute overland flow, interflow, and direct precipitation. For most drainage basins, direct precipitation adds only a modest amount of water to the stream. Interflow is a factor

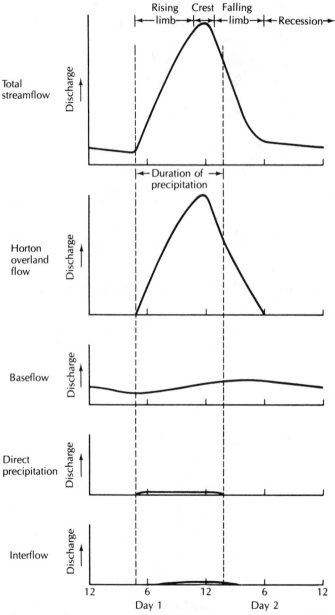

FIGURE 3.6. Hypothetical storm hydrograph for a period of evenly distributed precipitation, separated into Horton overland flow, direct precipitation, and interflow.

which can be highly variable, depending upon the geology of the drainage basin. A deep, sandy soil might not induce any interflow; on the other hand, a lava landscape covered by loose rubble might have no overland flow but great amounts of interflow below the base of the rubble and a hard, low-permeability lava flow. Steeply sloping land also promotes interflow. The most consistent factor in the storm hydrograph is overland flow. Figure 3.6 shows a hypothetical storm hydrograph broken down into overland flow, interflow, direct precipitation, and baseflow recession. The baseflow component is given for a stream that continues to receive groundwater discharge through the duration of the overland-flow peak.

One of the tasks in analyzing a storm hydrograph is to separate the overland-flow component from the baseflow. Generally, it is first assumed that both the direct precipitation and the interflow components are inconsequential; however, the hydrogeologist should be aware of the general geology and surface slope of the drainage area before assuming the inconsequence of the latter component. The overland flow is assumed to end some fixed time after the storm peak. As a general rule of thumb, this can be approximated by the formula (14)

$$D = 1.25\, K^{0.2} \qquad\qquad (3\text{-}2)$$

where

D is the number of days between the storm peak
and the end of overland flow

K is the drainage basin area in square kilometers*

The exponential constant of 0.2 is somewhat arbitrary; thus, blind use of the preceding formula could result in error. The value will depend upon many drainage basin characteristics, such as mean slope, vegetation, drainage density, roughness, etc.

The baseflow recession that existed prior to the storm peak is extended until it is approximately under the storm peak. It is then drawn so as to rise to meet the stream hydrograph at a point D days after the peak. In Figure

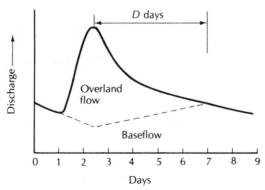

FIGURE 3.7. Hydrograph separation into overland-flow component and baseflow component for a stream receiving Horton overland flow.

*If K is in square miles, the equation is $D = K^{0.2}$

3.7, a storm hydrograph for a drainage basin of 680 square kilometers has been separated. For the given basin, D is equal to 4.6 days.

3.2.3 GAINING AND LOSING STREAMS

The typical stream of a humid region receives groundwater discharge; therefore, as one goes downstream the baseflow increases, even if no tributaries enter. This is a **gaining**, or **effluent**, stream. The water table slopes toward the stream, so that the hydraulic gradient of the aquifer is toward the stream. Figure 3.8B shows a cross section through a gaining reach of a stream.

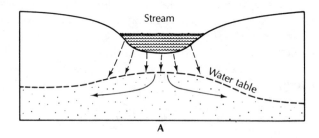

A

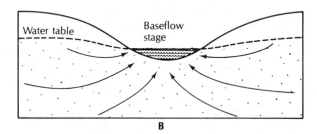

B

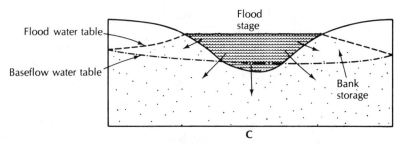

C

FIGURE 3.8. Cross sections of gaining and losing streams. **A.** A losing stream; **B.** A gaining stream; **C.** A stream which is gaining during low flow periods but which may temporarily become a losing stream during flood stage.

In arid regions, many rivers are fed by overland flow, interflow, and baseflow at high altitudes. As they wind their way to lower elevation, the local precipitation amounts decrease; consequently, there is less infiltration and a lower water table. There may also be a dramatic change in the depth to groundwater when a stream draining a high-altitude basin of lower permeability material flows out onto coarse alluvial materials. For whatever reason, if the bottom of the stream channel is higher than the local water table, water may drain from the stream into the ground (Figure 3.8A). As one goes downstream, less and less water will be found in the channel. The stream is **losing**, or **influent**. The rate of water loss is a function of the depth of water and the hydraulic conductivity of the underlying alluvium. Fine-grained deposits on the channel bottom will retard the rate of loss to the groundwater.

A stream that is normally a gaining stream during baseflow recessions may temporarily become a losing stream during floods. If the flood-crest depth in the channel is greater than the local water-table elevation, the hydraulic gradient in the aquifer next to the stream is reversed. Water flows from the stream into the ground (Figure 3.8C). The result is a temporary storage of flood water in the aquifer next to the stream. When the flood crest passes, the hydraulic gradient again reverses, and the stream is once again gaining (Figure 3.9).

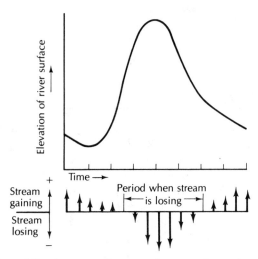

FIGURE 3.9. Effect of flood state on the groundwater regime adjacent to the river. As the flood peak passes, the normal direction of groundwater flow into the stream is reversed.

Heavy groundwater pumping near a stream can lower the water table to an elevation below the level of the stream bottom. The reach of the stream affected by the lowered water table will become a losing stream, while upstream and downstream reaches can still be gaining. Figure 3.10 is an idealization of this phenomenon, based on the behavior of the well field along the Fenton River of the University of Connecticut at Storrs.

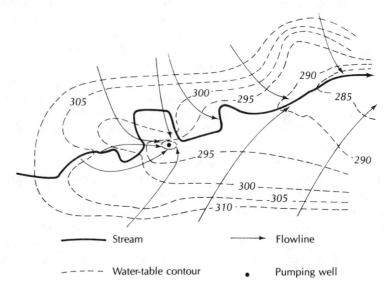

FIGURE 3.10. Induced stream-bed infiltration caused by a pumping well. SOURCE: P. Rahn, *Ground Water*, 6, no. 3 (1968): 21-32.

3.3 STREAM ORDER

Drainage basins may be likened to a complex jigsaw puzzle. For a given basin, there are a number of smaller interlocking subbasins. In turn, the basin may be but a part of a still larger drainage basin. All of these basins are interconnected by a network of rivers and streams. Stream order may be used to classify a stream (15). A small, unbranched tributary is a first-order stream; two first-order streams join to make a second-order one. A third-order stream has only first- and second-order tributaries (Figure 3.11). This classification system is often based on maps. If this is the case, large-scale maps will not show the few smallest orders of streams. These primary streams are typically ephemeral drainage channels. Stream-order classifications based on maps are thus several orders smaller than the same classifications based on field studies.

3.4 RATIONAL EQUATION

A relatively simple method of computing the rainfall-runoff relationship is known as the **rational method**. It is used to predict peak runoff rates from data

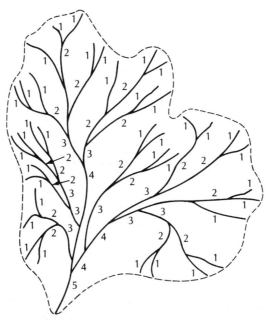

FIGURE 3.11. Stream order based on classification of tributaries. The given drainage basin has a fifth-order stream.

on the rainfall intensity and a knowledge of land use in the drainage basin. The rational method is of greatest validity when used in analysis of small drainage basins of 200 acres (100 hectares) or less. The rational equation is

$$Q = 0.278 \, CIA \qquad\qquad \text{(3-3)}$$

where

Q is the peak runoff rate (cubic meters per second)

I is the average rainfall intensity (millimeters per hour)

A is the drainage area (hectares)

C is the runoff coefficient from Table 3.1

In English units,

$$Q \text{ (cu ft/sec)} = CI \text{ (in./hr)} \, A \text{ (acres)}$$

The rational equation assumes that the rainfall event lasts long enough for the maximum discharge of the drainage basin to occur. In order for such a simple relationship to hold, the rate of infiltration must also be constant during the storm. In Table 3.1, a range of C values is given for many land uses. The lower values are used for storms of low intensity; storms of greater intensity will have proportionally more runoff, justifying the use of higher C factors. Models of the runoff process, which are much more accurate than the rational equation, are available. They are based on continuous simulation using a digital computer (17).

45

TABLE 3.1

Description of Area	C
Business	
Downtown	0.70–0.95
Neighborhood	0.50–0.70
Residential	
Single-family	0.30–0.50
Multiunits, detached	0.40–0.60
Multiunits, attached	0.60–0.75
Residential suburban	0.25–0.40
Apartment	0.50–0.70
Industrial	
Light	0.50–0.80
Heavy	0.60–0.90
Parks, cemeteries	0.10–0.25
Playgrounds	0.20–0.35
Railroad yard	0.20–0.35
Unimproved	0.10–0.30
Character of surface	
Pavement	
Asphalt and concrete	0.70–0.95
Brick	0.70–0.85
Roofs	0.75–0.95
Lawns, sandy soil	
Flat, up to 2% grade	0.05–0.10
Average, 2%–7% grade	0.10–0.15
Steep, over 7%	0.15–0.20
Lawns, heavy soil	
Flat, up to 2% grade	0.13–0.17
Average, 2%–7% grade	0.18–0.22
Steep, over 7%	0.25–0.35

SOURCE: American Society of Civil Engineers, *Manuals and Reports of Engineering Practice No. 37*, 1970.

The time of concentration of a drainage basin is the time it takes for water from the most distant part of the basin to reach the outlet. It can be approximated by the following equation (16):

$$tc = \frac{(0.305\ L)^{1.15}}{7700(0.305\ H)^{0.38}} \tag{3-4}$$

where

        tc is the time of concentration (hours)

        L is the length of the mainstream of the basin (meters)

        H is the difference in elevation from the most distant drainage divide to the outlet (meters)

If *L* and *H* are in feet,

$$tc = \frac{L^{1.15}}{7700\,H^{0.38}}$$

   If the rainfall duration is less than the time of concentration, the rational equation will yield a peak discharge that is too great.

FLOODWAVES 3.5

The hydrograph shows the rise and fall of water at a single gauging station on a river. What happens to that water it if passes downstream as a single-pulse wave? If the velocity of flow is very high and there is little if any storage of flood water in the channel, the wave can pass as a uniform flow with no change in peak height or hydrograph shape (Figure 3.12A). On the other hand, most stream channels tend to widen and have a more pronounced flood plain in the downstream direction. Storage that takes place in the channel and on the flood plain reduces the peak height and slows the progress of the floodwave as it passes a downstream point (Figure 3.12B).

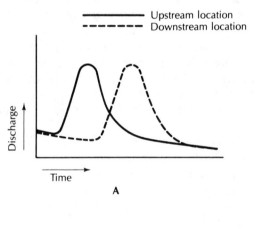

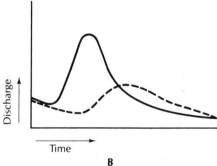

FIGURE 3.12.   Floodwave passing two gauging stations. **A.** On a stream with no available channel storage; **B.** On a stream where flood water is stored in the channel and on the flood plain.

On small catchment areas, the hydrograph may show the effect of variations in rainfall intensity. As the floodwave passes downstream or over large catchment areas, the hydrographs are smoother. This reflects the integrating effects of overland flow and return flow over a large drainage basin or the smoothing effects of channel storage. As progressively larger segments of the drainage basin are considered, the actual flood discharge will increase due to more water flowing. However, smaller areas with higher rainfall intensity will have more runoff per unit area than the drainage basin as a whole.

The movement of a floodwave through a reservoir or down a river can be analyzed to determine the peak discharge and duration of flooding at a downstream location. The process is known as **streamflow routing.** A mathematical treatment of stream routing is beyond the scope of this book, but the topic is covered in many advanced hydrology texts (14, 18). For complex river systems, the use of numerical methods of analysis is necessary (19, 20).

3.6 DURATION CURVES

For design or regulatory purposes, it may be necessary to know how often the discharge of a stream may be less than or greater than a given value. As an example, if a river is considered for a water-supply source, it is necessary to know how much water can be obtained. The average flow is not a particularly useful value, in that possibly 50 percent of the time the river would carry less than the average discharge. Depending upon the available storage and other sources of supply, some flow duration is selected as the reliable flow. For example, the 90 percent duration is the flow that will be equaled or exceeded 90 percent of the time.

Duration curves are generally constructed for either daily flow or annual flow, although other time periods could also be considered. Data are ranked from greatest to least flow values. They are then assigned a serial rank, m, starting with 1 for the greatest flow and going to n, the number of data values. If two or more data values are equal, each should receive a different serial rank. The probability, P, in percentage, that a given flow will be equaled or exceeded may be found by the equation

$$P = 100 \frac{m}{n + 1} \tag{3-5}$$

A plot of P as a function of flow will yield a duration curve showing the percentage of time a given flow is equaled or exceeded. The curve can be plotted on a type of graph paper known as probability paper. This paper is constructed with a special abscissa and ordinates that may be either arithmetic or logarithmic.

Figure 3.13 shows duration curves of daily flow for three rivers in Wisconsin. In order to compare the three directly, the discharge has been computed as cubic feet per second per square mile (cfs/mi²) of drainage basin.

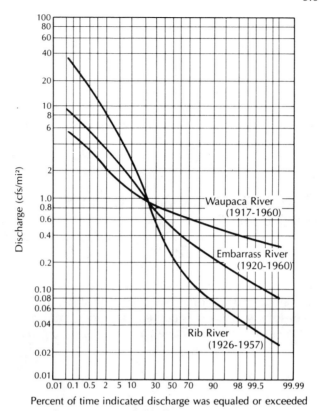

FIGURE 3.13. Daily duration curves for three streams having different runoff characteristics due to the differing geology of the drainage basins. SOURCE: U.S. Geological Survey.

This makes the flow independent of the size of the drainage basin. The three streams are all in central Wisconsin and the annual runoff (precipitation less evaporation) is about 28 centimeters (11 inches) per year for all three. Examination of Figure 3.13 reveals a great variability in the distribution of this annual runoff.

The Rib River has high flood values; 1.0 percent of the time flow equals or exceeds 12 cfs/mi². On the other hand, the 1 percent value for the Waupaca River is 2.7 cfs/mi² with the Embarrass River intermediate. All three rivers have the same 20 percent flow value: 0.9 cfs/mi². Whereas the Rib River had the greatest flood flows, it has the smallest low flows. One percent of the time the Rib River discharge is less than 0.04 cfs/mi². (The graph shows this as 99 percent of the time the flow equals or exceeds 0.04 cfs/mi².) The Waupaca River has low flows an entire magnitude greater; the flow is less than 0.39 cfs/mi² for 1 percent of the time. Again, the Embarrass River falls about evenly between the two.

This distribution of runoff is caused by the geology of the drainage basins. The Rib River is located in an area of crystalline bedrock, which has a very low hydraulic conductivity. Part of the drainage basin is in the driftless

49

area where surficial glacial deposits are lacking, overland flow and return flow are high, and baseflow is scant. The soils are thin, with little water-retaining capacity. The drainage basin of the Waupaca River has thick deposits of unconsolidated sand. Most of the potential overland flow is absorbed by the sand; hence, there are small flood peaks. This water can drain slowly and provide high baseflows. The Embarrass River has thick deposits of glacial drift—but it is till and lake clay—so the hydraulic response of the watershed is intermediate.

3.7 DETERMINING GROUNDWATER RECHARGE FROM BASEFLOW

The baseflow-recession equation (3-1) indicates that Q_0 varies logarithmically with time, t. A plot of a stream hydrograph with time on an arithmetic scale and discharge on a logarithmic scale will therefore yield a straight line for the baseflow recession. Figure 3.14 shows hypothetical stream hydrographs. The baseflow recessions are shown as dashed lines; they were considered to start when the summer stream level dropped below the adjacent water table and to

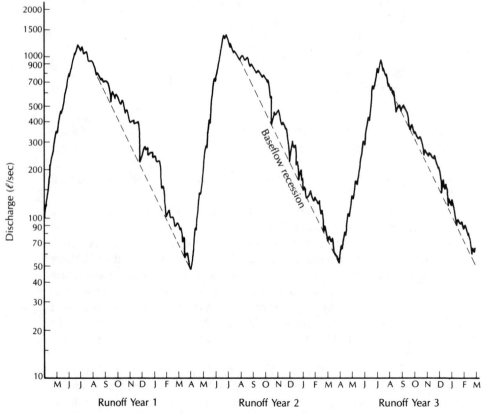

FIGURE 3.14. Semilogarithmic stream hydrographs showing baseflow recessions.

end when the first spring flood occurred. The total potential groundwater discharge, Q_{tp}, is the amount that would be discharged by a complete groundwater recession (22). The value of Q_{tp} can be found from

$$Q_{tp} = \frac{Q_0 t_1}{2.3} \tag{3-6}$$

where

Q_0 is the baseflow at the start of the recession

t_1 is the time it takes for the baseflow to go from Q_0 to $0.1 Q_0$.

If one determines the remaining potential groundwater discharge at the end of a recession and then the total potential groundwater discharge at the beginning of the next recession, the difference between the two is the groundwater recharge that has taken place between recessions. The amount of potential baseflow, Q_t, remaining some time, t, after the start of a baseflow recession is given by

$$Q_t = \frac{Q_{tp}}{10^{(t/t_1)}} \tag{3-7}$$

or

$$Q_t = \frac{(Q_0 t_1)/2.3}{10^{(t/t_1)}} \tag{3-8}$$

EXAMPLE PROBLEM

Refer to Figure 3.14. Determine the amount of groundwater recharge that takes place from the end of the baseflow recession of Runoff Year 1 to the start of the baseflow recession of Runoff Year 2.

The value of Q_0 for the first recession is 760 liters per second and it takes 6.3 months for the discharge to reach $0.1 Q_0$:

$$Q_{tp} = \frac{Q_0 t_1}{2.3}$$

$$Q_{tp} = \frac{760 \ \ell/sec \times 6.3 \ mon \times 30 \ days/mon \times 1440 \ min/day \times 60 \ sec/min}{2.3}$$

$$Q_{tp} = 5.4 \times 10^9 \ \ell$$

The value of Q_t at the end of the recession, which lasts 7.5 months, is

$$Q_t = \frac{Q_{tp}}{10^{(t/t_1)}} = \frac{5.4 \times 10^9}{10^{(7.5/6.3)}} = \frac{5.4 \times 10^9}{15.5} = 3.5 \times 10^8 \ \ell$$

For the next year's recession, the value of Q_0 is 1000 liters per second, and t is again 6.3 months. Therefore,

$$Q_{tp} = \frac{1000 \; \ell/\text{sec} \times 6.3 \times 30 \times 1440 \times 60 \; \text{sec}}{2.3}$$

$$= 7.1 \times 10^9 \; \ell$$

The amount of recharge is equal to the total potential baseflow remaining at the end of the first baseflow recession subtracted from Q_{tp} for the beginning of the next recession:

$$\text{Recharge} = 7.1 \times 10^9 \; \ell - 3.5 \times 10^8 \; \ell$$

$$= 6.75 \times 10^9 \; \ell$$

3.8 MEASUREMENT OF STREAMFLOW

3.8.1 STREAM GAUGING

Water flowing in an open channel is subject to friction as it comes in contact with the channel bottom and sides. As a result, the fastest current is at the surface in the center of the channel. If a series of careful measurements of flow velocity from the surface downward are made, a parabolic profile will emerge (Figure 3.15). Field studies have shown that the velocity at a depth equal to 0.6 times the total depth is very close to the average velocity for the entire section. The average of measurements made at 0.2 times depth and 0.8 times depth is also used to represent the average velocity of the entire profile.

The flow, q, in an open channel with a cross-sectional area, a, and average velocity, v, can be found from the equation

$$q = va \qquad \qquad \text{(3-9)}$$

The velocity of flow can be measured by using a **current meter.** The United States Geological Survey has standard specifications for two types of meters. Each has a horizontal wheel with sets of small, cone-shaped cups attached. The wheel turns in the current and a cam attached to the spindle of the wheel makes an electrical contact once every revolution. The meter wheel is wired to either a set of headphones and a battery, or to a direct readout meter. The operator with a headphone meter counts the clicks over a measured time period, usually thirty to sixty seconds, and uses a calibration curve furnished by the manufacturer to find the velocity.

The **Price-type meter** has a wheel about 13 centimeters (5 inches) in diameter; it is equipped with a vane to orient the meter perpendicular to the flow. The Price-type meter is usually suspended on a cable and lowered into a river, with a streamlined weight to pull it down. A **pygmy-Price meter** is smaller and is usually mounted on a graduated wading rod. The operator takes

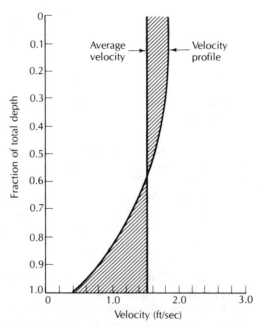

FIGURE 3.15. Typical parabolic velocity profile for a natural stream.

to the river with rubber boots and places the rod on the bottom of a stream in order to make the measurements.

In a typical stream, velocity will vary from bank to bank, necessitating a number of measurements. A straight reach of stream with a smooth shoreline, no brush hanging in the water, and no weeds or large rocks should be chosen. Places with back-eddies should be avoided; they will overestimate the total discharge, as the current meter will not distinguish the direction of flow. If a wading rod is to be used, the water must not be too deep or too swift for a person to wade. This is especially important to check when measuring peak flows.

A tape is stretched perpendicularly across the stream, or along the bridge. The channel is subdivided into fifteen to thirty segments. At the midpoint of each segment, the depth, d_i, is measured and recorded. The meter is then raised to 0.6 times the depth and the average velocity, v_i, for that segment is measured. If the water is deep, the average velocity at 0.8 depth and 0.2 depth should be used. The discharge, q_i, for a segment of width, w_i, is given by

$$q_i = v_i d_i w_i \qquad \text{(3-10)}$$

The process is repeated for each segment of the cross section. The total discharge, Q, for the river is the sum of the discharge for each segment. For a measurement with m segments,

$$Q = \sum_{i=1}^{m} q_i \qquad \text{(3-11)}$$

53

If measurements are made from a bridge using a cable-suspended meter, a swift current may draw the meter downstream. The amount of line let out to measure the depth is thus too great. A correction must be made based on the angle of the cable from the vertical (22).

Current measurements may be made through ice. A series of holes are cut in the ice across the river and the current measured by the preceding method. Because of friction between the ice and the underlying water, the velocity should be measured at 0.2 depth, 0.6 depth, and 0.8 depth and the results averaged.

It is possible to develop an empirical relationship between stream stages (elevation of the water surface above a datum) and discharge. As discharge measurements are slow and costly to make, knowledge of the preceding relationship is useful. A **rating curve** for a stream is made by simultaneously measuring the discharge of a stream and its stage, and then repeating the measurements for a number of different stage heights. Stage versus discharge is then plotted as co-ordinates on graph paper to produce the rating curve (Figure 3.16). If the stream channel does not scour during flooding, and if the stage is not affected by such factors as tributary flow, a simple rating curve is sufficient. Otherwise, a rating curve must also include a factor for water-surface slope (23). Automated stream-gauging stations employ a float device that measures the stage of a river by means of a stilling well connected to the stream. The stage data are transformed into discharge data by using either the rating curve or a rating table based on the curve. In the United States, most stream-stage measurements are recorded in digital form for automatic data processing.

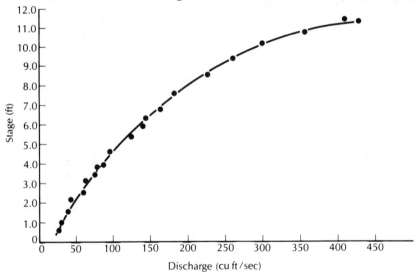

FIGURE 3.16. Typical stage-discharge rating curve.

3.8.2 WEIRS

The discharge of small streams can be conveniently measured by use of a **weir.** This is a small dam with a spillway opening of specified shape. A simple weir can be constructed from a sheet of ¾-inch-thick plywood. A spillway is cut out and rimmed with a metal strip that extends out into the flow (Figure 3.17). There are a number of standard shapes for sharp-crested weirs, the most common being a 90-degree V-notch or a rectangular cut-out. A small earthen or concrete dam is built and the weir set into it. The dam will impound a small amount of water that should free fall over the weir crest, or lowest point of the spillway. The elevation of the backwater above the weir crest, H, is measured. The discharge over the weir can be found from the following formulas:

Rectangular weir:
$$Q = \tfrac{1}{3}(L - 0.2H) H^{3/2} \qquad \text{(3-12)}$$

Ninety-degree V-notch weir:
$$Q = 2.5H^{5/2} \qquad \text{(3-13)}$$

where

Q is the discharge (cubic feet per second)
L is the length of the weir crest (feet)
H is the head of the backwater above the weir crest (feet)

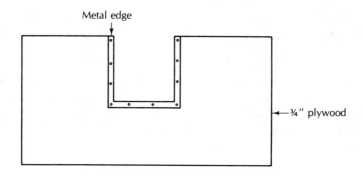

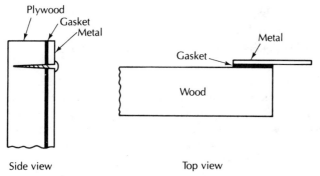

FIGURE 3.17. Construction details for a sharp-crested weir made of plywood and metal strips.

3.9 MANNING EQUATION

In open-channel hydraulics, the average velocity of flow of water may be found from the **Manning equation:**

$$V = \frac{1.49R^{2/3}S^{1/2}}{n} \qquad (3\text{-}14)$$

where

V is the average velocity in feet per second

R is the hydraulic radius, or the ratio of the cross-sectional area of flow to the wetted perimeter (see Figure 3.18)

S is the energy gradient, which is the slope of the water surface

n is the Manning roughness coefficient

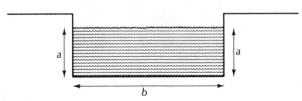

FIGURE 3.18. The cross-sectional area of flow is $a \times b$. The wetted perimeter is $a + b + a$.

The velocity of flow is dependent upon the amount of friction between the water and the stream channel. Smoother channels will have less friction and, hence, faster flow. Channel roughness contributes to turbulence, which dissipates energy and reduces flow velocity. The following values for n are typical:

mountain streams with rocky beds:	0.04–0.05
winding natural streams with weeds:	0.035
natural streams with little vegetation:	0.025
straight, unlined earth canals:	0.020
smoothed concrete:	0.012

The U.S. Geological Survey has published a series of photographs of rivers for which the value of the Manning roughness coefficient has been computed (24). Field measurements of velocity, slope, area and wetted perimeter were made and the value of n computed from Equation (3-14). Careful study of the photographs can be used to obtain an estimate for the value of n for a given river under study.

The Manning equation can be used to determine flow in situations that preclude direct measurements. For example, if the current is changing rapidly (during a rising or falling flood peak, for example), conventional streamflow measurements would take too long. It might take the better part of an hour to make a discharge measurement, and flow velocity and discharge could change substantially during the period of measurement. Under these conditions, an instant computation can be made using river cross sections and a measured slope.

References

1. BROWN, J. H. and A. C. BARKER. "An Analysis of Throughflow and Stemflow in Mixed Oak Stands." *Water Resources Research*, 6 (1970):316–23.

2. ROGERSON, T. L. and W. R. BYRNES. "Net Rainfall under Hardwoods and Red Pine in Central Pennsylvania." *Water Resources Research*, 4 (1968):55–58.

3. HELVEY, J. D. "Interception by Eastern White Pine." *Water Resources Research*, 3 (1967):723–30.

4. DUNNE, T. and L. B. LEOPOLD. *Water in Environmental Planning*. San Francisco: W. H. Freeman and Company, 1978, p. 88.

5. TRIMBLE, G. R., JR. and S. WEITZMAN. "Effect of a Hardwood Forest Canopy on Rainfall Intensities." *Transactions, American Geophysical Union*, 35 (1954):226–34.

6. THORUD, D. B. "The Effect of Applied Interception on Transpiration Rates of Potted Ponderosa Pine." *Water Resources Research*, 3 (1967):443–50.

7. HORTON, R. E. "The Role of Infiltration in the Hydrologic Cycle." *Transactions, American Geophysical Union*, 14 (1933):446–60.

8. HORTON, R. E. "An Approach toward a Physical Interpretation of Infiltration Capacity." *Soil Science Society of America, Proceedings*, 4 (1940):399–417.

9. International Great Lakes Levels Board. "Regulation of Great Lakes Water Levels," Appendix A, *Hydrology and Hydraulics*. Report to the International Joint Commission, December 7, 1973.

10. CHORLEY, R. J. "The Hillslope Hydrological Cycle." In *Hillslope Hydrology*, ed. M. J. Kirkby. Chichester, Sussex, England: John Wiley & Sons, 1978, pp. 1–42.

11. KIRKBY, M. J. and R. J. CHORLEY. "Throughflow, Overland Flow, and Erosion." *Bulletin, International Association Scientific Hydrology*, 12 (1967):5–21.

12. DUNNE, T. "Field Studies of Hillslope Flow Processes." In *Hillslope Hydrology*, ed. M. J. Kirkby. Chichester, Sussex, England: John Wiley & Sons, 1978, pp. 227–94.

13. DUNNE, T. and R. D. BLACK. "An Experimental Investigation of Runoff Production in Permeable Soils." *Water Resources Research,* 6 (1970):478–90.

14. LINSLEY, R. K., JR., M. A. KOHLER, and J. L. H. PAULHUS. *Hydrology for Engineers.* New York: McGraw-Hill Book Company, 1975, 230 pp.

15. HORTON, R. E. "Erosional Development of Streams." *Bulletin, Geological Society of America,* 56 (1945):281–83.

16. U.S. Soil Conservation Service. *National Engineering Handbook.* Washington, D.C.: U.S. Government Printing Office, 1972.

17. LINSLEY, R. K., JR. and N. CRAWFORD. "Continuous Simulation Models in Urban Hydrology." *Geophysical Research Letters,* 1, no. 1 (1974):59–62.

18. HJELMFELT, A. T., JR. and J. J. CASSIDY. *Hydrology for Engineers and Planners.* Ames, Iowa: Iowa State University Press, 1975, pp. 120–36.

19. FREAD, D. L. "Technique for Implicit Dynamic Routing in Rivers with Tributaries." *Water Resources Research,* 9 (1973):918–26.

20. BALTZER, R. A. and C. LAI. "Computer Simulation of Unsteady Flows in Waterways." *Journal of the Hydraulic Division, American Society of Civil Engineers,* 94 (1968):1083–117.

21. MEYBOOM, P. "Estimating Groundwater Recharge from Stream Hydrographs." *Journal of Geophysical Research,* 66 (1961):1203–14.

22. CORBETT, D. M. et al. *Stream-Gauging Procedure.* U.S. Geological Survey Water Supply Paper 888, 1945, pp. 43–51.

23. MITCHELL, W. D. *Stage-Fall-Discharge Relations for Steady Flow in Prismatic Channels.* U.S. Geological Survey Water Supply Paper 1164, 1954, 112 pp.

24. BARNS, H. H. *Roughness Characteristics of Natural Channels.* U.S. Geological Survey Water Supply Paper 1849, 1967, 213 pp.

Soil Moisture and Groundwater

chapter

4.1 POROSITY OF EARTH MATERIALS

At the time they are formed, some rocks contain void spaces while others are solid. Those rocks occurring near the surface of the earth are not totally solid. The physical and chemical weathering processes there continually decompose and disaggregate rock, thus creating voids. Slight movements of rock masses near the surface can cause rocks to crack or fracture. This also results in openings between rocks.

Sediments are assemblages of individual grains that were deposited by water, wind, ice, or gravity. There are openings called **pore spaces** between the sediment grains, so that sediments are not solid.

The cracks, voids, and pore spaces in earth materials are of great importance to hydrogeology. Groundwater and soil moisture occur in the voids in otherwise solid earth materials.

4.1.1 DEFINITION OF POROSITY

The **porosity** of earth materials is the percentage of the rock or soil that is void of material. It is defined mathematically by the equation

$$n = \frac{100V_v}{V} \tag{4-1}$$

where

n is the porosity (percentage)

V_v is the volume of void space in a unit volume of earth material

V is the unit volume of earth material, including both voids and solids

Laboratory porosity is determined by taking a sample of known volume (V). The sample is dried in an oven at 105° C until it reaches a constant weight. This expels moisture clinging to surfaces in the sample, but not water that is hydrated as a part of certain minerals. The dried sample is then submerged in a known volume of water and allowed to remain in a sealed

60

chamber until it is saturated. The volume of the voids (V_v) is equal to the original water volume less the volume in the chamber after the saturated sample is removed. This method excludes pores not large enough to contain water molecules and those which are not interconnected.

Under field conditions, some of the water in the pore spaces is tightly held to the surfaces of the mineral grains by surface tension. This water will not move through the sample when pulled by gravitational forces. The **effective porosity** is the ratio of the void space through which flow can occur to the total volume. It can be found by taking a saturated sample, prepared as previously described, and allowing it to drain. The sample is weighed before and after saturation. Gravity drainage is a slow process: periods up to a year are necessary for complete drainage. During the drainage process, the sample must be kept in an atmosphere with 100 percent relative humidity to prevent evaporation. Following drainage, the sample is again weighed. The effective porosity is found from the equation

$$n_e = 100\frac{W_s - W_r}{W_s - W_o} \times \frac{V_v}{V}$$ (4-2)

where

n_e is the effective porosity (percentage)

W_o is the weight of the air-dried sample

W_s is the weight of the saturated sample

W_r is the weight of the sample after gravity drainage

4.1.2 POROSITY OF SEDIMENTS

The porosity of sediments pertains to the void spaces between solid fragments. If the fragments are solid spheres of equal diameters, they can be put together in such a manner that each sphere sits directly on the crest of the underlying sphere (Figure 4.1). This is called **cubic packing**, with an associated porosity of 47.65 percent (1). If the spheres lie in the hollows formed by four adjacent spheres of the underlying layer, the result is **rhombohedral packing**, with a porosity of 25.95 percent (1).

These two configurations represent the extremes of porosity for arrangements of equidimensional spheres with each sphere touching all neighboring spheres. The diameter of the sphere does not influence the porosity. Thus, a room full of bowling balls in cubic packing would have the same porosity as a room full of 1-millimeter ball bearings. The volume of an individual pore would be much larger for the bowling balls. The porosity of well-rounded sediments, which have been sorted so that they are all about the same size, is independent of the particle size, and falls in the range of about 26 to 48 percent, depending upon the packing.

If a sediment contains a mixture of grain sizes, the porosity will be lowered. The smaller particles can fill the void spaces between the larger

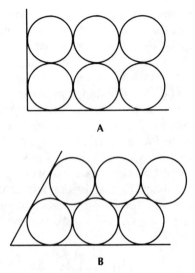

FIGURE 4.1. **A.** Cubic packing of spheres with a porosity of 47.65 percent; **B.** Rhombohedral packing of spheres with a porosity of 25.95 percent.

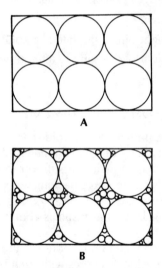

FIGURE 4.2. **A.** Cubic packing of spheres of equal diameter with a porosity of 47.65 percent; **B.** Cubic packing of spheres with void spaces occupied by grains of smaller diameter, resulting in a much lower overall porosity.

ones. The wider the range of grain sizes, the lower the resulting porosity (Figure 4.2). Geologic agents can sort sediments into layers of similar sizes. Wind, running water, and wave action tend to create well-sorted sediments. Other processes, such as glacial action and landslides, result in sediments with a wide range of grain sizes. These poorly sorted sediments have low porosities.

In addition to grain-size sorting, the porosity of sediments is affected by the shape of the grains. Well-rounded grains may be almost perfect spheres, but many grains are very irregular. They can be shaped like rods, disks, or books. Sphere-shaped grains will pack more tightly and have less porosity than particles of other shapes. The fabric or orientation of the particles, if they are not spheres, also influences porosity.

Sediments are classified on the basis of the size (diameter) of the individual grains. Clay particles are 2 microns (2×10^{-3} millimeter) or less in diameter. Silt falls in the range of 2 to 62 microns. Sand is no greater than 2 millimeters and no less than 62 microns. Pebbles or gravel are from 2 to 64 millimeters in diameter, while cobbles are in the range of 64 to 256 millimeters (Figure 4.3).

Clays and some clay-rich or organic soils can have very high porosities. Organic materials do not pack very closely because of their irregular

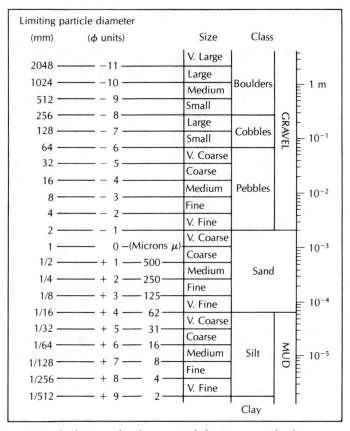

FIGURE 4.3. Standard sizes of sediments with limiting particle diameters and the ϕ scale of sediment size, in which ϕ is equal to $\log_2 s$ (the particle diameter). SOURCE: G. M. Friedman and J. E. Sanders, *Principles of Sedimentology* (New York: John Wiley & Sons, 1978). Used with permission.

shapes. The dispersive effect of the electrostatic charge present on the surfaces of certain book-shaped clay minerals causes clay particles to be repelled by each other. The result is a relatively large proportion of void space.

The general range of porosity that can be expected for some typical sediments is listed in Table 4.1.

TABLE 4.1. Porosity ranges for sediments (1, 2, 8, 9)

Well sorted sand or gravel	25–50%
Sand and gravel, mixed	20–35%
Glacial till	10–20%
Silt	35–50%
Clay	33–60%

4.1.3 POROSITY OF SEDIMENTARY ROCKS

Sedimentary rocks are formed from sediments through a process known as **diagenesis**. A sediment, which may be either a product of weathering or a chemically precipitated material, is buried. The weight of overlying materials and physiochemical reactions with fluids in the pore spaces induce changes in the sediment. This includes compaction, removal of material, addition of material, and transformation of minerals by replacement or change in mineral phase. Compaction reduces pore volume by rearranging the grains and reshaping them. The deposition of cementing materials such as calcite, dolomite, or silica will reduce porosity, although the dissolution of material that is dissolved by the pore fluid will increase porosity. The primary structures of the sediment may be preserved in the sedimentary rock. The porosity of a sandstone, for instance, will be influenced by the grain size, size sorting, grain shape, and fabric of the original sediment. Diagenesis is a complex process, but in general the primary porosity of a sedimentary rock will be less than that of the original sediment. This is especially true of fine-grained sediments (silts and clays) (Figure 4.4).

Rocks at the earth's surface are usually fractured to some degree. The fracturing may be mild, resulting in widely spaced joints. At the other extreme, violent fracturing may completely shatter the rock, resulting in fault breccias. Fractures create secondary porosity in the rock. Groundwater can be found in fractured sedimentary rocks in the pores between grains (**primary porosity**) as well as in fractures (**secondary porosity**). Groundwater flowing through fractures may enlarge them by solution of material. Bedding planes in the sedimentary rocks may have primary porosity formed during deposition of the sediments and secondary porosity if the rock has moved along a bedding plane.

Some cohesive sediments (those rich in silt and/or clay) are also subject to fracturing. In some cases, this is merely from shrinkage cracks that develop when the sediment dries. However, slumping, loading, or tectonic activity can also cause fracturing in nonplastic cohesive sediments. This fracturing can be a significant source of secondary porosity in such deposits.

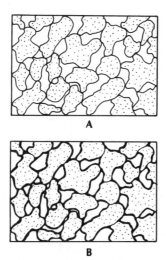

FIGURE 4.4. **A**. A clastic sediment with porosity between the grains; **B**. Reduction in porosity of the clastic sediment due to deposition of cementing materials in the pore spaces.

Limestones and dolomites are well-known and widespread examples of sedimentary rocks of a chemical or biochemical origin. They are formed of calcium carbonate and calcium-magnesium carbonate, respectively. Gypsum, a calcium sulfate, and halite or rock salt (sodium chloride) are also widely distributed common examples of chemical precipitates.

The materials that formed these rocks were originally part of an aqueous solution. Inasmuch as the precipitation process is reversible, the rock can be redissolved. When these rock types are in a zone of circulating groundwater, the rock may be removed by solution. Groundwater moves initially through pore spaces, as well as along fractures, joints, and bedding planes. As more water moves through the bedding planes, they are preferentially dissolved and enlarged, causing the rock to become very porous. Some limestone formations have openings large enough to permit thousands of tourists a day to pass through. The caverns at Carlsbad, New Mexico, and Ljubljana, Yugoslavia, exemplify such massive porosity. Gypsum and salt may also be cavernous (1).

The percent porosity of sedimentary rocks is highly variable. In clastic rocks, it can range from 3 to 30 percent (2, 5, 6, 7). Reported values for limestones and dolomites range from less than 1 to 30 percent (2, 3, 4, 7).

4.1.4 POROSITY OF PLUTONIC AND METAMORPHIC ROCKS

Plutonic rocks (those formed by intrusive igneous processes) and **metamorphic rocks** typically have a very low porosity (2). These rocks are formed of interlocking crystals; hence, there is virtually no void space in the inchoate rock. The porosity of newly crystallized igneous rocks at depth in the earth ap-

65

proaches zero. Rock resulting from high-grade metamorphism also exhibits interlocking crystalline structure, with a resultant low original porosity.

Two geologic processes, weathering and fracturing, increase overall rock porosity. Rock at depth is under pressure due to the weight of overlying materials. This rock may be fractured by expansion as the overlying weight is removed by erosion. Tectonic stresses in the earth can cause folding and faulting. Rock in a fault shear zone may be extensively fractured. Expansion cracks can form at the crest of a fold. Joint sets in crystalline rock are usually found in three mutually perpendicular directions (10). Fracturing increases porosity of crystalline rocks by about 2 to 5 percent (2, 11). Weathering due to chemical decomposition and physical disintegration operates with greater efficacy with increasing rock porosity. Weathered igneous and metamorphic rocks can have porosities in the range of 30 to 60 percent (12). Due to the sheetlike structure of some weathering minerals, such as the micas, porosities can exceed that of loosely packed spheres.

Porosity due to fracturing is concentrated in the rock along the sets of joints, and is a function of the width of the openings in the joints. Weathered rock has the pore spaces distributed throughout the rock, although weathering may be more intense along joint or weathering planes.

4.1.5 POROSITY OF VOLCANIC ROCKS

Volcanic rocks (those formed by extrusive igneous activity) are similar in chemical composition to plutonic rocks, as both are formed by the cooling of molten rock (magma). However, extrusive rocks are formed in a surficial environment, which results in radically different porosity characteristics. Volcanic rocks include sills, which are injected between layers of rock; dikes, which are injected in rock, but cut across any bedding planes; lava flows, which are at the surface; and unconsolidated deposits of ash and cinders thrown from the volcano. Sills and dikes can cool slowly; lava typically cools more quickly.

Lava cooling rapidly at the surface will trap degassing products, resulting in holes in the rock (vesicular texture). The holes create porosity, although they may not be interconnected. Shrinkage cracks that develop in the lava as it cools create joints. Flowing lava can form a crust, which then breaks apart to form a rubbly structure. The broken surface of buried lava flows, the remains of natural lava tubes and tunnels through which molten lava once poured, and stream gravels trapped between lava flows all produce a high porosity in some extrusive rocks. Porosity of basalt, a crystalline extrusive rock that is formed from magma with a low gas content, generally ranges from 1 to 12 percent (13). Pumice, a glassy rock that is formed from a magma with a very high gas content, can have a porosity of as high as 87 percent (2), although the vesicles are not well interconnected.

Pyroclastic deposits are formed by volcanic material thrown into the air when molten. They can have high porosities. Values of porosity of tuff ranging from 14 to 40 percent have been reported (14). Recent volcanic ash may have a porosity of 50 percent. Weathering of volcanic deposits can increase the porosity to values in excess of 60 percent (2).

SPECIFIC YIELD 4.2

Specific yield (S_y) is the ratio of the volume of water that drains from a saturated rock due to the attraction of gravity to the total volume of the rock (15) (Figure 4.5).

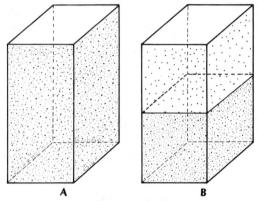

A   **B**

FIGURE 4.5. **A.** A volume of rock saturated with water; **B.** After gravity drainage, 1 unit volume of the rock has been dewatered with a corresponding lowering of the level of saturation. Specific yield is the ratio of the volume of water that drained from the rock due to gravity to the total rock volume.

Water molecules cling to surfaces due to the surface tension of the water (Figure 4.6). If gravity exerts a stress on a film of water surrounding a mineral grain, some of the film will pull away and drip downward. The remain-

FIGURE 4.6. Hygroscopic water clinging to spheres due to surface tension. Gravity attraction is pulling the water downward.

67

ing film will be thinner with a greater surface tension so that, eventually, the stress of gravity will be exactly balanced by the surface tension. **Hygroscopic water** is the moisture clinging to the soil particles due to surface tension. At that moisture content, gravity drainage will cease. The specific yield is approximately equal to the effective porosity.

If two samples are equivalent with regard to porosity, but the average grain size of one is much smaller than the other, the surface area of the finer sample will be larger. As a result, more water can be held as hygroscopic moisture by the finer grains.

The **specific retention** of a rock or soil is the ratio of the volume of water a rock can retain against gravity drainage to the total volume of the rock (15). Since the specific yield represents the volume of water that a rock will yield by gravity drainage, with specific retention the remainder, the sum of the two is obviously equal to porosity. The specific retention increases with decreasing grain size, so that a clay may have a porosity of 50 percent with a specific retention of 48 percent.

Table 4.2 lists the specific yield, in percent, for a number of sediment textures. The data for this table were compiled from a large number of samples in various geographic locations. Maximum specific yield occurs in sediments in the medium-to-coarse sand size range (0.5 to 1.0 millimeter). This is shown graphically in Figure 4.7, which plots specific yield as a function of grain size for several hundred samples from the Humboldt River Valley of Nevada.

TABLE 4.2. Specific yields in percent (16)

| Material | Specific Yield | | |
	Maximum	Minimum	Average
Clay	5	0	2
Sandy clay	12	3	7
Silt	19	3	18
Fine sand	28	10	21
Medium sand	32	15	26
Coarse sand	35	20	27
Gravelly sand	35	20	25
Fine gravel	35	21	25
Medium gravel	26	13	23
Coarse gravel	26	12	22

Both soil formed by weathering processes at the surface and sediments that are depositional generally contain a mixture of clay, silt, and sand. Figure 4.8 is a soil classification triangle showing lines of equal specific yield (16). It is apparent that the specific yield increases rapidly as the percentage of sand increases, and as the percentages of silt, and especially clay, decrease.

Specific yield may be determined by laboratory methods. A sample of sediment of known volume is fully saturated. This is usually done in a soil column that is flooded slowly from the bottom, allowing air to escape upward.

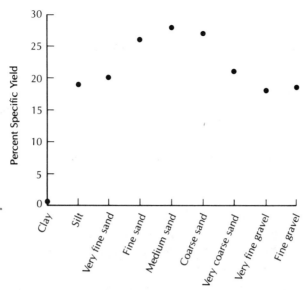

FIGURE 4.7. Specific yield of sediments from the Humboldt River Valley of Nevada as a function of the median grain size. SOURCE: Data from P. Cohen, U.S. Geological Survey Water Supply Paper 1975, 1965.

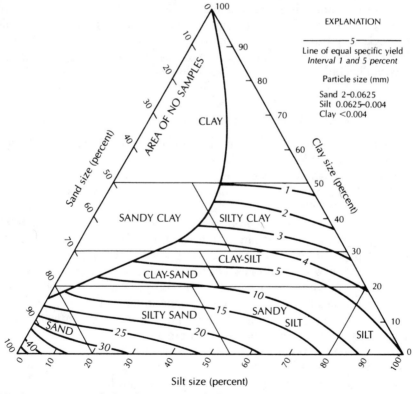

FIGURE 4.8. Textural classification triangle for unconsolidated materials showing the relation between particle size and specific yield. SOURCE: A. I. Johnson, U.S. Geological Survey Water Supply Paper 1662-D, 1967.

Water is then allowed to drain from the column (17). Care must be taken to avoid evaporation losses; even for sand-sized grains, columns must be allowed to drain for very long time periods (months) before equilibrium is reached (18). The ratio of the volume of water drained to the volume of the soil column is the specific yield (multiplied by 100 to express the value as a percentage).

The specific yield of sediment and rock can also be determined in the field. Water wells are pumped, and the rate at which the water level falls in nearby wells is measured (19, 20, 21). Chapter 8 includes a discussion of such pumping-test methods.

4.3 HYDRAULIC CONDUCTIVITY OF EARTH MATERIALS

We have seen that earth materials near the surface generally contain some void space and thus exhibit porosity. Moreover, in most cases, these voids are interconnected to some degree. Water contained in the voids is capable of moving from one void to another, thus circulating through the soil, sediment, and rock. It is the ability of a rock to transmit water which, together with its ability to hold water, constitute the most significant hydrogeologic properties. There are some rocks that exhibit porosity but lack interconnected voids, e.g., vesicular basalt. These rocks cannot convey water from one void to another. Some sediments and rocks have porosity, but the pores are so small that water flows through the rock with difficulty. Clay and shale are examples.

4.3.1 DARCY'S EXPERIMENT

In the mid-nineteenth century, a French engineer, Henry Darcy, made the first systematic study of the movement of water through a porous medium (22). He studied the movement of water through beds of sand used for water filtration. Darcy found that the rate of water flow through a bed of a "given nature" is proportional to the difference in the height of the water between the two ends of the filter beds and inversely proportional to the length of the flow path. He also determined that the quantity of flow is proportional to a coefficient, K, which is dependent upon the nature of the porous medium.

Figure 4.9 illustrates a horizontal pipe filled with sand. Water is applied under pressure through end A. The pressure can be measured and observed by means of a thin vertical pipe open in the sand at point A. Water flows through the pipe and discharges at point B. Another vertical pipe or piezometer is present to measure the pressure at B.

Darcy found experimentally that the discharge, Q, is proportional to the difference in the height of the water, h (hydraulic head), between the ends and inversely proportional to the flow length, L:

$$Q \propto h_A - h_B \qquad Q \propto 1/L$$

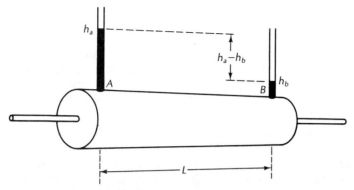

FIGURE 4.9.   Horizontal pipe filled with sand to demonstrate Darcy's experiment. (Darcy's original equipment was actually vertically oriented.)

The flow is also obviously proportional to the cross-sectional area of the pipe, A. When combined with the proportionality constant, K, the result is the expression known as **Darcy's law:**

$$Q = KA\left(\frac{h_A - h_B}{L}\right) \tag{4-3}$$

This may be expressed in more general terms as

$$Q = -KA\left(\frac{dh}{dl}\right) \tag{4-4}$$

where dh/dl is known as the **hydraulic gradient.** The quantity dh represents the change in head between two points that are very close together, and dl is the small distance between these two points. The negative sign indicates that flow is in the direction of decreasing hydraulic head. The use of the negative sign necessitates careful determination of the sign of the gradient. If the value of h_2 at point X_2 is greater than h_1 at point X_1, then flow is from point X_2 to X_1. If $h_1 > h_2$, then flow is from X_1 to X_2.

4.3.2   HYDRAULIC CONDUCTIVITY

Equation (4-4) can be rearranged to show that the coefficient K has the dimensions of length/time (L/T), or velocity. This coefficient has been termed the **hydraulic conductivity.** In older works, it may be referred to as the coefficient of permeability:

$$K = \frac{Q}{A(dh/dL)} \tag{4-5}$$

Discharge has the dimensions volume/time (L^3/T), area (L^2), and gradient (L/L). Substituting these dimensions into Equation (4-5), the dimensions of K are determined:

$$K = \frac{(L^3/T)}{(L^2)(L/L)} = (L/T)$$

Darcy did not address the fact that the value of K is a function of properties of both the porous medium and the fluid passing through it. It is intuitively obvious that a viscous fluid (one which is thick), such as crude oil, will move at a slower rate than water, which is thinner and has a lower viscosity. The hydraulic conductivity is directly proportional to the **specific weight**, γ, of the fluid. The specific weight is the force exerted by gravity on a unit volume of the fluid. This represents the driving force of the fluid. Hydraulic conductivity is also inversely proportional to the **dynamic viscosity** of the fluid, μ, which is a measure of the resistance of the fluid to the shearing that is necessary for fluid flow. A proportionality expression for K can be written as

$$K = K_i \left(\frac{\gamma}{\mu}\right) = K_i \left(\frac{\rho g}{\mu}\right) \tag{4-6}$$

where g is the acceleration of gravity and ρ is the density.

The new constant, K_i, is representative of the properties of the porous medium alone. It is termed the **intrinsic permeability**. This is basically a function of the size of the openings through which the fluid moves. The larger the square of the mean pore diameter, d, the lower the flow resistance. The cross-sectional area of a pore is also a function of the shape of the opening. A constant can be used to describe the overall effect of the shape of the pore spaces. Using this dimensionless constant, called the **shape factor**, C, the intrinsic permeability is given by the expression

$$K_i = Cd^2 \tag{4-7}$$

The dimensions of K_i are (L^2), or area.

Units for K_i can be in square feet or square centimeters. In the petroleum industry, the **darcy** is used as a unit of intrinsic permeability. (The petroleum engineer is similarly concerned with the occurrence and movement of fluids through porous media.) The darcy is defined as

$$1 \text{ darcy} = \frac{\dfrac{1 \text{ centipoise} \times 1 \text{ cm}^3/\text{sec}}{1 \text{ cm}^2}}{1 \text{ atmosphere}/1 \text{ cm}}$$

This expression can be converted to square centimeters, since

$$1 \text{ centipoise} = 0.01 \text{ dyne-sec/cm}^2$$

and

$$1 \text{ atmosphere} = 1.0132 \times 10^6 \text{ dynes/cm}^2$$

Substituting into the definition of the darcy, it may be seen that

$$1 \text{ darcy} = 9.87 \times 10^{-9} \text{ cm}^2$$

Both the viscosity and the density of a fluid are functions of its temperature. The colder the fluid, the more viscous it is (Figure 4.10). There is also a more complex relationship between temperature and density, as the density of water decreases with temperature to 4° C, at which temperature it is at a maximum. The hydraulic conductivity of a rock or sediment will vary with the temperature of the water. As solutions become saline, this may also affect the values of specific gravity and viscosity, which will also cause the hydraulic conductivity to vary.

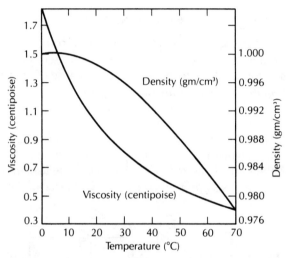

FIGURE 4.10. Variation of viscosity and density of pure water with temperature.

The laboratory or standard value of hydraulic conductivity is defined for pure water at a temperature of 15.6° C. The most logical units are those of distance/time, such as centimeters per second or feet per day. In the United States, a derived unit of gallons per day per square foot is often used. This is defined as the flow in gallons per day through a cross-sectional area of 1

TABLE 4.3. Conversion values for hydraulic conductivity

1 gal/day/ft^2	= 0.0408 m/day
1 gal/day/ft^2	= 0.134 ft/day
1 gal/day/ft^2	= 4.72 × 10^{-5} cm/sec
1 ft/day	= 0.305 m/day
1 ft/day	= 7.48 gal/day/ft^2
1 ft/day	= 3.53 × 10^{-4} cm/sec
1 cm/sec	= 864 m/day
1 cm/sec	= 2835 ft/day
1 cm/sec	= 21,200 gal/day/ft^2
1 m/day	= 24.5 gal/day/ft^2
1 m/day	= 3.28 ft/day
1 m/day	= 0.00116 cm/sec

73

square foot under a gradient of 1 at 60° F. The unit is named the **meinzer** after O. E. Meinzer, a pioneering groundwater geologist with the U.S. Geological Survey. However, the name meinzer is rarely used in practice. For pure water at 15.6° C, a porous medium with an intrinsic permeability of 1 darcy would have a hydraulic conductivity of 18.2 meinzers or 8.61×10^{-4} centimeter per second. In the SI system of units, conductivity is in meters per day. Units of centimeters per second are widely used in soil mechanics.

4.3.3 PERMEABILITY OF SEDIMENTS

Unconsolidated coarse-grained sediments represent some of the most prolific producers of groundwater. Likewise, clays are often used for engineering purposes, such as lining solid waste disposal sites, because of their extremely low intrinsic permeability. There is obviously a wide-ranging continuum of permeability values for unconsolidated sediments.

The intrinsic permeability is a function of the size of the pore opening. The smaller the size of the sediment grains, the larger the surface area the water contacts (Figure 4.11). This increases the frictional resistance to flow, which reduces the intrinsic permeability. For well-sorted sediments, the intrinsic permeability is inversely proportional to the grain size of the sediment (23).

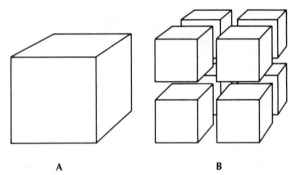

A   B

FIGURE 4.11. Relationship of sediment grain size to surface area of pore spaces. **A.** A cube of sediment with a surface area of six square units; **B.** The cube has been broken into eight pieces, each with a diameter of one-half of the cube in Part A. The surface area has increased to twelve square units—an increase of 100 percent.

For sand-sized alluvial deposits, several factors relating intrinsic permeability to grain size have been noted (24). These observations would hold true for all sedimentary deposits, regardless of origin of deposition.

1. As the median grain size increases, so does permeability. This is due to larger pore openings.

2. Permeability will decrease for a given median diameter as the standard deviation of particle size increases. The increase in standard deviation indicates a more poorly sorted

sample, so that the finer material can fill the voids between larger fragments.

3. Coarser samples show a greater decrease in permeability with an increase in standard deviation than fine samples.

4. Unimodal (one dominant size) samples have a greater permeability than bimodal (two dominant sizes) samples. This is again a result of poorer sorting of the sediment sizes, as the bimodal distribution indicates.

TABLE 4.4.   Ranges of intrinsic permeabilities and conductivities for unconsolidated sediments

Material	Intrinsic Permeability (darcys)	Conductivity (cm/sec)
Clay	$10^{-6} - 10^{-3}$	$10^{-9} - 10^{-6}$
Silt, sandy silts, clayey sands, till	$10^{-3} - 10^{-1}$	$10^{-6} - 10^{-4}$
Silty sands, fine sands	$10^{-2} - 1$	$10^{-5} - 10^{-3}$
Well-sorted sands, glacial outwash	$1 - 10^{2}$	$10^{-3} - 10^{-1}$
Well-sorted gravel	$10 - 10^{3}$	$10^{-2} - 1$

4.3.4   PERMEABILITY OF ROCKS

The intrinsic permeability of rocks is due to primary openings formed with the rock and secondary openings created after the rock was formed. The size of openings, the degree of interconnection, and the amount of open space are all significant.

Clastic sedimentary rocks have primary permeability characteristics similar to unconsolidated sediments. However, diagenesis can reduce the size of the throats which connect adjacent pores through cementation and compaction. This could reduce permeability substantially without a large impact on primary porosity. Primary permeability may also be due to sedimentary structures, such as bedding planes.

Crystalline rocks, whether of igneous, metamorphic, or chemical origin, typically have a low primary permeability, in addition to a low porosity. The intergrown crystal structure contains very few openings, so fluids cannot pass through as readily. The exceptions to this are volcanic rocks, which can have a high primary porosity. If the openings are large and well connected, then high permeability may also be present.

Secondary permeability can develop in rocks through fracturing. The increase in permeability is initially due to the number and size of the fracture openings. As water moves through the fractures, minerals may be dissolved from the rock and the fracture enlarged. This increases the permeability. Chemically precipitated rocks (limestone, dolomite, gypsum, halite) are most susceptible to solution enlargement, although even igneous rocks may be so affected. Bedding-plane openings of sedimentary rocks may also be enlarged by solution.

Weathering is another process which can result in an increase in permeability. As the rock is decomposed or disintegrated, the number and size of pore spaces, cracks, and joints can increase.

EXAMPLE PROBLEM

The intrinsic permeability of a consolidated rock is 2.7×10^{-3} darcy. What is the conductivity for water at $15°$ C?

At $15°$ C for water,

$$\rho = 0.999099 \text{ gm/cm}^3$$
$$\mu = 0.011404 \text{ poise}$$

The acceleration of gravity is given as

$$g = 980 \text{ cm/sec}^2$$

As 1 darcy $= 9.87 \times 10^{-9}$ square centimeter, the intrinsic permeability is 2.66×10^{-11} square centimeter:

$$K = K_i\left(\frac{\rho g}{\mu}\right) = 2.66 \times 10^{-11} \text{ cm}^2 \times \frac{0.999099 \text{ gm/cm}^3 \times 980 \text{ cm/sec}^2}{0.011404 \text{ poise}}$$

$$1 \text{ poise} = \frac{\text{dyne-sec}}{\text{cm}^2}$$

$$1 \text{ dyne} = \frac{\text{gm-cm}}{\text{sec}^2}$$

$$1 \text{ poise} = \frac{\text{gm}}{\text{sec-cm}}$$

$$K = 2.28 \times 10^{-6} \frac{\text{gm/cm}^3 \times \text{cm/sec}^2}{\text{gm/sec-cm}}$$

$$= 2.28 \times 10^{-6} \text{ cm/sec}$$

 FORCES ACTING ON GROUNDWATER

There are three outside forces acting on the water contained in the ground. The most obvious of these is **gravity**, which pulls water downward. The second force is **external pressure**. Above the zone of saturation, atmospheric pressure is acting. The combination of atmospheric pressure and the weight of overlying water creates pressures in the zone of saturation. The third force is **molecular attraction**, which causes water to adhere to solid surfaces. It also creates surface tension in water when the water is exposed to air. The combination of these two processes is responsible for the phenomenon of capillarity.

When water in the ground is flowing through a porous medium, there are forces resisting the fluid movement. These consist of the **shear stresses** acting tangentially to the surface of the solid and **normal stresses** acting perpendicularly to the surface (29). We can think of these forces collectively as "friction." The internal molecular attraction of the fluid, itself, resists the movement of fluid molecules past each other. This shearing resistance is known as the viscosity of the fluid.

WATER TABLE 4.5

Water may be present beneath the earth's surface as a liquid, solid, or vapor. Other gases may also be present, either in vapor phase or dissolved in water. In the lower zone of porosity, generally all that is present is mineral matter and liquid water. The rock is saturated with water, and the water may also contain dissolved gas. The fluid pressure is greater than atmospheric pressure due to the weight of overlying water. As the surface is approached, the fluid pressure decreases as the thickness of fluid above it decreases. At some depth, which varies from place to place, the pressure of the fluid in the pores is equal to atmospheric pressure. The undulating surface at which pore water pressure is equal to atmospheric pressure is called the **water table** (Figure 4.12).

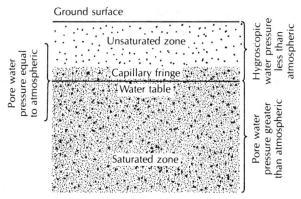

FIGURE 4.12.   Distribution of fluid pressures in the ground with respect to the water table.

Water in a shallow well (a meter or less below the water table) will rise to the elevation of the water table at that location. The position of the water table often follows the general shape of the topography, although the

77

water-table relief is not as great as the topographic relief. At all depths below the water table, the rock is generally saturated with water.*

A hypothetical experiment can serve to illustrate the formation of the water table. A box made of clear plastic is filled with sand. A notch is cut in one side of the plastic, and the surface of the sand is smoothed to model a valley draining toward the notch. A fine mist of water is then spread evenly over the surface of the sand, simulating rainfall. The precipitation rate is sufficiently low to preclude any overland flow. The water will move downward through the sand, so that, eventually, a zone of saturation will develop at the bottom. As shown in Figure 4.13A, this zone will have a level surface. As more rainfall is simulated, the water table will rise, continuing to be perfectly flat. It will follow this pattern until the water table reaches the lowest point in the valley.

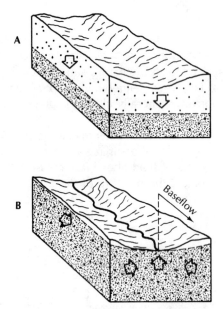

FIGURE 4.13. **A**. Diagram of a flat-lying water table in an aquifer where there is downward movement of water through the unsaturated zone but no lateral groundwater movement; **B**. Diagram of the water table in a region where water is moving downward through the unsaturated zone to the water table and moving as groundwater flow through the zone of saturation toward a discharge zone along the stream. Net discharge from the aquifer is occurring as baseflow from the stream.

Continuing rainfall will cause further increases in the height of the water table. In the valley, the water level will be above the surface, so that water will now flow through the notch. Elsewhere, the water table will be higher than the elevation of the notch, and groundwater will begin to flow

---

*There are exceptions. The rocks may contain trapped liquid and gaseous hydrocarbons, for example. Or there may be isolated voids, which cannot fill with any fluid.

laterally due to the hydraulic gradient. Now, water will flow through the saturated zone toward the point of discharge (Figure 4.13B).

We can make the following observations which pertain primarily to humid regions:

1. In the absence of groundwater flow, the water table will be flat.

2. A sloping water table indicates the groundwater is flowing.

3. Groundwater discharge zones are in topographical low spots.

4. The water table has the same general shape as the surface topography.

5. Groundwater generally flows away from topographical high spots and toward topographic lows.

CAPILLARITY AND THE CAPILLARY FRINGE

If fluid pressures are measured above the water table, they will be found to be negative with respect to local atmospheric pressure. This is called a **tension**. Air may also be present in the voids above the water table. Air pressure is equal to atmospheric pressure above the water table. Water vapor is also present in the voids above the water table.

Water molecules at the water table are subject to an upward attraction due to surface tension of the air-water interface and the molecular attraction of the liquid and solid phases. This is known as **capillarity**—a phenomenon well studied in classical physics.

In a tube of small diameter, the free-water surface will assume a shape with the minimum surface area. The attraction of the solid for the liquid will draw the liquid up into the tube. The upward force will eventually be offset by the weight of the column of water. The water, itself, is under tension; thus, the pressure is less than atmospheric. For pure water at 20° C rising in a glass tube, the height of the capillary rise is given by the equation

$$h_c = \frac{0.153}{r} \qquad \text{(4-8)}$$

where h_c is the height of the capillary rise above the water surface (centimeters) and r is the radius of the capillary tube (centimeters) (Figure 4.14).

In fine-grained soils and sediments, and in rocks with very small pores and cracks, the openings are of small enough diameter to act as capillary tubes. The classical formula for capillary rise suggests that for a pore radius of 0.02 millimeter there will be a capillary rise of 0.076 meter. In fine- to medium-grained sediments and soils there is a **capillary fringe** above the water table. The lower part of the capillary fringe may actually be saturated, although

79

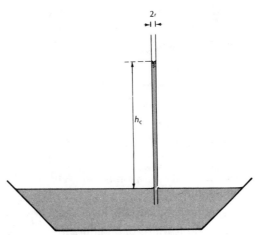

FIGURE 4.14. Capillary rise of water in a thin tube.

the pore water is under tension (negative pressure). The capillary formula can be used to estimate the height of the capillary fringe.

Because of irregularities in the size of the openings, capillary water does not rise to an even height above the water table; rather, it forms an irregular fringe.

TABLE 4.5. Typical height of capillary fringe

Material	Typical Pore Radius		Capillary Height	
Gravel, coarse	0.4	mm	0.38	cm
Sand, coarse	0.05	mm	3.0	cm
Sand, fine	0.02	mm	7.7	cm
Silt	0.001	mm	1.5	m
Clay	0.0005	mm	3.0	m

Above the capillary fringe, there is moisture coating the solid surfaces of the fragment or rock particles. This can move in any direction due to surface tension. If the liquid coating becomes too thick to be held by surface tension, a droplet will pull away and be drawn downward by gravity. The fluid can also evaporate and move through the air space in the pores as water vapor.

The amount of vapor movement in the unsaturated zone is much less important than transport in the liquid form (30). However, this might not hold true if the water content of the soil is very low or if there is a strong temperature gradient. The movement of vapor through the unsaturated zone is a function of the temperature and humidity gradients in the soil and molecular diffusion coefficients for water vapor in the soil (31).

INFILTRATION 4.7

Rainfall reaching the earth's surface can either form puddles, run across the surface, or infiltrate into the ground. Soil has a finite capacity to absorb water. The infiltration capacity varies not only from soil to soil, but is also different for dry versus moist conditions in the same soil.

If a soil is initially dry, the infiltration capacity is high. Surface effects between the soil particles and the water exert a tension that draws the moisture downward into the soil through labyrinthine capillary passages. As the capillary forces diminish with increased soil-moisture content, the infiltration capacity drops (Figure 4.15). In addition, colloidal particles in the soil swell as the moisture content increases. Eventually, the infiltration capacity reaches a more or less constant value determined by the unsaturated permeability of the soil.

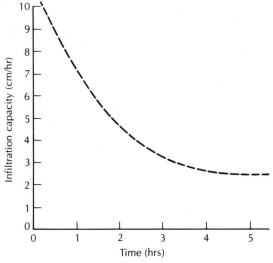

FIGURE 4.15. Decreasing infiltration capacity of an initially dry soil, as the soil-moisture content of the surface layer increases.

If the precipitation rate is lower than the initial infiltration capacity of the soil, there will be no overland flow (Figure 4.16A). But should the infiltration capacity decrease to a rate lower than the precipitation rate, depression storage will fill, and then overland flow can occur. Part B of the figure shows an idealized version of a rainfall chart: initially there is no overland flow, but depression storage fills, and then an overland-flow component develops as the infiltration capacity decreases with time. If the uniform precipitation rate is

81

initially higher than the infiltration capacity, depression storage starts to fill immediately, with overland flow starting when depression storage is filled (Part C).

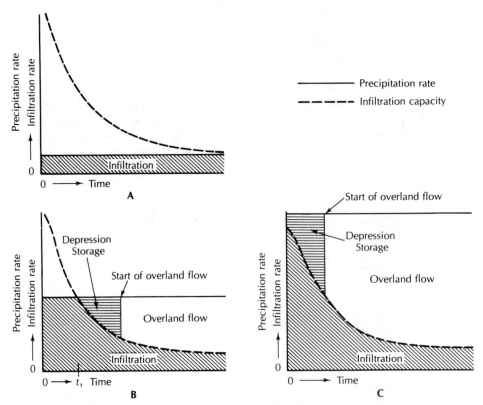

FIGURE 4.16. Relationship of precipitation rate and infiltration capacity in three different situations: **A.** No depression storage or overland flow; **B.** Filling of depression storage after time, t, has elapsed; **C.** Filling of depression storage starting immediately.

Conditions that encourage a high infiltration rate include coarse soils, well-vegetated land, low soil moisture, and a topsoil layer made porous by insects and other burrowing animals, in addition to land-use practices that avoid soil compaction. Once the final infiltration rate is reached, the depth of ponded water also promotes high infiltration.

Moisture flowing downward through the unsaturated zone from a rainstorm moves as a front. Waves of water pass through a given soil layer, each front displaced by one from a more recent rain (32). Radioactive tracer studies have shown that water infiltrated from a single storm can be identified for its entire journey through the unsaturated zone to the water table (33).

SOIL MOISTURE

The uppermost soil layers may have a three-phase system of soil, air, and water. The water is in both the liquid and vapor phases. Larger pore spaces contain a film of water coating the mineral grains, while the smaller and capillary pores may be saturated. Water movement takes place through the saturated pores, while water vapor passes through the air-filled pores.

The **water content** of the soil is the ratio of the volume of the contained water to that of the entire soil. The volume of the soil, if it has a high shrink-swell potential, may change with a change in volume of contained water (34). The percent of saturation is the ratio of the volume of contained water to the volume of pore space in the soil.

The pressure in the unsaturated zone **(vadose zone)** is negative, due to tension of the soil–surface-water contact. The negative pressure head, ψ, is measured in the field with a **tensiometer**. This device consists of a tube that is closed at the top, with a ceramic cup at the bottom to provide a porous membrane (Figure 4.17). When the tensiometer is inserted into soil, water

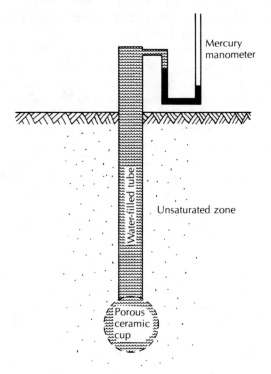

FIGURE 4.17. Porous-cup tensiometer with mercury manometer to measure soil-moisture tension.

83

within it is in contact with the soil moisture through the porous membrane. The suction exerted on the water in the tensiometer can be measured with a mercury manometer, vacuum gauge, or pressure transducer (35).

Capillary pores in the **zone of aeration** can draw water up from the zone of saturation beneath the water table. In very fine-grained soils, this capillary fringe can saturate the soil above the water table. However, tensiometer readings will reveal that the head is negative, indicating that the capillary fringe is part of the vadose zone. The zone of aeration is best defined as the zone where the soil moisture is under tension.

The moisture content of a soil sample can be measured directly. A sample of the moist material of known volume, V, is weighed, W_w. The sample is oven dried at 105° C to obtain the dry weight, W_d. The gravimetric moisture content is $(W_w - W_d)/W_d$. On a volume basis, the volumetric moisture content, θ, is $(W_w - W_d)/V$. In order to make this direct measurement, the soil must be excavated to obtain a sample each time a measurement is to be made.

Soil moisture can also be measured indirectly by nondestructive means. One method involves burying small blocks of either nylon or gypsum in which electrodes are imbedded, and then passing an electrical current through the wire. The electrical resistance of the block is proportional to the moisture it contains, which, in turn, is dependent upon the soil moisture. The meter can be calibrated for the soil type and, once calibrated, can be used for repeated measurements at the same location. This type of apparatus is relatively inexpensive, but the buried resistance blocks must be left in the soil.

A more expensive device uses a source of fast neutrons enclosed in a probe. The probe is lowered into a tube in the soil. When the fast neutrons encounter hydrogen atoms in water, they become slow neutrons. The density of slow neutrons produced is a function of the amount of soil moisture. A slow neutron counter is also a part of the probe (Figure 4.18). The neutron meter must be calibrated against the known water content of one part of the soil profile. The method can be used for repeated soil-moisture measurements in the same access tube. The water content in a volume of 15-centimeter radius is measured by this method.

Soil moisture at a location varies with changes in the amount of precipitation and evapotranspiration. Figure 4.19 shows a soil-moisture budget for a humid area based on measured precipitation and computed potential and actual evapotranspiration. During the period of water surplus, there is moisture available for groundwater recharge and runoff. Major fluctuations are seasonal in nature. In the spring, soil moisture is high, as snowmelt and spring rains have created large amounts of water available for infiltration. At times, the top layer of soil may be completely saturated, even though lower layers are unsaturated. During these periods of very high soil moisture, groundwater recharge can occur. Moisture moves downward by gravity flow. As water is withdrawn from the soil by evapotranspiration, the soil-moisture content drops. When the soil-moisture content of a layer reaches the point at which the force of gravity acting on the water equals the surface tension, gravity drainage ceases. This soil-moisture content is the **field capacity** of the soil.

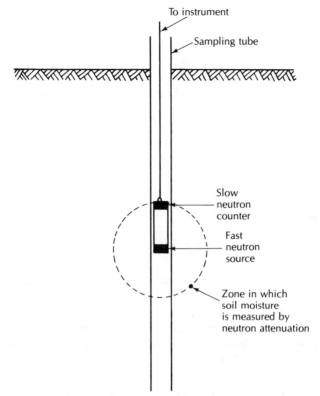

FIGURE 4.18. Neutron soil-moisture meter based on scattering of fast neutrons by hydrogen atoms, which are primarily found in the water molecules of the soil moisture.

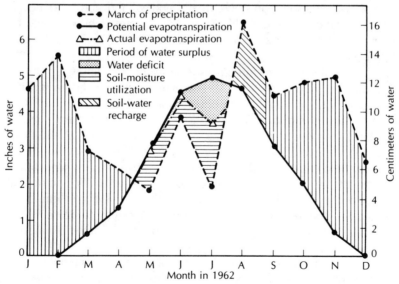

FIGURE 4.19. Soil-moisture budget for Bridgehampton, New York. The diagram is based on measured precipitation and computed potential and actual evapotranspiration. The Thornthwaite method was used for evapotranspiration computations. SOURCE: C. W. Fetter, Jr., *Bulletin, Geological Society of America,* 87 (1976):401-6.

The concept of field capacity is somewhat vague. Gravity drainage may take a long period of time to occur. The amount of moisture retained for a few days is much more than that retained for a long period. The following table shows the soil moisture of a silt-loam (fine-textured) soil as a function of time (36):

TABLE 4.6. Moisture content of a silt loam as a function of time since saturation (36)

Time	θ (%)
1 day	20.2
7 days	17.5
30 days	15.9
60 days	14.7
156 days	13.6

Field capacity, then, is not a single value, but a time-dependent parameter.

Soil moisture becomes lower than the field capacity as evapotranspiration removes still more water. During summer periods, the soil often dries. Occasional rainstorms may cause short-term rises in soil-moisture content, but there is generally no groundwater recharge. Exceptionally heavy summer rains, which replenish the depleted soil moisture and raise the water content above the field capacity, can create a wave of infiltrated water that courses downward through the soil-moisture zone and past the roots of thirsty plants, recharging the groundwater reservoir and thus causing the water table to rise. After fall frosts kill plants and cause deciduous trees to lose their leaves, the rate of evapotranspiration slows greatly, and soil moisture increases if rainfall and infiltration continues. Figure 4.20 illustrates a hypothetical annual cycle of soil moisture for a moderately humid area (50 to 75 centimeters per year precipitation).

If the soil moisture drops too low, the remaining moisture is too tightly bound to the soil particles for the plant roots to withdraw it. The soil-moisture content at which this first occurs is the **wilting point**. Plants wilt and may die for lack of moisture. The wilting-point moisture content is greater for fine-textured soils due to their greater surface area. Figure 4.21 shows typical wilting-point moisture content values for various soil types. Also shown are generalized field capacity and porosity values. The available water capacity of a soil is the difference between the field capacity and the wilting point. Water in a soil above the field will drain downward if the soil is permeable, or will waterlog a fine-grained, slowly permeable soil. Brief study of Figure 4.21 shows that the available water capacity is greatest for soils of intermediate texture, e.g., loams and silt loams.

It should be stressed that the soil-moisture zone only extends to the depth of plant roots. In agricultural soils, this might be a few meters or less. Some trees have tap roots extending tens of meters downward. The unsaturated zone may extend beyond the zone of soil moisture. In this case, there is an intermediate zone between the soil-moisture zone and the capillary fringe. In other cases, the soil-moisture zone may extend to the water table—an especially likely occurrence if the water table is within a few meters of the surface.

86

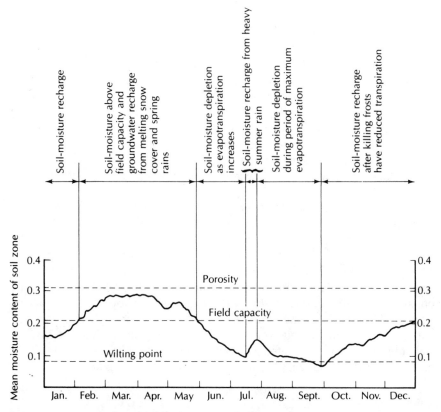

FIGURE 4.20. Hypothetical fluctuation of soil moisture for a sandy loam soil through an annual cycle in a region with a moderate amount of rainfall (20 to 30 inches [50 to 75 centimeters] per year) and heavy rains in the spring.

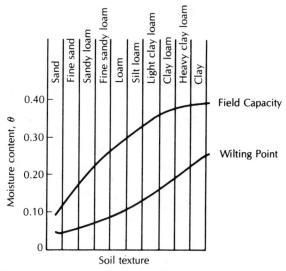

FIGURE 4.21. Water-holding properties of soils based on texture. The available water supply for a soil is the difference between field capacity and wilting point.
SOURCE: U.S. Department of Agriculture, *Yearbook*, 1955.

4.9 THEORY OF UNSATURATED FLOW

The infiltration of water into the unsaturated zone can be considered mathematically in terms of both a **gravity potential**, Z, and a **moisture potential**, ψ (34). The moisture potential is a negative pressure due to the soil-water attraction. The moisture potential increases with decreasing amounts of soil moisture. Depending upon the soil-moisture content, either the moisture potential or the gravity potential may predominate. At moisture contents close to the specific retention, the gravity potential is greater; but when the soil is very dry, the moisture potential may be several orders of magnitude greater than the gravity potential.

When the soil is not saturated, soil moisture flows downward by gravity flow through interconnected pores that are filled with water. With increasing water content, more pores fill, and the rate of downward water movement increases. Darcy's law is valid for flow in the unsaturated zone, although the **unsaturated hydraulic conductivity**, $K(\theta)$, is not a constant. The unsaturated hydraulic conductivity is a function of the volumetric moisture content, θ. As θ increases, so does $K(\theta)$. Figure 4.22 shows the relationship between $K(\theta)$ and θ for a clay soil. The value of the moisture potential, ψ, is also a function of θ, often ranging over many orders of magnitude (Figure 4.23). The **total potential**, ϕ, in unsaturated flow is the sum of the moisture potential, $\psi(\theta)$, and the elevation head, Z:

$$\phi = \psi(\theta) + Z \tag{4-9}$$

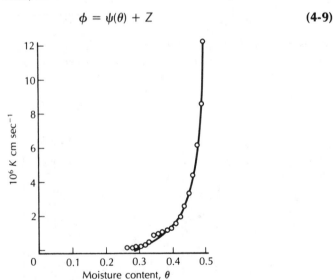

FIGURE 4.22. The relationship between hydraulic conductivity, K, and moisture content, θ, for a clay. SOURCE: J. R. Philip, "Theory of Infiltration," in *Advances in Hydroscience*, vol. 5, ed. V. T. Chow (New York: Academic Press, 1969). Used with permission.

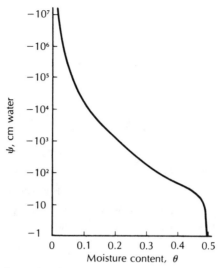

FIGURE 4.23.   The relationship between moisture potential, ψ, and moisture content, θ, for the clay soil of Figure 4.22. SOURCE: J. R. Philip, "Theory of Infiltration," in *Advances in Hydroscience*, vol. 5, ed. V. T. Chow (New York: Academic Press, 1969). Used with permission.

The relationship of unsaturated hydraulic conductivity and moisture content is determined experimentally. A sample of the rock or sediment is placed in a container. The moisture content is kept constant, and the rate at which water moves through the soil is measured. The value of $K(\theta)$ can be determined by Darcy's law. Figure 4.22 is based on a number of different measurements at differing values of θ.

The moisture potential, ψ, is measured by a suction applied to a soil. The curve of ψ versus water content is a plot of the water content starting with a saturated sample to which increased suction is gradually applied. A rather substantial problem occurs because of hysteresis in the soil-moisture content–moisture potential relationship. The curve of ψ as a function of θ is different if it is determined for decreasing as opposed to increasing values of θ. Thus, two curves such as that of Figure 4.23 are needed: one if the sediment has a particular θ as a result of an increase in soil-water content, and another if the soil is drying. Figure 4.24 shows the impact of hysteresis for an idealized sandy soil. Therefore, to use a value of ψ, one must know the prior moisture history of the sample.

Flow in the unsaturated zone is further complicated by the fact that both K and ψ may change as θ varies, and, by its very nature, unsaturated flow involves many changes in moisture content as waves of infiltrated water pass. The flow equations are nonlinear and not subject to easy solutions (37, 38).

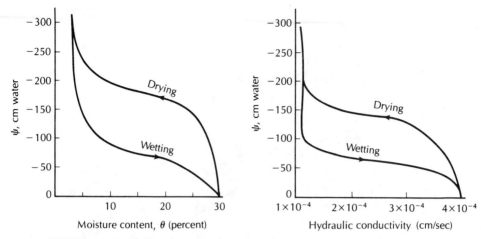

FIGURE 4.24. Idealized curves showing relationships of soil moisture, θ, hydraulic conductivity, K, and soil-moisture tension head, ψ. The effect of hysteresis is included for wetting and drying cycles.

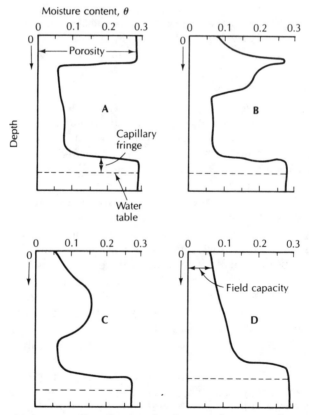

FIGURE 4.25. Moisture profiles showing the downward passage of a wave of infiltrated water. The soil is saturated at a moisture content of 0.29 and has a field capacity moisture content of 0.06. SOURCE: Modified from Agronomy Monograph 17, "Drainage for Agriculture," 1974, pp. 359–405. Used with permission of the American Society of Agronomy.

The downward movement of a slug of infiltrated water is shown in Figure 4.25. During infiltration (Part A), the topmost soil layer is brought up to the saturated water content of almost 0.3. The rainfall rate is exceeding the infiltration capacity, so the vertical hydraulic conductivity of the soil controls the rate of downward movement. After infiltration ceases (Part B), the slug of infiltrated water begins to move downward, although only a small layer remains at the saturated moisture content. Eventually, all of the downward-moving slug is at an unsaturated state, and $K(\theta)$ controls the downward movement (Part C). Finally, the slug reaches the water table and raises it (Part D). By this time, gravity drainage has reduced the upper soil layer to field capacity.

One important consideration in unsaturated flow is that at low moisture contents, the relations which hold true in saturated flow may be invalid. The best example is the fact that for coarse materials, such as sand and gravel, the pores are large and drain quickly. At lower moisture contents, there may be very few saturated pores. On the other hand, finer-grained soils may have most of the pores still saturated. Thus, at lower values of θ, the unsaturated hydraulic conductivity of a clay may be greater than that of a sand. A layer of sand in a fine-textured, unsaturated soil may retard downward movement of infiltrating water due to its low unsaturated hydraulic conductivity.

WATER-TABLE RECHARGE

When the front of infiltrating water reaches the capillary fringe, it displaces air in the pore spaces and causes the water table to rise. The capillary fringe is also higher, and the latest arriving recharge is actually found at the top of the capillary fringe. The time of movement of infiltrating water is a function of the thickness of the unsaturated zone and the vertical unsaturated hydraulic conductivity. The presence of layers of low-permeability material, such as silts and clays, can retard the rate of recharge, even if the layers are thin. The time lag may be only a few hours in the humid regions for very coarse soils with a water table close to the surface. In arid environments, with very infrequent recharge and great depths to the water table, water may take years to pass through the unsaturated zone.

In the arid Hualapai Plateau area of northwestern Arizona, the annual recharge rate is only 0.25 centimeter per year. Although rainfall averages 23 to 33 centimeters per year, potential evapotranspiration is in the range of 183 to 193 centimeters per year (39). Grassland regions have more precipitation and, consequently, more recharge. In the Sand Hills region of South Dakota and Nebraska, the mean annual precipitation is 46.3 centimeters, and groundwater recharge amounts to 6.9 centimeters per year (40). On eastern Long Island, New York, precipitation averages 116.8 centimeters per year, and evapotranspiration, as computed by the Thornthwaite method, averages 57.4 centimeters per year. The soils are permeable, and much of the remaining 59.4 centimeters per year of water recharges the water table (41).

On eastern Long Island, the response of the water table to recharge is fairly rapid. Figure 4.26 is a hydrograph for 1962 of a water-table well with an unsaturated zone on the order of 3 to 4 meters. This well is a few miles from the precipitation station that yielded the data for Figure 4.19 (page 85). As the water table is being drained by stream and spring discharge, it will decline (a) if there is no recharge or (b) if the amount of recharge is less than groundwater outflow. The hydrograph peaks in March and then begins a recession which lasts until October. Periods of surplus water are indicated in Figure 4.19 for January through April, and then September through December. The peak in March corresponds to snowmelt and spring rains, and the upturn in water levels by November is due to fall recharge. The response time of this water-table aquifer is more rapid than for those cases in which the unsaturated zone is many tens of meters thick.

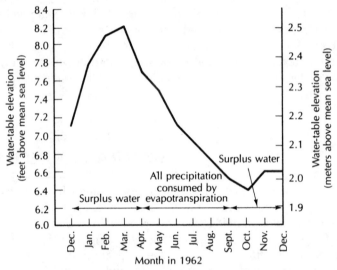

FIGURE 4.26. Hydrograph of a shallow well in a water-table aquifer in Long Island. The period of record is the same as for the soil-moisture budget diagram of Figure 4.19.

4.11 AQUIFERS

Natural earth materials have a very wide range of hydraulic conductivities. Near the earth's surface there are very few, if any, geologic formations that are absolutely impermeable. Weathering, fracturing, and solution have affected most rocks to some degree. However, the rate of groundwater movement can be exceedingly slow in units of low hydraulic conductivity.

An **aquifer** is a geologic unit that can store and transmit water at rates fast enough to supply reasonable amounts to wells. The intrinsic permeability of aquifers would range from about 10^{-2} darcy upward. Unconsolidated sands and gravels, sandstones, limestones and dolomites, basalt flows, and fractured plutonic and metamorphic rocks are examples of rock units known to be aquifers.

A **confining layer** is a geologic unit having little or no intrinsic permeability—less than about 10^{-2} darcy. This is a somewhat arbitrary limit, and depends upon local conditions. In areas of clay, with intrinsic permeabilities of 10^{-4} darcy, a silt of 10^{-2} darcy might be used to supply water to a small well. On the other hand, the same silt might be considered a confining layer if it were found in an area of coarse gravels with intrinsic permeabilities of 100 darcys. Groundwater moves through most confining layers, although the rate of movement is very slow.

Confining layers are somtimes subdivided into aquitards, aquicludes and aquifuges. An **aquifuge** is an absolutely impermeable unit that will not transmit any water. An **aquitard** is a layer of low permeability that can store groundwater and also transmit it slowly from one aquifer to another. The term **leaky confining layer** is also applied to such a unit. An **aquiclude** is also a unit of low permeability, but is located so that it forms an upper or lower boundary to a groundwater flow system. Most authors now use the terms confining layer and leaky confining layer.

Confining layers can be important elements of regional flow systems and leaky confining layers can transmit significant amounts of water if the cross-sectional area is large. On Long Island, New York, there is a deep aquifer, the Lloyd Sand Member of the Raritan Formation. The recharge to this aquifer passes downward across a confining layer 60 to 100 meters thick, the Clay Member of the Raritan Formation. Over an areal extent measured in hundreds of square kilometers, a very low rate of vertical seepage provides recharge to the underlying aquifer. Some of the recharge to the principal artesian aquifer of the southeastern United States takes place by downward leakage through overlying confining layers, although much of the recharge is carried through sinkholes.

Aquifers can be close to the land surface, with continuous layers of materials of high intrinsic permeability extending from the land surface to the base of the aquifer. Such an aquifer is termed a **water-table aquifer** or **unconfined aquifer**. Recharge to the aquifer can be from downward seepage through the unsaturated zone (Figure 4.27A). Recharge can also occur through lateral groundwater flow or upward seepage from underlying strata.

Some aquifers, called **confined** or **artesian aquifers**, are overlain by a confining layer. Recharge to them can occur either in a recharge area, where the aquifer crops out, or by slow downward leakage through a leaky confining layer. If a tightly-cased well is placed through the confining layer, water from the aquifer may rise considerable distances above the top of the aquifer (Figure 4.27B). This indicates that the water in the aquifer is under pressure. The **potentiometric surface** for a confined aquifer is the surface representative of the level to which water will rise in a well cased to the aquifer. (The term piezometric was used in the past, but it has now been replaced by potentiometric.) If the potentiometric surface of an aquifer is above the land surface, a flowing artesian well may occur. Water will flow from the well casing without need for a pump. Of course, if a pump were installed, the amount of water obtained from the well could be increased.

93

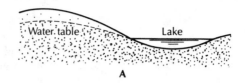

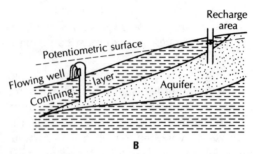

FIGURE 4.27. A. Water-table aquifer; **B.** Confined aquifer.

In some cases, a layer of low-permeability material will be found as a lens in more permeable materials. Water moving downward through the unsaturated zone will be intercepted by this layer, and will accumulate on top of the lens. A layer of saturated soil will form above the main water table. This is termed a **perched aquifer** (Figure 4.28). Water moves laterally above the low-permeability layer up to the edge, and then seeps downward toward the main water table. Perched aquifers are common in glacial outwash, where lenses of clay formed in small glacial ponds are present. They are also often present in volcanic terrains, where weathered ash zones of low permeability can occur sandwiched between high-permeability basalt layers.

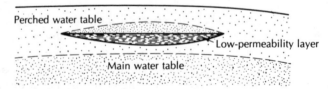

FIGURE 4.28. A perched aquifer.

Perched aquifers are usually not very large; most would only supply enough water for household use. Some lakes are perched on low-permeability sediments. Such ponds are especially vulnerable to widely fluctuating lake-stage levels with changes in the amount of rainfall.

4.12 AQUIFER CHARACTERISTICS

We have thus far considered the intrinsic permeability of earth materials and their hydraulic conductivity when transmitting water. A useful concept in many studies is aquifer **transmissivity**. This is a measure of the amount of water that can be transmitted horizontally by the full saturated thickness of the aquifer under a hydraulic gradient of 1. The transmissivity, T, is the product of the hydraulic conductivity and the saturated thickness of the aquifer, b:

$$T = bK \tag{4-10}$$

For a multilayer aquifer, the total transmissivity is the sum of the transmissivity of each of the layers:

$$T = \sum_{i=1}^{n} T_i \tag{4-11}$$

The dimensions of transmissivity are (L^2/T). Common units are square meters per day or gallons per day per foot. Aquifer transmissivity is a concept that assumes flow through the aquifer to be horizontal. In some cases, this is a valid assumption; in others, it is not.

When the head in a saturated aquifer or confining unit changes, water will be either stored or expelled. The **storage coefficient** or **storativity** (S) is the volume of water that a permeable unit will absorb or expel from storage per unit surface area per unit change in head. It is a dimensionless quantity.

In the saturated zone, the head creates pressure, affecting the arrangement of mineral grains as well as the density of the water in the voids. If the pressure increases, the mineral skeleton will expand; if it drops, the mineral skeleton will contract. This is known as **elasticity**. Likewise, water will contract with an increase in pressure and expand if the pressure drops. When the head in an aquifer or confining bed declines, the aquifer skeleton compresses, which reduces the effective porosity and expels water. Additional water is released as the pore water expands due to lower pressure.

The **specific storage** (S_s) is the amount of water per unit volume of a saturated formation that is stored or expelled from storage due to compressibility of the mineral skeleton and the pore water. This is also called the **elastic storage coefficient**. The concept can be applied both to aquifers and confining units.

The specific storage is given by the expression (42, 43, 44)

$$S_s = \rho g (\alpha + n\beta) \tag{4-12}$$

where

ρ is the density of the water

g is the acceleration of gravity

α is the compressibility of the aquifer skeleton

n is the porosity

β is the compressibility of the water

Specific storage has dimensions of $1/L$. The value of specific storage is very small, generally 0.0001/meter or less.

95

In a confined aquifer, the head may decline—yet the potentiometric surface remains above the unit (Figure 4.29). Although water is released from storage, the aquifer remains saturated. The storativity (S) of a confined aquifer is the product of the specific storage (S_s) and the aquifer thickness (b):

$$S = bS_s \qquad (4\text{-}13)$$

All of the water released is accounted for by the compressibility of the mineral skeleton and the pore water. The water comes from the entire thickness of the aquifer. The value of the storativity of confined aquifers is on the order of 0.005 or less.

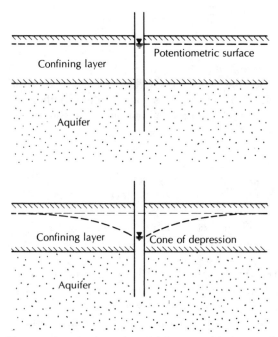

FIGURE 4.29. Diagram showing lowering of the water table in a confined aquifer with the resultant water level still above the aquifer materials. In this circumstance, the aquifer remains saturated.

In an unconfined unit, the level of saturation rises or falls with changes in the amount of water in storage. As the water level falls, water drains from the pore spaces. This storage or release is due to the **specific yield** (S_y) of the unit. Water is also stored or expelled due to the specific storage of the unit. For an unconfined unit, the storativity is found by the formula

$$S = S_y + hS_s \qquad (4\text{-}14)$$

where h is the thickness of the saturated zone.

The value of S_y is several orders of magnitude greater than hS_s for an unconfined aquifer, and the storativity is usually taken to be equal to the specific yield. For a fine-grained unit, the specific yield may be very small, approaching the same order of magnitude as hS_s. Storativity of unconfined aquifers ranges from 0.02 to 0.30.

The volume of water drained from an aquifer, V_w, may be found from the equation

$$V_w = SA\Delta h \qquad\qquad (4\text{-}15)$$

where A is the area and Δh is the decline in head.

EXAMPLE PROBLEM

An unconfined aquifer with a storativity of 0.13 has an area of 123 square kilometers. The water falls 0.23 meter during a drought. How much water is lost from storage?

$V_w = SA\Delta h$

$\quad = 0.13 \times 123 \text{ km}^2 \times 10^6 \text{ m}^2/\text{km}^2 \times 0.23 \text{ m}$

$\quad = 3.68 \times 10^6 \text{ m}^3$

If the same aquifer had been confined with a storativity of 0.0005, what change in the amount of water in storage would have resulted?

$V_w = 0.0005 \times 123 \text{ km}^2 \times 10^6 \text{ m}^2/\text{km}^2 \times 0.23 \text{ m}$

$\quad = 1.41 \times 10^4 \text{ m}^3$

HOMOGENEITY AND ISOTROPY

Hydrogeologists are interested in two key properties of geologic formations: hydraulic conductivity and specific storage or specific yield. A third property, the thickness, is also important, since the overall hydrogeologic response of a unit is a function of the product of the hydraulic parameters and the thickness.

A **homogeneous** unit is one that has the same properties at all locations. For a sandstone, this would indicate that the grain-size distribution, porosity, degree of cementation, and thickness are variable only within small limits. The value of the transmissivity and storativity of the unit would be about the same wherever present. A plutonic or metamorphic rock would have the same amount of fracturing everywhere, including the strike and dip of the joint sets. A limestone would have the same amount of jointing and solution openings at all locations.

In **heterogeneous** formations, hydraulic properties change spatially. One example would be a change in thickness. A sandstone that thickens as a wedge is nonhomogeneous, even if porosity, hydraulic conductivity, and storativity remain constant. The change in thickness results in a change in the hydraulic properties of the unit. Layered units may also be nonhomogeneous. Most sedimentary units were deposited as successive layers of sediments with intervals of nondeposition. These layers are known to vary in thickness, from microscopic layers to those measured in meters. If the hydraulic properties of the layers are different, the entire unit is heterogeneous. The individual beds may be homogeneous. The third type of heterogeneity in a sedimentary unit occurs when a facies change in the unit involves a transformation of hydraulic characteristics as well as lithologic features. Figure 4.30 illustrates these examples of heterogeneity in clastic sedimentary units.

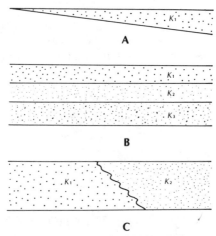

FIGURE 4.30. **A.** Heterogeneous formation consisting of a sediment which thickens in a wedge; **B.** Heterogeneous formation consisting of three layers of sediments of differing hydraulic conductivity; **C.** Heterogeneous formation consisting of sediments with different hydraulic conductivities lying next to each other.

Carbonate units may be heterogeneous (a) if there is a change in thickness, or (b) if the degree of solution openings of fractures varies. The formation of solution passageways by moving groundwater is typically concentrated along preferred fractures or bedding planes; thus, limestone formations are often heterogeneous. Plutonic rocks may have uneven fracturing or sporadic shear zones which render them heterogeneous. Basalt flows are virtually always heterogeneous by the very nature of the way they are formed. As might be expected, it is a very unusual geologic formation that is perfectly homogeneous. Geologic processes operate at varying rates and over uneven terrain, resulting in heterogeneity.

In a porous medium made of spheres of the same diameter packed uniformly, the geometry of the voids is the same in all directions. Thus, the intrinsic permeability of the unit is the same in all directions, and the unit is said to be **isotropic**. On the other hand, if the geometry of the voids is not uniform, there may be a direction in which the intrinsic permeability is greater. The medium is thus **anisotropic**. For example, a porous medium composed of book-shaped grains arranged in a subparallel manner would have a greater permeability parallel to the grains than crossing the grain orientation.

In fractured rock units, the direction of groundwater flow is completely constrained by the direction of the fractures. There may be zero intrinsic permeability in directions not parallel to a set of fractures (Figure 4.31). Basalt flows are highly anisotropic, as flow parallels the dip of the flow as it moves in the interflow zones. Shrinkage cracks in the basalt are vertical, yielding some vertical permeability.

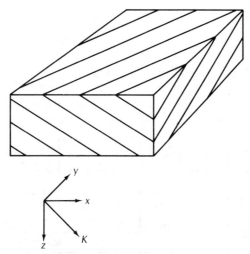

FIGURE 4.31.   Anisotropy of fractured rock units due to directional nature of fracturing.

In sedimentary units, there may be several layers, each of which is homogeneous. The equivalent vertical and horizontal hydraulic conductivity can be easily computed. Figure 4.32 shows a three-layered unit, each unit having a different horizontal and vertical hydraulic conductivity (K_h and K_v).

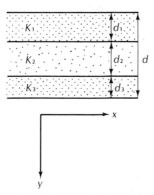

FIGURE 4.32.   Heterogeneous formation consisting of three layers of differing hydraulic conductivity.

99

The average horizontal conductivity is found from the summation

$$K_h \text{ avg} = \sum_{m=1}^{n} \frac{K_{h_m} d_m}{d} \qquad (4\text{-}16)$$

where d is the total thickness and d_m is the thickness of each layer. The overall vertical hydraulic conductivity is given by:

$$K_v \text{ avg} = \frac{d}{\sum_{m=1}^{n} \frac{d_m}{K_{v_m}}} \qquad (4\text{-}17)$$

4.14 HYDROSTRATIGRAPHIC UNITS

Geological formations are grouped or separated on the basis of mineralogy, lithology, fossil assemblages, and color. Some, but not all, of these characteris-

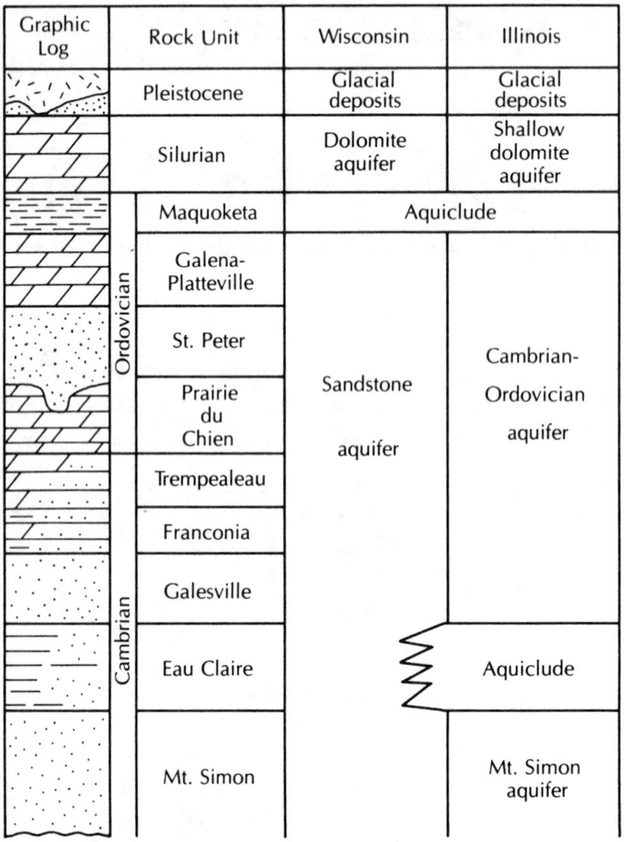

Graphic Log	Rock Unit		Wisconsin	Illinois
	Pleistocene		Glacial deposits	Glacial deposits
	Silurian		Dolomite aquifer	Shallow dolomite aquifer
	Maquoketa	Ordovician	Aquiclude	
	Galena-Platteville	Ordovician		
	St. Peter	Ordovician		Cambrian-Ordovician aquifer
	Prairie du Chien	Ordovician	Sandstone aquifer	
	Trempealeau	Ordovician		
	Franconia	Ordovician		
	Galesville	Cambrian		
	Eau Claire	Cambrian		Aquiclude
	Mt. Simon	Cambrian	Mt. Simon	Mt. Simon aquifer

FIGURE 4.33. Geologic formations and hydrostratigraphic units of southeastern Wisconsin and northeastern Illinois.

tics are related to hydrologic properties of the rocks. Lithology and grain-size distributions are very closely correlated with hydraulic conductivity, but fossil assemblages may have no correlation.

In hydrogeologic studies, it is more convenient to group units on the basis of similar hydraulic conductivity. Units so grouped are called **hydrostratigraphic units**, and often they will cross stratigraphic boundaries. Frequently, several formations are grouped into a single aquifer. Such is the case with the sandstone aquifer of central and eastern Wisconsin. Figure 4.33 shows rock units and hydrostratigraphic units of eastern Wisconsin and northeastern Illinois. As many as eight formations are included in the sandstone aquifer. The Eau Claire Formation is included in the sandstone aquifer in Wisconsin. However, due to differing hydrogeologic characteristics, it is considered a leaky confining layer in Illinois. A geologic formation may also be subdivided into different hydrostratigraphic units. On Long Island, the Raritan Formation has a Clay Member (a leaky confining layer) and the Lloyd Sand Member (an aquifer). The naming of aquifers does not adhere to the strict rules used in stratigraphic nomenclature. Aquifer names are informal; hence, they are not capitalized.

References

1. MEINZER, O. E. *The Occurrence of Ground Water in the United States.* U.S. Geological Survey Water Supply Paper 489, 1923, 321 pp.

2. DAVIS, S. N. "Porosity and Permeability in Natural Materials." In *Flow through Porous Media,* ed. R. J. M. DeWeist. New York: Academic Press, 1969, pp. 53–89.

3. ARCHIE, G. E. "Classification of Carbonate Reservoir Rocks and Petro-physical Considerations." *American Association of Petroleum Geologists Bulletin,* 36 (1950):943–61.

4. MURRAY, R. C. "Origin of Porosity in Carbonate Rocks." *Journal of Sedimentary Petrology,* 30 (1960):59–84.

5. WINSAUER, W. O. et al. "Resistivity of Brine-Saturated Sands in Relation to Pore Geometry." *American Association of Petroleum Geologists Bulletin,* 36 (1952):253–77.

6. WYLLIE, M. R. J. and M. B. SPANGLER. "Application of Electrical Resistivity Measurements to Problems of Fluid Flow in Porous Media." *American Association of Petroleum Geologists Bulletin,* 36 (1952):359–403.

7. MANGER, G. E. *Porosity and Bulk Density of Sedimentary Rocks.* U.S. Geological Survey Bulletin 1144-E, 1963.

8. COHEN, P. *Water Resources of the Humboldt River Valley Near Winnemucca, Nevada.* U.S. Geological Survey Water Supply Paper 1975, 1965.

9. MAC GARY, L. M. and T. W. LAMBERT. *Reconnaissance of Ground-Water Resources of the Jackson Purchase Region, Kentucky.* U.S. Geological Survey Hydrologic Atlas HA-13, 1962, 9 pp.

10. KRYNINE, D. P. and W. R. JUDD. *Principles of Engineering Geology and Geotechnics*. New York: McGraw-Hill Book Company, 1957, 730 pp.

11. BRACE, W. F., B. W. PAULDING, JR., and C. SCHOLZ. "Dilatancy in the Fracture of Crystalline Rock." *Journal of Geophysical Research*, 71 (1966):3939–53.

12. STEWART, J. W. *Infiltration and Permeability of Weathered Crystalline Rocks, Georgia Nuclear Laboratory, Dawson County, Georgia*. U.S. Geological Survey Bulletin 1133-D, 1964.

13. SCHOELLER, H. *Les Eaux souterraines*. Paris: Mason et Cie, 1962.

14. KELLER, G. V. *Physical Properties of Tuffs in the Oak Spring Formation, Nevada*. U.S. Geological Survey Professional Paper 400-B, 1960.

15. MEINZER, O. E. *Outline of Groundwater Hydrology, with Definitions*. U.S. Geological Survey Water Supply Paper 494, 1923, 71 pp.

16. JOHNSON, A. I. *Specific Yield—Compilation of Specific Yields for Various Materials*. U.S. Geological Survey Water Supply Paper 1662-D, 1967, 74 pp.

17. JOHNSON, A. I., R. C. PRILL, and D. A. MORRIS. *Specific Yield—Column Drainage and Centrifuge Moisture Content*. U.S. Geological Survey Water Supply Paper 1662-A, 1963, 60 pp.

18. PRILL, R. C., A. I. JOHNSON, and D. A. MORRIS. *Specific Yield—Laboratory Experiments Showing the Effect of Time on Column Drainage*. U.S. Geological Survey Water Supply Paper 1662-B, 1965, 55 pp.

19. WENZEL, L. K. *Methods of Determining Permeability of Water-Bearing Materials*. U.S. Geological Survey Water Supply Paper 887, 1942, 192 pp.

20. FERRIS, J. G., D. B. KNOWLES, R. H. BROWN, and R. W. STALLMAN. *Theory of Aquifer Tests*. U.S. Geological Survey Water Supply Paper 1536-E, 1962, 174 pp.

21. PRICKETT, T. A. "Type-Curve Solution to Aquifer Tests under Water-Table Conditions." *Ground Water*, 3, no. 5 (1965).

22. DARCY, H. *Les Fontaines publiques de la ville de Dijon*. Paris: Victor Dalmont, 1856, 647 pp.

23. NORRIS, S. E. and R. E. FIDLER. *Relation of Permeability to Grain Size in a Glacial-Outwash Aquifer at Piketown, Ohio*. U.S. Geological Survey Professional Paper 525-D, 1965, pp. 203–6.

24. MASCH, F. E. and K. J. DENNY. "Grain-Size Distribution and Its Effect on the Permeability of Unconsolidated Sands." *Water Resources Research*, 2 (1966):665–77.

25. NORRIS, S. E. *Permeability of Glacial Till*. U.S. Geological Survey Professional Paper 450-E, 1963, pp. 150–51.

26. FOLEY, F. C., W. C. WALTON, and W. J. DRESCHER. *Ground-Water Conditions in the Milwaukee-Waukesha Area, Wisconsin*. U.S. Geological Survey Water Supply Paper 1229, 1953, 96 pp.

27. MUSKAT, M. *The Flow of Homogeneous Fluids through Porous Media*. New York: McGraw-Hill Book Company, 1937.

28. VISHNER, F. N. and J. F. MINK. *Ground-Water Resources of Southern Oahu, Hawaii*. U.S. Geological Survey Water Supply Paper 1778, 1964, 133 pp.

102

29. RUMER, R. R., JR. "Resistance to Flow through Porous Media." In *Flow through Porous Media,* ed. R. J. M. DeWeist. New York: Academic Press, 1969, pp. 91–108.

30. SWARTZENDRUBER, D. "The Flow of Water in Unsaturated Soils." In *Flow through Porous Media,* ed. R. J. M. DeWeist. New York: Academic Press, 1969, pp. 215–92.

31. RIPPLE, C. D., J. RUBIN, and T. E. A. VAN HYLCKAMA. *Estimating Steady-State Evaporation Rates from Bare Soils under Conditions of High Water Table.* U.S. Geological Survey Water Supply Paper 2019-A, 1972, 39 pp.

32. SMITH, W. O. "Infiltration in Sands and Its Relation to Groundwater Recharge." *Water Resources Research,* 3 (1967):539–55.

33. ZIMMERMAN, U. et al. "Tracers Determine Movement of Soil Moisture and Evapotranspiration." *Science,* 152 (1960):346–47.

34. CHILDS, E. C. "Soil Moisture Theory." In *Advances in Hydroscience,* vol. 4, ed. V. T. Chow. New York: Academic Press, 1967, pp. 73–117.

35. WATSON, K. K. "A Recording Field Tensiometer with Rapid Response Characteristics." *Journal of Hydrology,* 5 (1967):33–39.

36. HILLEL, D. *Soil and Water.* New York: Academic Press, 1971, 288 pp.

37. PHILIP, J. R. "Theory of Infiltration." In *Advances in Hydroscience,* vol. 5, ed. V. T. Chow. New York: Academic Press, 1969, pp. 215–96.

38. SWARTZENDRUBER, D. "The Flow of Water in Unsaturated Soils." In *Flow through Porous Media,* ed. R. J. M. DeWeist. New York: Academic Press, 1969, pp. 215–92.

39. HUNTOON, P. "Cambrian Stratigraphic Nomenclature and Ground-Water Prospecting Failures on the Hualapai Plateau, Arizona." *Ground Water,* 15 (1977):426–33.

40. RAHN, P. H. and H. A. PAUL. "Hydrogeology of a Portion of the Sand Hills and Ogallala Aquifer, South Dakota and Nebraska." *Ground Water,* 13 (1975):428–37.

41. FETTER, C. W., JR. "Hydrogeology of the South Fork of Long Island, New York." *Bulletin, Geological Society of America,* 87 (1976):401–6.

42. JACOB, C. E. "On the Flow of Water in an Elastic Artesian Aquifer." *Transactions, American Geophysical Union,* 21 (1940):574–86.

43. JACOB, C. E. "Flow of Groundwater." In *Engineering Hydraulics,* ed. H. Rouse. New York: John Wiley & Sons, 1950, pp. 321–86.

44. COOPER, H. H. "The Equation of Groundwater Flow in Fixed and Deforming Coordinates." *Journal of Geophysical Research,* 71 (1966):4785–90.

103

Principles of
Groundwater Flow

chapter

Groundwater possesses energy in mechanical, thermal, and chemical forms. Because the amounts of energy vary spatially, groundwater is forced to move from one region to another in nature's attempt to eliminate these energy differentials. The flow of groundwater is thus controlled by the laws of physics and thermodynamics. To enable a separate examination of mechanical energy, we will make the simplifying assumption that the water is of nearly constant temperature. Thermal energy must be considered, however, in such applications as geothermal flow systems and burial of radioactive heat sources.

Energy is the capacity to do work, which implies that some resistance to change in movement must be overcome. **Work** is done when a force is applied to a fluid while the fluid is moving. Work is equal to the product of the net force exerted and the distance through which the force moves:

$$W = Fd \qquad \text{(5-1)}$$

where

> W is the work
> F is the force
> d is the distance

The **force** acting on a body is equal to the product of mass of the body and its acceleration (Newton's second law of motion):

$$F = ma \qquad \text{(5-2)}$$

where

> F is the force
> m is the mass
> a is the acceleration

The unit of **mass** is a kilogram, and if acceleration is expressed in meters per second per second (m/sec^2), then the unit of force is the newton, or $Kg\text{-}m/sec^2$.

The **weight** of a body is the gravitational force exerted on it by the earth. The gravitational acceleration, g, varies from place to place, but is approximately 9.8 m/sec^2. The weight of a body is given by

$$w = mg \qquad\qquad (5\text{-}3)$$

where

w is the weight

g is the acceleration of gravity

Weight has the same units as force. The mass of a body, the weight of which is 1 newton, at a place where $g = 9.80$ m/sec^2, is

$$m = \frac{w}{g} = \frac{1\ N}{9.80\ \text{m/sec}^2} = 0.102\ \text{kg}$$

In the English system, the pound is a unit of force; hence, a unit of weight. The unit of mass is the slug. These definitions may seem somewhat confusing, since the pound as a unit of weight is often compared to the kilogram—a unit of mass. Balances and scales are calibrated in kilograms and grams. If we say that a sample weighs 0.13 kilogram, we are saying that the sample has a mass of 0.13 kilogram in the earth's gravitational field. The weight would be 1.27 newtons (0.29 pound). A kilogram has a mass of 1000 grams.

The **density** of a fluid is its mass per unit volume. Density is usually referred to by the Greek letter rho, ρ. It has units of gm/cm^3 or kg/m^3:

$$\rho = m/V \qquad\qquad (5\text{-}4)$$

where V is the volume.

The **specific weight** of a substance is its weight per unit volume, indicated by the Greek letter gamma, γ. The units are newtons/m^3 or newtons/cm^3, and it changes with location on the earth, since the value of g changes:

$$\gamma = w/V \qquad\qquad (5\text{-}5)$$

EXAMPLE PROBLEM

A fluid has a density of 1.085 gm/cm^3. If the acceleration of gravity is 9.81 m/sec^2, what is the specific weight of the fluid?

$\gamma = w/V \qquad w = mg \qquad \rho = m/V$

$\gamma = \rho g$

$\rho = 1.085\ \text{gm/cm}^3 \times 1/1000\ \text{kg/gm} \times 10^6\ \text{cm}^3/\text{m}^3$
   $= 1.085 \times 10^3\ \text{kg/m}^3$

$\gamma = 1.085 \times 10^3\ \text{kg/m}^3 \times 9.81\ \text{m/sec}^2$
   $= 1.064 \times 10^4\ \text{N/m}^3$

107

5.2 MECHANICAL ENERGY

There are a number of different types of mechanical energy recognized in classical physics. Of these, we will consider kinetic energy, gravitational potential energy, and energy of fluid pressures.

A moving body or fluid tends to remain in motion, according to Newtonian physics. This is because it possesses energy due to its motion—**kinetic energy**. This energy is equal to one half the product of its mass and the square of the magnitude of the velocity:

$$E_k = \tfrac{1}{2}mv^2 \qquad (5\text{-}6)$$

where

E_k is the kinetic energy

v is the velocity

If m is in kilograms and v in m/sec, then E_k has the units of kg • m²/sec² or newton-meters. The unit of energy is the joule, which is one newton-meter. The joule is also the unit of work.

Imagine that a weightless container filled with water of mass m is moved vertically upward a distance, z, from some reference surface (a datum). Work is done in moving the mass of water upward. This work is equal to

$$W = Fd = (mg)z \qquad (5\text{-}7)$$

where

z is the elevation of the center of gravity of the fluid
above the reference elevation

m is the mass

g is the acceleration of gravity

The units are kg × m/sec² × m, or newton-meters, with dimensions of $[ML^2T^{-2}]$.

The mass of water has now acquired energy equal to the work done in lifting the mass. This is a potential energy, due to the position of the fluid mass with respect to the datum. E_g is **gravitational potential energy**:

$$W = E_g = mgz \qquad (5\text{-}8)$$

A fluid mass has another source of potential energy due to the **pressure** of the surrounding fluid acting upon it. Pressure is the force per unit area acting on a body:

$$P = F/A \qquad (5\text{-}9)$$

where

P is the pressure

A is the cross-sectional area

The units of pressure are pascals, or newtons/m². A newton/m² is equal to a N-m/m³ or joule/m³. Pressure may thus be thought of as potential energy per unit volume of fluid.

For a unit volume of fluid, the mass, m, is numerically equal to the density, ρ, since density is defined as mass per unit volume. The total energy of the unit volume of fluid is the sum of the three components—kinetic, gravitational, and fluid-pressure energy:

$$E_{tv} = \tfrac{1}{2}\rho v^2 + \rho gz + P \tag{5-10}$$

where E_{tv} is the total energy per unit volume.

If Equation (5-10) is divided by ρ, the result is total energy per unit mass, E_{tm}:

$$E_{tm} = \frac{v^2}{2} + gz + \frac{P}{\rho} \tag{5-11}$$

which is known as the **Bernoulli equation**. The derivation of the Bernoulli equation may be found in textbooks on fluid mechanics (1).

For steady flow of a frictionless, incompressible fluid along a smooth line of flow, the sum of the three components is a constant. Each term of Equation (5-11) has the units of $(L/T)^2$:

$$\frac{v^2}{2} + gz + \frac{P}{\rho} = \text{constant} \tag{5-12}$$

Steady flow indicates that the conditions do not change with time. The density of an incompressible fluid would not change with changes in pressure. A frictionless fluid would not require energy to overcome resistance to flow. An ideal fluid would have both of these characteristics; real fluids have neither one. Real fluids are compressible and do suffer frictional flow losses; however, Equation (5-12) is useful for purposes of comparing the components of mechanical energy.

If each term of Equation (5-12) is divided by g, the following expression results:

$$\frac{v^2}{2g} + z + \frac{P}{\rho g} = \text{constant} \tag{5-13}$$

This equation expresses all terms in units of energy per unit weight. These are joules/newtons, or m. Thus, Equation (5-13) has the advantage of having all units in length dimensions (L). The first term of $v^2/2g$ is (m/sec)²/(m/sec²), or m; the second term, z, is already in m; and the third term, $P/\rho g$, is pascals/(kg/m³)(m/sec²), or (N/m²)/(kg/m³)(m/sec²), which reduces to m. The sum of these three factors is the total mechanical energy, known as the **hydraulic head**, h. This is usually measured in the field or laboratory in units of length.

109

5.3 HYDRAULIC HEAD

A **piezometer** is used to measure the total energy of the fluid flowing through a pipe packed with sand, as shown in Figure 5.1. The piezometer is open at the top and bottom, and water rises in it in direct proportion to the total fluid energy at the point at which the bottom of the piezometer is open in the sand. At point A, which is at an elevation, z, above a datum, there is a fluid pressure, P. The fluid is flowing at a velocity, v. The total energy per unit mass can be found from Equation (5-11).

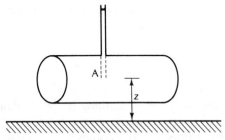

FIGURE 5.1. Piezometer measuring fluid pressure and the elevation of water.

EXAMPLE PROBLEM

At a place where $g = 9.80$ m/sec^2 the fluid pressure is 1500 N/m^2; the distance above a reference elevation is 0.75 m; and the fluid density is 1.02×10^3 kg/m^3. The fluid is moving at a velocity of 10^{-6} m/sec. Find E_{tm}.

$$E_{tm} = gz + \frac{P}{\rho} + \frac{v^2}{2}$$

$$= 9.80 \text{ m/sec}^2 \times 0.75 \text{ m} + \frac{1500 \text{ N/m}^2}{1.02 \times 10^3 \text{ kg/m}^3} + \frac{(10^{-6})^2}{2} \frac{\text{m}^2}{\text{sec}^2}$$

$$= 7.35 \text{ m}^2/\text{sec}^2 + 1.47 \text{ m}^2/\text{sec}^2 + 5.0 \times 10^{-13} \text{ m}^2/\text{sec}^2$$

The total energy per unit mass is 8.82 m^2/sec^2. The energy is almost exclusively in the pressure and gravitational potential energy terms, which are thirteen orders of magnitude greater than the value of kinetic energy.

The preceding problem shows that the amount of energy developed as kinetic energy by flowing groundwater is small. The velocity of groundwater flowing in porous media under natural hydraulic gradients is very low. The example velocity of 10^{-6} meter per second results in a movement of 30 meters per year, which is typical for groundwater.

Velocity components of energy may be safely ignored in groundwater flow because they are so much smaller than the other two terms. By dropping $v^2/2g$ from Equation (5-13), the total hydraulic head, h, is given by the formula

$$h = z + \frac{P}{\rho g} \tag{5-14}$$

Figure 5.2 shows the components of head. The head is the total potential energy per unit weight of water. For a fluid at rest, the pressure at a point is equal to the weight of the overlying water per unit cross-sectional area:

$$P = \rho g h_p \tag{5-15}$$

where h_p is the height of the water column. Substituting into Equation (5-14), we see that

$$h = z + h_p \tag{5-16}$$

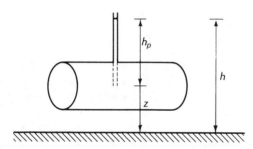

FIGURE 5.2.   Total head, h, elevation head, z, and pressure head, h_p.

The total hydraulic head is equal to the sum of the elevation head and the pressure head. The elevation and pressure heads, when used in the form of Equation (5-16) correlate with energy per unit weight of water with dimensions (L).

EXAMPLE PROBLEM

An aquifer has a total saturated depth of 50 meters. At some point in the aquifer, there are no vertical components of flow. What are the total elevation and pressure heads at the top and bottom of the aquifer?

The bottom of the aquifer is assumed to be the reference elevation. At the bottom of the aquifer, the elevation head is

---

> zero and the pressure head is 50 meters. At the top of the aquifer, the elevation head is 50 meters, but the pressure head is zero. (We are using gauge pressure relating atmospheric pressure to an assumed zero.) The total head in both cases is 50 meters. At an elevation of 25 meters above the bottom, the pressure head is 25 meters, the elevation head 25 meters, and the total head also 50 meters.

5.4 FORCE POTENTIAL AND HYDRAULIC HEAD

In Equation (5-11) we showed the total potential energy per unit mass to be equal to the sum of the kinetic energy, elevation energy, and pressure. This total potential energy has been termed the **force potential**, and is indicated by the capital Greek letter phi, Φ (2):

$$\Phi = gz + \frac{P}{\rho} = gz + \frac{\rho g h_p}{\rho} = g(z + h_p) \tag{5-17}$$

Since $z + h_p = h$, the hydraulic head,

$$\Phi = gh \tag{5-18}$$

The force potential is the driving force behind groundwater flow and is equal to the product of hydraulic head and the acceleration of gravity. Both force potential and hydraulic head are potentials. Hydraulic head is energy per unit weight and force potential is energy per unit mass.

Figure 5.3 shows a pipe filled with sand with water flowing through it from left to right. The pipe can be rotated to any inclination, with the discharge of water remaining constant. In Part A of the figure, the water flows from point 1 (of elevation z_1) to point 2 (of elevation z_2), z_2 being somewhat greater than z_1. In Part B, the slope has been reversed: rather than flowing uphill, the water now flows downhill. However, the fluid pressure head at point 2 (h_{p2}) is greater than at point 1 (h_{p1}). The fluid is thus moving from a region of low pressure to one of higher pressure. Clearly, neither elevation head nor pressure head alone controls groundwater motion. Part C of Figure 5.3 shows equal elevation heads with pressure head declining in the direction of flow. Part D has equal pressure heads, but elevation head declines in the direction of flow.

In this example the total hydraulic head showed the same decrease in the direction of flow. This would be true no matter what the inclination of the pipe, so long as other factors remained constant. Since the force potential is the controlling force in groundwater flow, this demonstrates that the proportion of pressure and elevation head is not a factor.

From Figure 5.3, we see that the force potential and, hence, hydraulic head, decrease in the direction of flow. As groundwater moves, it

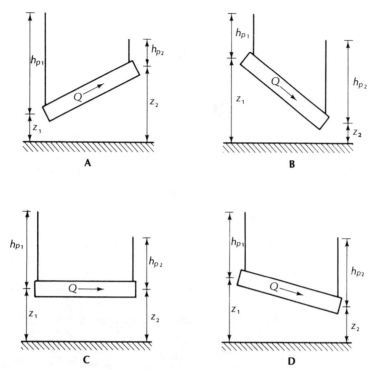

FIGURE 5.3. Apparatus to demonstrate how changing the slope of a pipe packed with sand will change the components of elevation, z, and pressure, h_p, heads. The direction of flow, Q, is indicated by the arrow.

encounters frictional resistance between the fluid and the porous media. The smaller the openings through which the fluid moves, the greater the friction. In overcoming the frictional resistance, some of the force potential is lost. It is transformed into heat (a lower form of energy). Thus, groundwater is warmed slightly as it flows, and potential energy is converted to thermal energy. Under most circumstances, the resulting change in temperature is not measurable.

DARCY'S LAW 5.5

5.5.1 DARCY'S LAW IN TERMS OF FORCE AND POTENTIAL

In Section 4.3 it was shown that flow through a pipe filled with sand is proportional to the decrease in hydraulic head divided by the length of the pipe. This ratio is called the **hydraulic gradient**. It should now be apparent that the hydraulic head is the sum of the pressure head and the elevation head. Expressed in terms of hydraulic head, Darcy's law is

113

$$Q = -KA \frac{dh}{dl} \tag{5-19}$$

Since the fluid potential, Φ is equal to gh, Darcy's law can also be expressed in terms of fluid potential as (2)

$$Q = -\frac{KA}{g} \frac{d\Phi}{dl} \tag{5-20}$$

As expressed above, Darcy's law is in a one-dimensional form, as water flows through the pipe in only one direction. In later sections, we will examine various forms of Darcy's law for two and three directions.

5.5.2 THE APPLICABILITY OF DARCY'S LAW

When a fluid at rest starts to move, it must overcome resistance to flow, due to the viscosity of the fluid. Slowly moving fluids are dominated by viscous forces. There is a low energy level and the resulting fluid flow is **laminar**. In laminar flow, molecules of water follow smooth lines, called **streamlines** (Figure 5.4A).

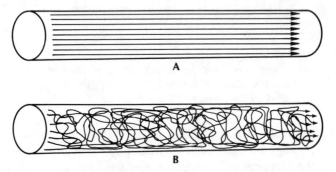

FIGURE 5.4. **A.** Flow paths of molecules of water in laminar flow; **B.** Flow paths of molecules of water in turbulent flow.

As the velocity of flow increases, the moving fluid gains kinetic energy. Eventually, the inertial forces due to movement are more influential than the viscous forces, and the fluid particles begin to rush past each other in an erratic fashion. The result is **turbulent flow**, in which the water molecules no longer move along parallel streamlines (Figure 5.4B).

The **Reynolds number** relates the four factors that determine whether the flow will be laminar or turbulent (1):

$$R = \frac{\rho v d}{\mu} \tag{5-21}$$

where

R is the Reynolds number

ρ is the fluid density

v is the fluid velocity

d is the diameter of the passageway through which the fluid moves

μ is the viscosity

For open-channel or pipe flow, d is simply the channel width or pipe diameter. In such cases, the transition from laminar to fluid flow occurs when the average velocity is such that R exceeds a value of 2000 (1). For a porous medium, however, it is not easy to determine the value of d. Rather than an average or characteristic pore diameter, the average grain diameter is often used.

Turbulence in groundwater flow is difficult to detect. The inception of fluid turbulent flow in groundwater has been reported at a Reynolds number ranging from 60 (3) to 600 (4). However, experimentation has shown that Darcy's law is valid only when conditions are such that the resistive forces of viscosity predominate. These conditions prevail when the Reynolds number is in the range of 1 to 10 (5, 6, 7). This means that Darcy's law only applies to very slowly moving groundwaters. It is possible to have laminar groundwater flow, but under conditions such that the Reynolds number is so great as to invalidate Darcy's law. Under most natural groundwater conditions, the velocity is sufficiently low for Darcy's law to be valid. Exceptions might be areas of rock with large openings, such as solution openings and basalt flows. Likewise, areas of steep hydraulic gradients, such as the vicinity of a pumping well, might result in high velocities with a correspondingly high Reynolds number.

EXAMPLE PROBLEM

A sand aquifer has a median grain diameter of 0.5 millimeter. For pure water at 15° C, what is the greatest velocity for which Darcy's law is valid?

At 15° C,

$$\rho = 0.999 \times 10^3 \text{ kg/m}^3$$

$$\mu = 1.15 \times 10^{-3} \text{ Pa-sec}$$

$$R = \frac{\rho v d}{\mu}$$

$$v = \frac{R\mu}{\rho d}$$

A pascal-second (Pa-sec) is a N-sec/m². If R cannot exceed 10,

$$v = \frac{10 \times 0.00115 \text{ N-sec/m}^2}{0.999 \times 10^3 \text{ kg/m}^3 \times 0.0005 \text{ m}}$$

$$= 0.023 \text{ m/sec}$$

Darcy's law will be valid for velocities equal to or less than 0.023 meters per second.

5.5.3 DISCHARGE AND SEEPAGE VELOCITIES

When water flows through an open channel or a pipe, the discharge, Q, is equal to the product of the velocity, v, and the cross-sectional area of flow, A:

$$Q = vA \qquad (5\text{-}22)$$

Rearrangement of Equation (5-22) yields an expression for velocity,

$$v = Q/A \qquad (5\text{-}23)$$

One can apply the same reasoning to Equation (5-19), Darcy's law, for flow through a porous medium:

$$v = \frac{Q}{A} = -K\frac{dh}{dl} \qquad (5\text{-}24)$$

A moment's reflection will reveal that this velocity is not quite the same as the velocity of water flowing through an open pipe. The discharge is measured as water coming from the pipe. In an open pipe, the cross-sectional area of flow inside the pipe is equivalent to the area of the end of the pipe. However, if the pipe is filled with sand, the open area through which water may flow is much smaller than the cross-sectional area of the pipe. The velocity of flow determined by Equation (5-24) is termed the **discharge velocity**.* It is an apparent velocity, representing the velocity at which water would move through an aquifer if the aquifer were an open conduit.

The cross-sectional area of flow for a porous medium is actually much smaller than the dimensions of the aquifer. It is equal to the product of the effective porosity of the aquifer material and the physical dimensions. Water can move only through the pore spaces. Moreover, part of the pore space is occupied by stagnant water, which clings to the rock material. The effective porosity is that portion of the pore space through which saturated flow occurs. To find the velocity at which water is actually moving, the discharge velocity is divided by the effective porosity to account for the actual open space available for flow. The result is the **seepage velocity**—a true velocity representing the rate at which water actually moves through the pore spaces. It is greater than the discharge velocity:

$$v_s = \frac{Q}{n_e A} = -\frac{Kdh}{n_e dl} \qquad (5\text{-}25)$$

where

v_s is the seepage velocity

n_e is the effective porosity

A rigorous approach would also consider the **sinuosity** of the flow paths. Water must move around sand grains to get from pore to pore, so the flow paths are longer than straight-line flow paths. However, values for

---

*The term specific discharge is also used.

sinuosity are not ordinarily known; they are generally assumed to be 1. This assumption is inherent in Equation (5-25).

EXAMPLE PROBLEM

Compute the discharge and seepage velocities for water flowing through a pipe filled with sand, with $K = 10^{-4}$ centimeter per second, $dh/dl = 0.01$, $A = 75$ square centimeters, and $n_e = 0.22$.

Discharge velocity:

$$v = \frac{Q}{A} = K\frac{dh}{dl}$$

$$= 10^{-4} \text{ cm/sec} \times 0.01$$

$$= 10^{-6} \text{ cm/sec}$$

Seepage velocity:

$$v_s = \frac{Q}{n_e A} = \frac{Kdh}{n_e dl}$$

$$= \frac{1}{0.22} \times 10^{-4} \times 0.01$$

$$= 4.54 \times 10^{-6} \text{ cm/sec}$$

At a rate of 4.54×10^{-6} centimeter per second, the water would move 1.43 meters in one year.

PERMEAMETERS

The value of the hydraulic conductivity of earth materials can be measured in the laboratory. Not surprisingly, the devices used to do this are called **permeameters**.

Permeameters all have some type of a chamber to hold a sample of rock or sediment. Rock permeameters hold a core of solid rock, usually cylindrical. Unconsolidated samples may be remolded into the permeameter chamber. It is also possible to make permeability analyses of "undisturbed" samples of unconsolidated materials if they are left in the field-sampling tubes, which become the permeameter sample chambers. If sediments are repacked into the permeameter, they will yield values of hydraulic conductivity that only approximate the value of K for undisturbed material.

The **constant-head permeameter** is used for noncohesive sediments, such as sand and rocks. A chamber with an overflow provides a supply

117

of water at a constant head. Water moves through the sample at a steady rate. The hydraulic conductivity is determined from a variation of Darcy's law:

$$K = \frac{VL}{Ath} \qquad (5\text{-}26)$$

where

V is the volume of water discharging in time, t
L is the length of the sample
A is the cross-sectional area of the sample
h is the hydraulic head
K is the hydraulic conductivity

A constant-head permeameter is illustrated in Figure 5.5. It is critical in such permeameters to have hydraulic gradients approaching those in the field. The head should never be more than about 0.5 of the sample length. Some commercial permeameters permit heads of up to ten times the sample length. Under such conditions, the Reynolds number may become so high that Darcy's law is invalidated. If the permeameter is designed for upward flow, too great an upward-flow velocity may result in quicksand conditions in the permeameter.

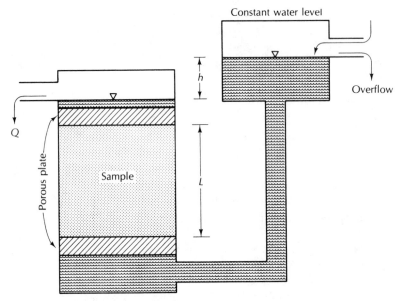

FIGURE 5.5. Constant-head permeameter apparatus.

For cohesive sediments with low conductivities, a **falling-head permeameter** is used (Figure 5.6). A much smaller volume of water moves through the sample. A falling-head tube is attached to the permeameter. The initial water level above the outlet in the falling-head tube, h_0, is noted. After

118

some time period, t (generally several hours), the water level is again measured, h. The inside diameter of the falling head tube, d_t, the length of the sample, L, and the diameter of the sample, d_c, must also be known. The conductivity, K, is found by the formula

$$K = \frac{d_t^2 L}{d_c^2 t} \ln \left(\frac{h_0}{h} \right) \qquad \text{(5-27)}$$

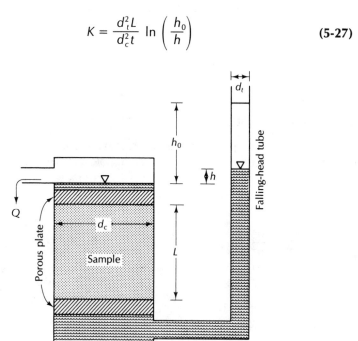

FIGURE 5.6. Falling-head permeameter apparatus.

In using any permeameter, it is critical that the sample be completely saturated. Air bubbles in the sample will reduce the cross-sectional flow area, resulting in lowered measurements of conductivity. The sample must also be tightly pressed against the sidewall of the chamber. If it is not, water may move along the sidewall, avoiding the porous medium. In this case, measurements of conductivity may be too great.

EXAMPLE PROBLEM

A constant-head permeameter has a sample of medium-grained sand 15 centimeters in length and 25 square centimeters in cross-sectional area. With a head of 5 centimeters, a total of 100 milliliters of water is collected in 12 minutes. Find the hydraulic conductivity.

$$K = \frac{QL}{Ath}$$

119

$$K = \frac{100 \text{ cm}^3 \times 15 \text{ cm}}{25 \text{ cm}^2 \times 12 \text{ min} \times 60 \text{ sec/min} \times 5 \text{ cm}}$$

$$= 1.67 \times 10^{-2} \text{ cm/sec} \quad \text{or} \quad 14.4 \text{ m/day}$$

NOTE: Units must be such that the resultant answer is in the desired length/time units.

EXAMPLE PROBLEM

A falling-head permeameter containing a silty, fine sand has a falling-head tube diameter of 2 centimeters, a sample diameter of 10 centimeters, and a flow length of 15 centimeters. The initial head is 5 centimeters. It falls to 0.5 centimeter over a period of 528 minutes. Find the hydraulic conductivity.

$$K = \frac{d_t^2 L}{d_c^2 t} \ln \left(\frac{h_o}{h} \right)$$

$$= \frac{2^2 \text{cm}^2}{10^2 \text{cm}^2} \times \frac{15 \text{ cm}}{528 \text{ min} \times 60 \text{ sec/min}} \times \ln \frac{5 \text{ cm}}{0.5 \text{ cm}}$$

$$= 4.36 \times 10^{-5} \text{ cm/sec} \quad \text{or} \quad 3.76 \times 10^{-2} \text{ m/day}$$

 EQUATIONS OF GROUNDWATER FLOW*

5.7.1 CONFINED AQUIFERS

The flow of fluids through porous media is governed by the laws of physics. As such, it can be described by differential equations. Since the flow is a function of several variables, it is usually described by partial differential equations in which the spatial coordinates, x, y, and z, and time, t, are the independent variables.

In deriving the equations, the laws of conservation for mass and energy are employed. The **law of mass conservation,** or **continuity principle,** states that there can be no net change in the mass of a fluid contained in a small

---

*The main equation of groundwater flow is derived in this section following a method used by Jacob (8, 9) and modified by Domenico (10). Those not familiar with differential calculus can skip this material without compromising practical understanding.

volume of an aquifer. Any change in mass flowing into the small volume of the aquifer must be balanced by a corresponding change in mass flux out of the volume, or a change in the mass stored in the volume, or both. The **law of conservation of energy** is also known as the **first law of thermodynamics**. It states that within any closed system there is a constant amount of energy, which can neither be lost nor increased. It can, however, change form. The **second law of thermodynamics** implies that when energy changes forms, it tends to go from a more useful form, such as mechanical energy, to a less useful form, such as heat. Based upon these principles and Darcy's law, the main equations of groundwater flow have been derived (8–11).

We will consider a very small part of the aquifer, called a **control volume**. The three sides are of lengths dx, dy, and dz, respectively. The area of the faces normal to the x-axis is $dydz$; the area of the faces normal to the y-axis is $dxdz$; and the area of the faces normal to the z-axis is $dxdy$ (Figure 5.7).

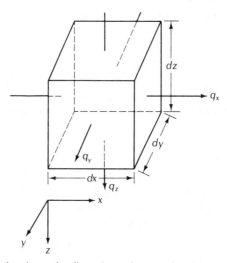

FIGURE 5.7. Control volume for flow through a confined aquifer.

Assume the aquifer is homogeneous and isotropic. The fluid moves in only one direction through the control volume. However, the actual fluid motion can be subdivided on the basis of the components of flow parallel to the three principal axes. If Q is the total flow, $\rho_w q_x$ is the portion parallel to the x-axis, etc., where ρ_w is the fluid density.

The mass flux into the control volume is $\rho_w q_x \, dydz$ along the x-axis. The mass flux out of the control volume is $\rho_w q_x \, dydz + \dfrac{\delta}{\delta x}(\rho_w q_x) \, dx \, dydz$. The net accumulation in the control volume due to movement parallel to the x-axis is equal to the inflow less the outflow, or $-\dfrac{\delta}{\delta x}(\rho_w q_x) \, dx \, dydz$. Since there are flow components along all three axes, similar terms can be determined for the other two directions: $-\dfrac{\delta}{\delta y}(\rho_w q_y)\,dy\,dxdz$ and $-\dfrac{\delta}{\delta z}(\rho_w q_z)\,dz\,dxdy$. Combining these three terms yields the net total accumulation of mass in the control volume:

$$-\left(\frac{\delta}{\delta x}\,\rho_w q_x + \frac{\delta}{\delta y}\,\rho_w q_y + \frac{\delta}{\delta z}\,\rho_w q_z\right)dxdydz \qquad (5\text{-}28)$$

The volume of water in the control volume is equal to $n\,dxdydz$, where n is the porosity. The initial mass of the water is thus $\rho_w n\,dxdydz$. The volume of solid material is $(1-n)\,dxdydz$. Any change in the mass of water, M, with respect to time is given by

$$\frac{\delta M}{\delta t} = \frac{\delta}{\delta t}(\rho_w n\,dxdydz) \qquad (5\text{-}29)$$

As the pressure in the control volume changes, the fluid density will change, as will the porosity of the aquifer. The compressibility of water, β, is defined as the rate of change in density with a change in pressure, P:

$$\beta dP = \frac{d\rho_w}{\rho_w} \qquad (5\text{-}30)$$

The aquifer also changes in volume with a change in pressure. We will assume the only change is vertical. The aquifer compressibility, α, is given by

$$\alpha dP = \frac{d(dz)}{dz} \qquad (5\text{-}31)$$

As the aquifer compresses or expands, n will change, but the volume of solids, V_s, will be constant. Likewise, if the only deformation is in the z-direction, $d(dx)$ and $d(dy)$ will equal zero:

$$dV_s = 0 = d[(1-n)\,dxdydz] \qquad (5\text{-}32)$$

Differentiation of Equation (5-32) yields

$$dzdn = (1-n)d(dz) \qquad (5\text{-}33)$$

and

$$dn = \frac{(1-n)d(dz)}{dz} \qquad (5\text{-}34)$$

The pressure, P, at a point in the aquifer is equal to $P_0 + \rho_w gh$, where P_0 is atmospheric pressure, a constant, and h is the height of a column of water above the point. Therefore, $dP = \rho_w g\,dh$, and Equations (5-30) and (5-31) become

$$d\rho_w = \rho_w \beta(\rho_w g\,dh) \qquad (5\text{-}35)$$

and

$$d(dz) = -dz\alpha(\rho_w g\ dh) \tag{5-36}$$

Equation (5-34) can be rearranged if $d(dz)$ is replaced by Equation (5-36):

$$dn = (n - 1)\alpha\rho_w g\ dh \tag{5-37}$$

If dx and dy are constant, the equation for change of mass with time in the control volume, Equation (5-29), can be expressed as

$$\frac{\delta M}{\delta t} = \left\{ \rho_w n \frac{\delta(dz)}{\delta t} + \rho_w dz \frac{\delta n}{\delta t} + ndz \frac{\delta\rho_w}{\delta t} \right\}\ dxdy \tag{5-38}$$

Substitution of Equations (5-35), (5-36), and (5-37) into Equation (5-38) yields, after minor manipulation,

$$\frac{\delta M}{\delta t} = (\alpha\rho_w g + n\beta\rho_w g)\,\rho_w\ dxdydz\ \frac{\delta h}{\delta t} \tag{5-39}$$

The net accumulation of material expressed as Equation (5-28) is equal to Equation (5-39), the change in mass with time:

$$-\left\{ \frac{\delta(q_x)}{\delta x} + \frac{\delta(q_y)}{\delta y} + \frac{\delta(q_z)}{\delta z} \right\}\ \rho_w\ dxdydz = (\alpha\rho_w g + n\beta\rho_w g)\rho_w\ dxdydz\ \frac{\delta h}{\delta t} \tag{5-40}$$

From Darcy's law,

$$q_x = -K\frac{\delta h}{\delta x} \tag{5-41}$$

$$q_y = -K\frac{\delta h}{\delta y} \tag{5-42}$$

and

$$q_z = -K\frac{\delta h}{\delta z} \tag{5-43}$$

Substituting these into Equation (5-40) yields the main equation of flow for a confined aquifer:

$$K\left(\frac{\delta^2 h}{\delta x^2} + \frac{\delta^2 h}{\delta y^2} + \frac{\delta^2 h}{\delta z^2} \right) = (\alpha\rho_w g + n\beta\rho_w g)\frac{\delta h}{\delta t} \tag{5-44}$$

which is a general equation for flow in three dimensions. For two-dimensional flow with no vertical components, the equation can be rearranged and terms introduced for the storativity, $S = m(\alpha\rho_w g + n\beta\rho_w g)$, and transmissivity, $T = Km$, where m is the aquifer thickness:

$$\frac{\delta^2 h}{\delta x^2} + \frac{\delta^2 h}{\delta y^2} = \frac{S}{T}\frac{\delta h}{\delta t} \tag{5-45}$$

123

In steady-state flow, there is no change in head with time; for example, in cases in which there is no change in the position or slope of the water table. Under such conditions, time is not one of the independent variables, and steady flow is described by the three-dimensional partial differential equation, known as the **Laplace equation**:

$$\frac{\delta^2 h}{\delta x^2} + \frac{\delta^2 h}{\delta y^2} + \frac{\delta^2 h}{\delta z^2} = 0 \qquad (5\text{-}46)$$

The preceding equations are based on the assumption that all flow comes from water stored in the aquifer. In the field, it is more often than not the case that significant flow is generated from leakage into the aquifer through overlying or underlying confining layers. We will consider the leakage to appear in the control volume as horizontal flow. This assumption is justified on the grounds that the conductivity of the aquifer is usually orders of magnitude greater than that of the confining layer. The law of refraction indicates that, for these conditions, flow in the confining layer will be nearly vertical if flow in the aquifer is horizontal.

The leakage rate, or rate of accumulation, is designated as e. The general equation of flow (in two dimensions, since horizontal flow was assumed) is given by

$$\frac{\delta^2 h}{\delta x^2} + \frac{\delta^2 h}{\delta y^2} + \frac{e}{T} = \frac{S}{T}\frac{\delta h}{\delta t} \qquad (5\text{-}47)$$

The rate of vertical movement, e, can be determined from Darcy's law. If the head at the top of the aquitard is h_0 and the head in the aquifer just below the aquitard is h, the aquitard has a thickness b' and a conductivity (vertical) of K':

$$e = K'\frac{(h_0 - h)}{b'} \qquad (5\text{-}48)$$

5.7.2   UNCONFINED AQUIFERS

Water is derived from storage in water-table aquifers by vertical drainage of water in the pores. This drainage results in a decline in the position of the water table near a pumping well as time progresses. In the case of a confined aquifer, although the potentiometric surface declined, the saturated thickness of the aquifer remained constant. In the case of an unconfined aquifer, the saturated thickness can change with time. Under such conditions, the ability of the aquifer to transmit water—the transmissivity—changes, as it is the product of the conductivity (K) and the saturated thickness (h).

The general flow equation for two-dimensional unconfined flow is known as the **Boussinesq equation** (14):

$$\frac{\delta}{\delta x}\left(h\,\frac{\delta h}{\delta x}\right) + \frac{\delta}{\delta y}\left(h\,\frac{\delta h}{\delta y}\right) = \frac{S_y}{K}\frac{\delta h}{\delta t} \qquad (5\text{-}49)$$

where S_y is specific yield. This equation is a type of differential equation that cannot be solved using calculus, except in some very specific cases. In mathematical terms, it is nonlinear.

   If the drawdown in the aquifer is very small compared with the saturated thickness, the variable thickness, h, can be replaced with an average thickness that is assumed to be constant over the aquifer (m). The Boussinesq equation can thus be linearized by this approximation to the form

$$\frac{\delta^2 h}{\delta x^2} + \frac{\delta^2 h}{\delta y^2} = \frac{S_y}{Km}\frac{\delta h}{\delta t} \qquad (5\text{-}50)$$

which has the same form as Equation (5-45).

SOLUTION OF FLOW EQUATIONS

The flow of water in an aquifer can be mathematically described by either Equation (5-44), (5-45), (5-46), (5-47), or (5-50). These are all partial differential equations in which the head, h, is described in terms of the variables x, y, z, and t. They are solved by means of a mathematical model consisting of the applicable governing flow equation, equations describing the hydraulic head at each of the boundaries of the aquifer, and equations describing the initial conditions of head in the aquifer.

   If the aquifer is homogeneous and isotropic, and the boundaries can be described with algebraic equations, then the mathematical model can be solved using an analytical solution based on integral calculus. However, if the aquifer does not correspond to those conditions, (e.g., a layered aquifer), then a numerical solution to the mathematical model is needed. Numerical solutions are based on the concept that the partial differential equation can be replaced by a similar equation that can be solved using arithmetic. Likewise, the equations governing initial and boundary conditions are replaced by numerical statements of these conditions. Numerical solutions are typically solved on digital computers. The use of digital-computer models is treated in Chapter 12.

GRADIENT OF HYDRAULIC HEAD

The potential energy, or force potential, Φ, of groundwater consists of two parts: elevation and pressure. It is equal to the product of the acceleration of gravity and the total head (2), and represents potential energy per unit mass:

$$\Phi = gh \qquad (5\text{-}51)$$

   Force potential is a physical quantity. To obtain it one only needs to measure the heads in an aquifer with piezometers and multiply the results by the acceleration of gravity. If a point in an aquifer has a head of 15.1 meters, and the value of g is 9.81 m/sec^2, then Φ is $15.1 \times 9.81 = 148.1$ m^2/sec^2. For practical purposes, as g is usually constant throughout an area, most field problems are solved in terms of hydraulic head, h.

If the value of h is variable in an aquifer, a contour map may be made showing lines of equal value of h (equipotential lines). Such a map is similar to a topographic map of the land-surface elevation. In three-dimensional cases, one deals with surfaces of equal value of h (equipotential surfaces).

Figure 5.8 shows a family of equipotential lines for a two-dimensional field. Also shown is a vector, known as the gradient of h (grad h). Remember, a vector is a directed line segment, so grad h has a value and a direction. It is roughly analogous to the maximum slope of the potential field. In the notation of differential calculus,

$$\text{grad } h = \frac{dh}{ds} \qquad (5\text{-}52)$$

Grad h has a direction perpendicular to the equipotential lines.

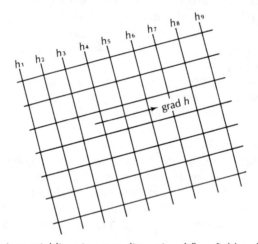

FIGURE 5.8. Equipotential lines in a two-dimensional flow field and the gradient of h.

For isotropic aquifers, the value of K is equal in all directions. In such aquifers, the fluid flow will be parallel to grad Φ, which means that it will also be perpendicular to the equipotential lines. By contrast, anisotropic media have differing values of K in different directions. Under such conditions, the flowlines will not be parallel, but rather oblique to the direction of grad h. They will thus cross the equipotential lines at other than a 90-degree angle; however they cannot be parallel to the equipotential lines.

If the potential is the same everywhere, it will be manifest in a condition such as a flat water table. In this case, grad h equals zero, since there is no slope to h. There will be no groundwater flow, since grad h must have a positive value before groundwater will move.

FLOW NETS 5.10

The Laplace equation (p. 124) may be solved for some types of problems by means of the construction of a **flow net** showing equipotential lines and streamlines (12, 13). It is an appropriate method for two-dimensional problems in an isotropic medium; especially adaptable to problems in which the flow boundaries are known before the solution is attempted. Thus, hydrogeological field work is necessary before a solution is undertaken.

Flow nets are bounded by either equipotential lines or streamlines. In Figure 5.9, a flow net is drawn for seepage through an earthen dam resting on an impervious surface. The water table in the dam is represented by Line A-B. It is a streamline. Line D-C is also a streamline. Line A-D is an equipotential line, with the head equal to the depth of the impounded water. Line B-C is also an equipotential line, the head being equal to the depth of the tailwater.

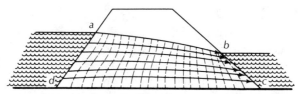

FIGURE 5.9.   A flow net for seepage through an earthen dam resting on an impervious surface.

A flow net is made by trial-and-error process. The boundaries of the flow region are carefully drawn to scale and a few streamlines sketched on the drawing. They should be evenly spaced across the width of the flow region. These are only a few of the infinite number of streamlines. In many flow nets, at least two of them will be boundaries of the flow region. The streamlines will begin and end at equipotential surfaces. They must intersect these equipotential surfaces at right angles.

The final step is to add the intermediate equipotential lines. They must intersect all streamlines at right angles, including the boundary streamlines. The equipotential lines and streamlines should bound areas that are approximately square. At certain odd angles in the boundaries, these areas may be three- or five-sided. Most beginners at the art of flow-net construction will find an ample supply of paper, pencils, and erasers essential. As a check on the quality of the flow net, the diagonals of the squares can be drawn. These should form smooth curves that intersect each other at right angles. This should, of course, be done on a copy of the final product.

In addition to presenting a graphic display of the groundwater flow, the completed flow net can be used to determine the quantity of water flowing by the following formula:

127

$$Q = \frac{Kmh}{n}$$ (5-53)

where

    Q is the total volume of flow

    K is the hydraulic conductivity

    m is the number of flow channels bounded by adjacent pairs of streamlines

    h is the total head loss over the length of the streamlines

    n is the number of squares bounded by any two adjacent streamlines and covering the entire length of flow

---

EXAMPLE PROBLEM

In Figure 5.9 on the preceding page there is an earthen dam 13 meters across and 7.5 meters high. The impounded water is 6.2 meters deep, while the tailwater is 2.2 meters deep. The dam is 72 meters long. If the hydraulic conductivity is 6.1×10^{-4} centimeter per second, what is the seepage through the dam?

$$K = 6.1 \times 10^{-4} \text{ cm/sec} = 0.527 \text{ m/day}$$

From the flow net, the total head loss, h, is $6.2 - 2.2 = 4.0$ meters. There are 6 flow channels (m) and 21 squares of equipotential drop along each flow path (n):

$$Q = \frac{Kmh}{n} \times \text{dam length}$$

$$= \frac{0.527 \text{ m/day} \times 6 \times 4 \text{ m}}{21} \times \text{dam length}$$

$= 0.60$ m³/day for each meter of dam length

$= 0.60 \times 72 = 43.2$ m³/day for the entire 72-meter length of the dam

---

The preceding problem illustrates one of the pitfalls of two-dimensional problem solutions. It must be recognized that two-dimensional problems imply a third dimension, with an axis of symmetry perpendicular to the two-dimensional representation. The length of this axis must be included to determine the total volume of flow. An alternative method is to state flow in terms of discharge per unit width. For an aquifer, flow might be stated in cubic meters per day per kilometer width of the aquifer (measured orthogonal to the direction of flow).

Further examples of flow nets are given in Figure 5.10. These represent flow beneath a dam with a cutoff and flow beneath a sheet pile.

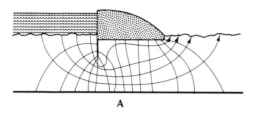

A

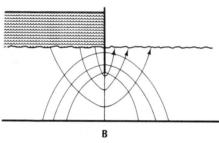

B

FIGURE 5.10. **A.** Flow net for flow beneath a dam with a cutoff; **B.** Flow beneath a vertical sheet piling.

REFRACTION OF STREAMLINES

When flow passes from one stratum to another stratum of different hydraulic conductivity, the direction of flow changes (2). The same type of refraction occurs when seismic waves pass through layers of the earth, or light waves pass from air into water. From the continuity principle and Darcy's law, when a fluid moves into a medium with higher permeability, less of the aquifer area is needed to transmit the fluid. Therefore, the flow channels can be narrower. If the flow moves from a region of higher to lower permeability, then the flow channels must be wider to accommodate the same volume of flow.

By analogy to the physical laws governing refraction of light, we know that the ratio of the hydraulic conductivities is equal to the ratio of the tangent of the angles made by the flow paths with a line perpendicular to the boundary (Figure 5.11A):

$$\frac{K_1}{K_2} = \frac{\tan \sigma_1}{\tan \sigma_2} \qquad (5\text{-}54)$$

129

As a consequence, the direction of refraction for flow going from a region of low conductivity to one of high conductivity will be different from that for flow going from high to low conductivity (Figure 5.11B,C). Likewise, if the streamlines are refracted, and they are perpendicular to the equipotential lines, then the equipotential lines must also be refracted. Figure 5.12 shows a portion of a flow net crossing a conductivity boundary.

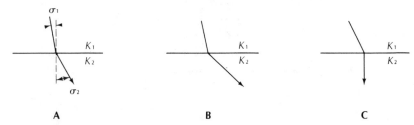

A   B   C

FIGURE 5.11. **A.** Refraction of a streamline crossing a conductivity boundary; **B.** Refracted streamline going from a region of low to high conductivity; **C.** Refracted streamline going from a region of high to low conductivity.

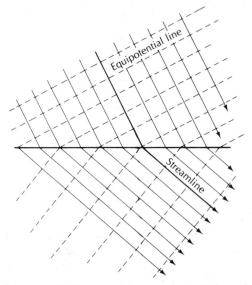

FIGURE 5.12. A flow net with flow crossing a conductivity boundary showing refraction of streamlines and equipotential lines.

5.12 STEADY FLOW IN A CONFINED AQUIFER

If there is the steady movement of groundwater in a confined aquifer, there will be a gradient or slope to the potentiometric surface of the aquifer. Likewise, we know that the water will be moving in the opposite direction of grad h. For flow of this type, Darcy's law may be used directly. In Figure 5.13, a portion of a confined aquifer of uniform thickness is shown. The potentiometric surface has a linear gradient; i.e., its two-directional projection is a straight line. There are two observation wells where the hydraulic head can be measured.

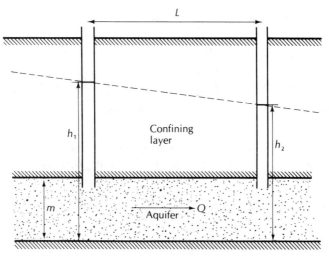

FIGURE 5.13. Steady flow through a confined aquifer of uniform thickness.

The quantity of flow per unit width, q, may be determined from Darcy's law:

$$q = -Km \frac{dh}{dl}$$ (5-55)

where

K is the hydraulic conductivity

m is the aquifer thickness

$\dfrac{dh}{dl}$ is the slope of potentiometric surface

q is the flow per unit width

One may wish to know the head, h, at some intermediate distance, a, between h_1 and h_2. This may be found from the equation

$$h = h_1 - \frac{Q}{Km} a$$ (5-56)

131

EXAMPLE PROBLEM

A confined aquifer is 33 meters thick and 7 kilometers wide. Two observation wells are located 1.2 kilometers apart in the direction of flow. The head in Well 1 is 97.5 meters and in Well 2 it is 89.0 meters. The hydraulic conductivity is 1.2 meters per day. What is the total daily flow of water through the aquifer?

$$Q = -Km \frac{dh}{dl} \text{ width}$$

$$= 1.2 \text{ m/day} \times 33 \text{ m} \times \frac{97.5 \text{ m} - 89.0 \text{ m}}{1200 \text{ m}} \times 7000 \text{ m}$$

$$= 1963.5 \text{ m}^3\text{/day}$$

What is the elevation of the piezometric surface at a point located 0.3 kilometer from Well h_1 and 0.9 kilometer from Well h_2?

$$h = h_1 - \frac{Q}{Km} a$$

Discharge per unit width is 1963.5 m³/day/7000 m = 0.2805 m³/day:

$$h = 97.5 \text{ m} - \frac{0.2805 \text{ m}^3\text{/day}}{1.2 \text{ m/day} \times 33 \text{ m}} \times 300 \text{ m}$$

$$= 97.5 - 2.125$$

$$= 95.375 \text{ m}$$

5.13 STEADY FLOW IN AN UNCONFINED AQUIFER*

In an unconfined aquifer, the fact that the water table is also the upper boundary of the region of flow complicates flow determinations. Figure 5.14 illustrates the problem. On the left side of the figure, the saturated flow region is h_1 feet thick. On the right side, it is h_2 feet thick, which is $h_1 - h_2$ feet thinner than the left side. If there is no recharge or evaporation as the flow traverses the region, the quantity of water flowing through the left side is equal to the right

---

*The equations in this section are derived following methods used by Polubarinova-Kochina (16) and Haar (17). Those not familiar with calculus may skip the material from Equation (5-57) to Equation (5-69) without compromising practical understanding.

side. From Darcy's law, it is obvious that since the cross-sectional area is smaller on the right side, the hydraulic gradient must be greater. Thus, the gradient of the water table in unconfined flow is not constant; it increases in the direction of flow.

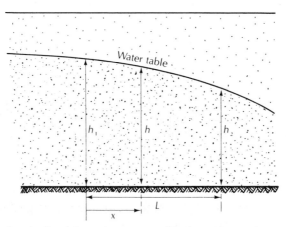

FIGURE 5.14.   Steady flow through an unconfined aquifer resting on a horizontal impervious surface.

This problem was solved by Dupuit (15), and his assumptions are known as **Dupuit flow**. The assumptions are that (a) the hydraulic gradient is equal to the slope of the water table and (b) for small water-table gradients, the streamlines are horizontal and the equipotential lines are vertical. Solutions based on these assumptions have proved to be very useful in many practical problems. However, the Dupuit assumptions do not allow for a seepage face above the outflow side.

From Darcy's law,

$$q = -Kh\frac{dh}{dx} \tag{5-57}$$

where h is the saturated thickness of the aquifer. At $x = 0$, $h = h_1$; at $x = L$, $h = h_2$.

Equation (5-57) may be set up for integration with the boundary conditions:

$$\int_0^L q\, dx = -K\int_{h_1}^{h_2} h\, dh$$

Integration of the above yields

$$qx\Big|_0^L = -K\frac{h^2}{2}\Big|_{h_1}^{h_2}$$

133

Substitution of the boundary conditions for x and h yields

$$qL = -\frac{K}{2}\left(\frac{h_2^2}{2} - \frac{h_1^2}{2}\right) \tag{5-58}$$

Rearrangement of Equation (5-58) yields the **Dupuit equation**:

$$q = -\frac{1}{2}K\left(\frac{h_2^2 - h_1^2}{L}\right) \tag{5-59}$$

If we consider a small prism of the unconfined aquifer, it will have the shape of Figure 5.15. On one side it is h units high and slopes in the x-direction. Given the Dupuit assumptions, there is no flow in the z-direction. The flow in the x-direction, per unit width, is q_x. From Darcy's law, the total flow in the x-direction through the left face of the prism is

$$q_x dy = -K\left(h\,\frac{\delta h}{\delta x}\right)_x dy \tag{5-60}$$

where dy is the width of the face of the prism. The discharge through the right face, q_{x+dx}, is

$$q_{x+dx}\,dy = -K\left(h\,\frac{\delta h}{\delta x}\right)_{x+dx} dy \tag{5-61}$$

Note that $\left(h\,\dfrac{\delta h}{\delta x}\right)$ has different values at each face. The change in flow rate in the x-direction between the two faces is given by

$$(q_{x+dx} - q_x)dy = -K\,\frac{\delta}{\delta x}\left(h\,\frac{\delta h}{\delta x}\right)dxdy \tag{5-62}$$

Through a similar process, it can be shown that the change in the flow rate in the y-direction is

$$(q_{y+dy} - q_y)dx = -K\,\frac{\delta}{\delta y}\left(h\,\frac{\delta h}{\delta y}\right)dydx \tag{5-63}$$

For steady flow, any change in flow through the prism must be equal to a gain or loss of water across the water table. This could be infiltration or evapotranspiration. The net addition or loss is at a rate of w, and the volume change within the initial volume is w dxdy where dxdy is the area of the surface. If w represents evapotranspiration, it will have a negative value. As the change in flow is equal to the net addition,

$$-K\,\frac{\delta}{\delta x}\left(h\,\frac{\delta h}{\delta x}\right)dxdy - K\,\frac{\delta}{\delta y}\left(h\,\frac{\delta h}{\delta y}\right)dydx = w\,dxdy \tag{5-64}$$

We can simplify Equation (5-64) by dropping out dxdy and combining the differentials:

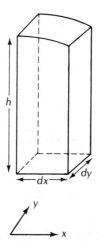

FIGURE 5.15. Control volume for flow through a prism of an unconfined aquifer with the bottom resting on a horizontal impervious surface and the top coinciding with the water table.

$$-K \left(\frac{\delta^2 h^2}{\delta x^2} + \frac{\delta^2 h^2}{\delta y^2} \right) = 2w \qquad \text{(5-65)}$$

If $w = 0$, then Equation (5-65) reduces to a form of the Laplace equation:

$$\frac{\delta^2 h^2}{\delta x^2} + \frac{\delta^2 h^2}{\delta y^2} = 0 \qquad \text{(5-66)}$$

If flow is in only one direction, and we align the x-axis parallel to the flow, then there is no flow in the y-direction, and Equation (5-65) becomes

$$\frac{d^2(h^2)}{dx^2} = -\frac{2w}{K} \qquad \text{(5-67)}$$

Integration of this equation yields the expression

$$h^2 = -\frac{wx^2}{K} + c_1 x + c_2 \qquad \text{(5-68)}$$

where c_1 and c_2 are constants of integration.

The following boundary conditions can be applied: at $x = 0$, $h = h_1$; at $x = L$, $h = h_2$ (Figure 5.16). By substituting these into Equation (5-68), the constants of integration can be evaluated with the following result:

$$h^2 = h_1^2 - \frac{(h_1^2 - h_2^2)x}{L} + \frac{w}{K}(L - x)x \qquad \text{(5-69)}$$

135

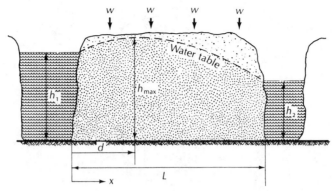

FIGURE 5.16. Unconfined flow, which is subject to infiltration or evaporation.

This equation can be used to find the elevation of the water table anywhere between two points located L distance apart, if the saturated thickness of the aquifer is known at the two end points.

For the case in which there is no infiltration or evaporation, $w = 0$ and Equation (5-69) reduces to

$$h^2 = h_1^2 - \frac{(h_1^2 - h_2^2)x}{L} \tag{5-70}$$

By integrating Equation (5-69), and because $q_x = -Kh(dh/dx)$, it may be shown that the discharge per unit width, q_x, at any section x distance from the origin is given by

$$q_x = \frac{K(h_1^2 - h_2^2)}{2L} - w\left(\frac{L}{2} - x\right) \tag{5-71}$$

If the water table is subject to infiltration, there may be a water divide with a crest in the water table. In this case, q_x will be zero at the water divide. If d is the distance from the origin to a water divide, then substituting $q_x = 0$ and $x = d$ into Equation (5-71) yields

$$d = \frac{L}{2} - \frac{K}{w}\frac{(h_1^2 - h_2^2)}{2L} \tag{5-72}$$

Once the distance from the origin to the water divide has been found, then the elevation of the water table at the divide may be determined by substituting d for x in Equation (5-69).

EXAMPLE PROBLEM

An unconfined aquifer has a hydraulic conductivity of 0.002 centimeter per second and a porosity of 0.27. The aquifer is in a bed of sand with a uniform thickness of 31 meters as measured from the land surface. At Well 1, the water table is 21 meters below land surface. At Well 2, located some 175 meters away, the water table is 23.5 meters from the surface. What is (a) the discharge per unit width, (b) the seepage velocity at Well 1, and (c) the water-table elevation midway between the two wells?

Part A:

$$q = K \frac{(h_1^2 - h_2^2)}{2L}$$

$$h_1 = 31 - 21 = 10 \text{ m}$$

$$h_2 = 31 - 23.5 = 7.5 \text{ m}$$

$$K = 0.002 \text{ cm/sec} = 1.7 \text{ m/day}$$

$$L = 175 \text{ m}$$

$$q = 1.7 \text{ m/day} \times \frac{10^2 - 7.5^2 \text{ m}^2}{2 \times 175}$$

$$= 0.21 \text{ m}^3/\text{day per unit width}$$

Part B:

$$v = q/nA \qquad A = h_1 \times \text{unit width}$$

$$v = \frac{0.21 \text{ m}^3/\text{day}}{0.27 \times 10 \text{ m} \times 1 \text{ m}} = 0.078 \text{ m/day}$$

Part C:

$$h = \sqrt{h_1^2 - (h_1^2 - h_2^2) \frac{x}{L}}$$

$$= \sqrt{10^2 - (10^2 - 7.5^2) \frac{87.5}{175}}$$

$$= 8.84 \text{ m}$$

EXAMPLE PROBLEM

A canal was constructed running parallel to a river 1.5 kilometers away. Both fully penetrate a sand aquifer with a hydraulic conductivity of 1.2 meters per day. The area is subject to rainfall of 1.8 meters per year and evaporation of 1.3 meters per year. The elevation of the water in the river is 31 meters and in the canal it is 27 meters. Determine (a) the water divide, (b) the maximum water-table elevation, (c) the daily discharge per kilometer into the canal, and (d) the daily discharge per kilometer into the river.

Part A:

$$d = \frac{L}{2} - \frac{K}{w}\frac{(h_1^2 - h_2^2)}{2L}$$

$h_1 = 31$ m

$h_2 = 27$ m

$L = 1500$ m

$K = 1.2$ m/day

$w = 1.8$ m/year infiltration $-$ 1.3 m/year evaporation
$\quad = 0.5$ m/year accretion
$\quad = 0.0014$ m/day

$$d = \frac{1500}{2} - \frac{1.2}{0.0014}\frac{(31^2 - 27^2)}{2 \times 1500}$$

$\quad = 684$ m

Part B:

$$h = \sqrt{h_1^2 - \frac{(h_1^2 - h_2^2)x}{L} + \frac{w}{K}(L - x)x}$$

$x = d = 684$ m

$h = h_{max}$

$$h_{max} = \sqrt{31^2 - \frac{(31^2 - 27^2)684}{1500} + \frac{0.0014}{1.2}(1500 - 684)684}$$

$\quad = 38.8$ m

Part C: q_x at $x = 0$:

$$q_x = \frac{K(h_1^2 - h_2^2)}{2L} - w\left(\frac{L}{2} - x\right)$$

$$= \frac{1.2 \times (31^2 - 27^2)}{2 \times 1500} - 0.0014\left(\frac{1500}{2} - 0\right)$$

$$= -0.957 \text{ m}^3/\text{day/m} \quad \text{and} \quad -957 \text{ m}^3/\text{day/km}$$

The negative sign indicates flow is going in a direction opposite to x, or into the river.

Part D: q_x at $x = L$:

$$q_x = \frac{K(h_1^2 - h_2^2)}{2L} - w\left(\frac{L}{2} - x\right)$$

$$= \frac{1.2 \times (31^2 - 27^2)}{2 \times 1500} - 0.0014\left(\frac{1500}{2} - 1500\right)$$

$$= 1.143 \text{ m}^3/\text{day/m} \quad \text{and} \quad 1143 \text{ m}^3/\text{day/km}$$

Flow is in the direction of x, or into the canal.

FRESHWATER-SALINE WATER RELATIONS

5.14.1 COASTAL AQUIFERS

We have assumed to this point that the content of dissolved solids of groundwater is so low that it does not affect the physics of flow. However, if fresh groundwater is adjacent to saline groundwater, the difference in density between the two fluids becomes very important. Due to the difference in dissolved solids, the density of the saline water, ρ_s, is greater than the density of freshwater, ρ_f. Salt water is found adjacent to fresh water in inland areas, often in the same aquifer, as well as in oceanic coastal areas and oceanic islands. Highly saline water in inland aquifers could be either trapped from the time of

139

formation of the rock unit (connate water), or occur through mineralization due to stagnant flow conditions. At coastal locations, the fresh groundwater beneath land is discharging near the coast and mixing with saline groundwater beneath the sea floor.

Fresh groundwater usually grades into saline water with a steady increase in the content of dissolved solids. In some situations, the contact may be quite sharp; that is, a very thin zone of mixed water. The mixture of fresh water and salt water yields the zone in which there is a salinity gradient. If the aquifer is subject to hydraulic head fluctuations caused by tides, the zone of mixed water will be enlarged. In unconfined coastal aquifers, there is groundwater flow occurring in both the fresh zone and the saline zone (18). Fresh water is flowing upward to discharge near the shoreline, and there is a cyclic flow in the salty water near the interface (Figure 5.17). We will make the simplifying assumption that there is a sharp interface between fresh water and saline water. Although the saltwater interface problem can be studied using dispersion and mass-transport theory (19, 20), the mathematical treatment is beyond the scope of this book. The zone of dispersion is often thin with respect to the overall thickness of the freshwater lens. Likewise, we will consider only the steady-state case. Solutions have been developed for moving interface problems (19–23), but they are too complex to be considered here.

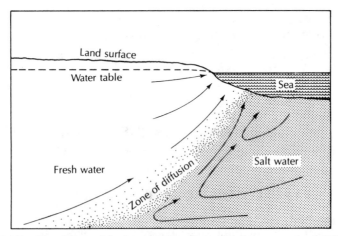

FIGURE 5.17. Circulation of fresh and saline groundwater at a zone of diffusion in a coastal aquifer. SOURCE: H. H. Cooper, Jr., U.S. Geological Survey Circular 1613-C, 1964.

A number of scientists have made significant contributions to the study of the saline water-freshwater interface in coastal aquifers. Studies by W. Baydon-Ghyben (24) and A. Herzberg (25) in the late nineteenth century have been widely cited and have given rise to the Ghyben-Herzberg principle, which we will now discuss. However, their work was antidated by more than half a century by an American, Joseph DuCommun (26), who clearly made the same observations in 1828. Unfortunately, DuCommun is not given just credit in the literature.

These early observers noted that in unconfined coastal aquifers the depth to which fresh water extends below sea level is approximately 40 times the height of the water table above sea level. The (misnamed) **Ghyben-Herzberg principle** states that

$$z_{(x,y)} = \frac{\rho_f}{\rho_s - \rho_f} h_{(x,y)} \qquad (5\text{-}73)$$

where

$z_{(x,y)}$ is the depth to the saltwater interface below sea level at location (x,y)

$h_{(x,y)}$ is the elevation of the water table above sea level at point (x,y)

ρ_f is the density of fresh water

ρ_s is the density of salt water

The application of this principle is limited to situations in which both the fresh water and salt water are static.

EXAMPLE PROBLEM

If $\rho_f = 1.000$ gram per cubic centimeter and $\rho_s = 1.025$ gram per cubic centimeter, what is the ratio of $z_{(x,y)}$ to $h_{(x,y)}$?

$$z_{(x,y)} = \frac{\rho_f}{\rho_s - \rho_f} h_{(x,y)}$$

$$= \frac{1.000}{1.025 - 1.000} h_{(x,y)}$$

$$= 40 \, h_{(x,y)}$$

Figure 5.18 illustrates the Ghyben-Herzberg principle for an unconfined coastal aquifer. Hubbert (2) pointed out that $h_{(x,y)}$ should actually be the hydraulic head at the interface at point (x,y). However, for thin aquifers with a large vertical extent, the Dupuit assumption that equipotential lines are vertical can be made, so that the hydraulic head at the saltwater interface is equal to the elevation of the water table at that location (27). In a study of the saltwater interface on eastern Long Island, it was shown that even at the coastline, where the greatest deviations from the Dupuit assumptions could be expected, there was almost no difference in the position of the interface as computed from potential theory versus Dupuit flow (Figure 5.19).

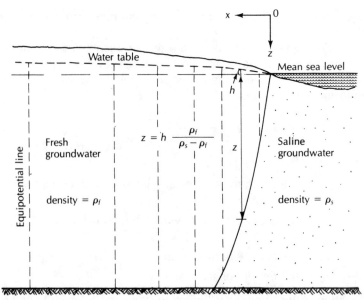

FIGURE 5.18. Relationship of freshwater head and depth to saltwater interface.

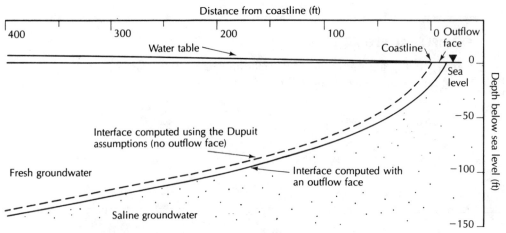

FIGURE 5.19. Comparison of the position of the interface on eastern Long Island as computed using the Dupuit assumptions with that computed using an outflow face. SOURCE: C. W. Fetter, Jr., *Water Resources Research,* 8 (1972):1307–14.

Flow in coastal aquifers can be described by means of the Dupuit equations in combination with the Ghyben-Herzberg principle. The steady flow of groundwater is given by the partial differential equation (27)

$$\frac{\delta^2 h}{\delta x^2} + \frac{\delta^2 h}{\delta y^2} = \frac{-2w}{K(1+G)} \qquad \text{(5-74)}$$

where

w is the recharge to the aquifer

K is the hydraulic conductivity

G is equal to $\dfrac{\rho_f}{\rho_s - \rho_f}$

Should the value of the depth to the saltwater interface, as computed by Equation (5-73), exceed the depth of the aquifer, then the saltwater wedge is missing. This is the case on the left side of Figure 5.18. In this region, the governing equation is (16)

$$\frac{\delta^2 h}{\delta x^2} + \frac{\delta^2 h}{\delta y^2} = \frac{-w}{K(m+h)} \qquad \text{(5-75)}$$

where m is the aquifer thickness below sea level. Both Equation (5-74) and Equation (5-75) can be solved for an infinite-strip coastline; that is, one with flow in only one direction. The x- and y-axes are shown on Figure 5.18.

The **Dupuit-Ghyben-Herzberg model** of one-dimensional flow in coastal aquifers yields the following expression for the x- and z-coordinates of the interface (28):

$$z = \sqrt{\frac{2qxG}{K}} \qquad \text{(5-76)}$$

where q is the discharge from the aquifer at the coastline, per unit width $[(L^3/T)/L]$.

One of the failings of the Dupuit-Ghyben-Herzberg model of coastal aquifers is that the saltwater interface intercepts the water table at the coastline. This does not allow for vertical components of flow and discharge of the fresh water into the sea floor. Therefore, a simple model has been developed in which the x- and z-coordinates of the interface are given by the following relation (29):

$$z = \frac{Gq}{K} + \sqrt{\frac{2Gqx}{K}} \qquad \text{(5-77)}$$

Note that Equation (5-77) is identical to (5-76), except that a constant, Gq/K, has been added. Thus, when $x = 0$, z will still have a value. The interface is shown in Figure 5.20.

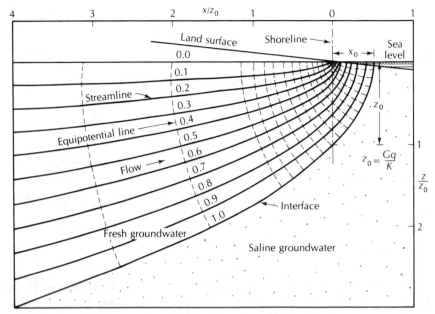

FIGURE 5.20. Flow pattern near a beach as computed using Equation (5-79). SOURCE: R. E. Glover, U.S. Geological Survey Water Supply Paper 1613-C, 1964.

The width of the outflow face, x_0, may be found from the expression

$$x_0 = -\frac{Gq}{2K} \tag{5-78}$$

and the height of the water table at any distance, x, from the coast is given by

$$h = \sqrt{\frac{2qx}{GK}} \tag{5-79}$$

5.14.2 OCEANIC ISLANDS

Oceanic islands are underlain by salty groundwater as well as being surrounded by sea water. The fresh water takes the form of a freshwater lens floating on the more dense salty water. Equation (5-74) also describes the flow of water in an oceanic island. It can be solved for islands of regular shape, such as an infinite strip or circle.

For an infinite-strip island receiving recharge at a rate, w, with a width equal to $2a$, the head of the water table, h, at any distance, x, from the shoreline is given by (27):

$$h^2 = \frac{w(a^2 - (a - x)^2)}{K(1 + G)} \tag{5-80}$$

A circular island with a radius of distance R can be evaluated using

$$h = \frac{w(R^2 - r^2)}{2K(1 + G)} \qquad \text{(5-81)}$$

where h is the head above sea level at some radial distance, r, from the center of the island.

EXAMPLE PROBLEM

Part A: An infinite-strip island has a width of 2 kilometers. The permeability of the sediments is 10^{-2} centimeter per second and there is a daily accretion of 0.13 centimeter per day. The density of fresh water is 1.000 and the density of salty groundwater is 1.025. Compute a water-table profile across the island using Equation (5-80). Then determine the interface depth using Equation (5-73).

$$G = \frac{1}{1.025 - 1} = 40$$

$$K = 10^{-2} \text{ cm/sec} = 8.64 \text{ m/day}$$

$$w = 0.0013 \text{ m/day}$$

$$a = 1000 \text{ m}$$

$$h^2 = \frac{w(a^2 - (a - x)^2)}{K(1 + G)}$$

$$z = Gh$$

x (m)	h (m)	z (m)
1000	1.92	76.8
900	1.91	76.4
800	1.88	75.2
700	1.83	73.2
600	1.76	70.4
500	1.66	66.4
400	1.53	61.2
300	1.36	54.4
200	1.15	46.0
100	0.84	33.6
0	0	0

Part B: From the computed profile, it can be seen that the freshwater lens thins very rapidly in the last 100 meters as the shoreline is approached. As the Dupuit assumptions may not be valid near the coastline, the profile of the interface close to the coast can be computed using Equation (5-77):

$$z = \frac{Gq}{K} + \sqrt{\frac{2Gqx}{K}}$$

x (m)	q ((m³/day)/m)	z (m)
0	1.3	6.0
20		21.5
40		28.0
60		32.9
80		37.0
100		40.7

Find the width of the outflow face:

$$x_0 = -\frac{Gq}{2K}$$

$$= -3.0 \text{ m}$$

Find the height of the water table at a distance from the coast of 100 meters:

$$h = \sqrt{\frac{2qx}{GK}}$$

$$= 0.87 \text{ m}$$

Comparison of the numerical solutions using Equations (5-80) and (5-73) versus those obtained by Equations (5-77), (5-78), and (5-79) will show that the results are similar. However, Equation (5-77) is only valid close to the coastline and yields a greater value for the head and depth to the interface.

5.15 TIDAL EFFECTS

Aquifers located next to tidal bodies are subjected to short-term fluctuations in the head, h, due to the tide. Water-level recorders located in coastal wells show a fluctuation in the hydraulic head, which parallels the rise and fall of the tide. The amplitude of the fluctuation is greatest at the coast and diminishes as one goes inland.

If we have a confined aquifer, as shown in Figure 5.21, water can enter at the subsurface outcrop. The governing flow equation in one dimension can be used to describe the flow of water into and out of the aquifer as the tide changes.

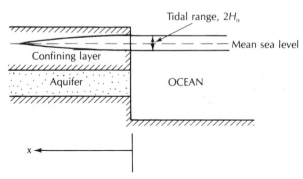

FIGURE 5.21. Coastal aquifer showing the tidal range, $2H_0$, and the effect of the tide on the potentiometric surface.

The amplitude of the tidal change is H_0 and the tidal period, or time for the tide to go from one extreme to the other, is t_0. At any distance, x, inland from the coast, the amplitude of the tidal fluctuation, H_x, is given by (9)

$$H_x = H_0 \exp(-x\sqrt{\pi S/t_0 T}) \tag{5-82}$$

where S and T are the aquifer storativity and transmissivity.

The time lag, t_r, between the high tide and the peak of the water level (or low tide and the low point in the water level) is given by

$$t_r = x\sqrt{t_0 S/4\pi T} \tag{5-83}$$

The preceding equations can also be applied to unconfined flow, as an approximation, if the range of tidal fluctuations is small compared with the saturated aquifer thickness.

References

1. STREETER, V. L. *Fluid Mechanics.* New York: McGraw-Hill Book Company, 1962, 555 pp.

2. HUBBERT, M. K. "The Theory of Ground-Water Motion." *Journal of Geology*, 48, no. 8 (1940):785–944.

3. SCHNEEBELI, G. "Expériences sur la limite de validité de la loi de Darcy et l'apparition de la turbulence dans un écoulement de filtration." *La Houille Blanche*, 10, no. 2 (1955):141–49.

4. HUBBERT, M. K. "Darcy's Law and the Field Equations of Flow of Underground Fluids." *Transactions, American Institute of Mining and Metallurgical Engineers,* 207 (1956):222–39.

5. LINDQUIST, E. *On the Flow of Water through Porous Soil.* Premier Congres des grands barrages (Stockholm), 1933, pp. 81–101.

6. ROSE, H. E. "An Investigation into the Laws of Flow of Fluids through Beds of Granular Materials." *Proceedings of the Institute of Mechanical Engineers,* 153 (1945):141–48.

7. ROSE, H. E. "On the Resistance Coefficient—Reynolds Number Relationship for Fluid Flow through a Bed of Granular Materials." *Proceedings of the Institute of Mechanical Engineers,* 153 (1945):154–68.

8. JACOB, C. E. "On the Flow of Water in an Elastic Artesian Aquifer." *Transactions, American Geophysical Union,* 21 (1940):574–86.

9. JACOB, C. E. "Flow of Groundwater." In *Engineering Hydraulics,* ed. H. Rouse. New York: John Wiley & Sons, 1950, pp. 321–86.

10. DOMENICO, P. A. *Concepts and Models in Groundwater Hydrology.* New York: McGraw-Hill Book Company, 1972, 405 pp.

11. COOPER, H. H. "The Equation of Groundwater Flow in Fixed and Deforming Coordinates." *Journal of Geophysical Research,* 71 (1966):4785–90.

12. CASAGRANDE, A. "Seepage through Dams." *Journal of Boston Society of Civil Engineers* (1940):295–337.

13. FORCHHEIMER, P. *Hydraulik.* Leipzig: B. G. Teubner, 1914, 566 pp.

14. BOUSSINESQ, J. "Recherches théoriques sur l'écoulement des nappes d'eau infiltrées dans le sol et sur le débit des sources." *Journal de Mathématiques Pures et Appliquées,* 10 (1904):5–78.

15. DUPUIT, J. *Études théoriques et pratiques sur le mouvement des eaux dans les canaux découverts et à travers les terrains perméables,* 2nd ed. Paris: Dunod, 1863, 300 pp.

16. POLUBARINOVA-KOCHINA, P. Y. *Theory of Ground Water Movement,* trans. J. M. R. DeWeist. Princeton, N.J.: Princeton University Press, 1962, 613 pp.

17. HARR, M. E. *Groundwater and Seepage.* New York: McGraw-Hill Book Company, 1962, 315 pp.

18. COOPER, H. H., Jr. A Hypothesis Concerning the Dynamic Balance of Fresh Water and Salt Water in a Coastal Aquifer. *Journal of Geophysical Research,* 64 (1959):461–67.

19. BREDEHOEFT, J. D. and G. F. PINDER. "Mass Transport in Flowing Groundwater." *Water Resources Research,* 9 (1973):194–210.

20. SEGOL, G., G. F. PINDER, and W. G. GRAY. "A Galerkin-Finite Element Technique for Calculating the Transient Position of the Saltwater Front." *Water Resources Research,* 11 (1975):343–47.

21. ANDERSON, M. P. "Unsteady Groundwater Flow beneath Strip Oceanic Islands." *Water Resources Research,* 12 (1976):640–44.

22. COLLINS, M. A. and L. W. GELHAR. "Seawater Intrusion in Layered Aquifers." *Water Resources Research,* 7 (1971):971–79.

23. PINDER, G. F. and H. H. COOPER, JR. "A Numerical Technique for Calculating the Transient Position of the Saltwater Front." *Water Resources Research,* 6 (1970):875–82.

24. GHYBEN, W. B. *Nota in verband met de voorgenomen putboring nabij Amsterdam.* Koninklyk Instituut Ingenieurs Tijdschrift (The Hague), 1888–1889, pp. 8–22.

25. HERZBERG, A. "Die Wasserversorgung einiger Nordseebader." *Journal Gasbeleuchtung und Wasserversorgung* (Munich) 44 (1901):815–19, 842–44.

26. DUCOMMUN, J. "On the Cause of Fresh Water Springs, Fountains, and c." *American Journal of Science,* 1st ser., vol. 14 (1828):174–76. As cited in S. N. Davis, "Flotation of Fresh Water on Sea Water, A Historical Note." *Ground Water,* 16, no. 6 (1978):444–45.

27. FETTER, C. W., Jr. "Position of the Saline Water Interface beneath Oceanic Islands." *Water Resources Research,* 8 (1972):1307–14.

28. TODD, D. K. "Sea Water Intrusion in Coastal Aquifers." *Transactions, American Geophysical Union,* 34 (1953):749–54.

29. GLOVER, R. E. "The Pattern of Fresh-Water Flow in a Coastal Aquifer." In *Sea Water in Coastal Aquifers,* U.S. Geological Survey Water Supply Paper 1613-C, 1964, pp. 32–35.

Regional Groundwater Flow

chapter

INTRODUCTION

In the zone of actively flowing groundwater, the water moves through the porous media under the influence of the fluid potential. This movement is a three-dimensional phenomenon, yet we are usually forced to represent it on a two-dimensional medium. In the diagrams in this chapter, the reader will have to imagine the implied third dimension. We will start by examining steady flow through isotropic, homogeneous media and then include the effects of nonhomogeneity and anisotropy.

Flow nets will be used to illustrate the various regional flow patterns. These are a means of portraying the solution to the Laplace equation which governs steady flow (see page 124). The various solutions will represent differing conditions of hydraulic conductivity and aquifer geometry. This type of flow net is constructed by drawing streamlines on a potential field. The potential fields are solutions to a mathematical model of the aquifer systems. Laplace's equation was solved either analytically (2, 3) or numerically (4, 5) with different boundary conditions. One of the most critical boundary conditions is the shape of the water table or potentiometric surface. The streamlines are drawn to illustrate some of the possible flow paths.

6.2 STEADY REGIONAL GROUNDWATER FLOW IN UNCONFINED AQUIFIERS

6.2.1 RECHARGE AND DISCHARGE AREAS

In unconfined aquifers, some characteristics are common to most **recharge areas**; likewise, most **discharge areas** have some common denominators. Recharge areas are usually in topographical high places; discharge areas are located in topographic lows. In the recharge areas, there is often a rather deep unsaturated zone between the water table and the land surface. Conversely, the water table is found either close to or at the land surface in discharge areas.

152

Streamlines on a flow net tend to diverge from recharge areas and converge toward discharge areas. This convergence will not occur if the discharge zone is large, such as a coastline. A water-table contour map can often be used to locate groundwater recharge and discharge areas. Streamlines can be drawn on the basis of groundwater contours, crossing them at right angles if the aquifer is isotropic.

In the field, vegetation and surface water can sometimes be used to locate discharge areas. There may be some physical manifestation of the discharging groundwater, which can take the form of a spring, seep, lake, or stream. The presence of vegetation common to wet soils may be indicative of discharge areas. In arid regions, groundwater may be discharged as evaporation or transpiration. In such cases, a thicker-than-normal cover of vegetation or a salt deposit may indicate a discharge area.

The aforementioned physical manifestations sometimes betoken a groundwater discharge area—but not always. In nonhomogeneous materials, a low-permeability layer may, for example, form a perched aquifer, which could result in a wetland or pond. A thorough evaluation of all hydrogeologic information should always be made.

Many field studies conducted in humid regions note that the water table in unconfined aquifers usually has the same general shape as the surface topography. This is not surprising, since recharge taking place in topographical high areas has a greater potential energy than recharge in lower areas. This greater energy is reflected in the higher elevations of the water table in those locations. This generalization may not be true in arid regions.

6.2.2 GROUNDWATER FLOW PATTERNS IN HOMOGENEOUS AQUIFERS

A descriptive model of regional steady-state groundwater flow in an unconfined aquifer was first presented by Hubbert (1). He demonstrated that the hydraulic head at a point in a potential field represents the elevation to which water will rise in a **piezometer*** which is open only at that point. At the point where an equipotential line intersects the water table, water in the piezometer will rise to the water table. Elsewhere, water in a piezometer intersecting the equipotential line may be above or below the water table, depending upon the relative hydraulic potential.

In Figure 6.1, the water levels in both Piezometer A and Piezometer B are equal, as they both end on the same equipotential line. Piezometer A is located on the water table, while Piezometer B is in an area where the water table has a higher elevation; hence, a greater potential, than A. The water level of Piezometer B is lower than the water table at that location. If a shallow piezometer were installed next to Piezometer B, the water level would be higher in the shallower piezometer. In this area, the value of the hydraulic potential decreases downward, indicating that the direction of flow is

---

*A piezometer is a small-diameter well with a very short well screen or section of slotted pipe at the end. It measures the hydraulic head at a point in the aquifer.

downward. Areas with this distribution of potential are recharge areas for a water-table aquifer.

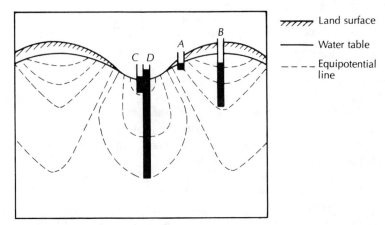

	Land surface
	Water table
	Equipotential line

FIGURE 6.1. Water level in a piezometer will rise to the elevation of the hydraulic head represented by the potential at the open end of the piezometer. The water level of Piezometers A and B is the same, since both terminate on the same equipotential line.

In certain places in the potential field, in adjacent piezometers of different depths, there will be a higher water level in the deeper one. Such is the case with Piezometers C and D. Here, the hydraulic potential increases with depth, indicating upward flow. This is a discharge area.

Having identified the direction of flow at various points in this potential field, we can draw flowlines (Figure 6.2). If the diagram had equal horizontal and vertical scales, and the medium were isotropic, the flowlines would cross the equipotential lines at right angles. However, due to the vertical exaggeration of the drawing, the lines will cross at less than a right angle, even though the flow field is in an isotropic medium.* Hubbert's model was for an unconfined aquifer of great depth. The crest of the water table represents a groundwater divide, with flow going in opposite directions on either side. The valley bottoms are areas of concentrated groundwater discharge into streams, with the streamlines converging toward them.

Not all drainage basins have concentrated discharge areas. If groundwater is discharged primarily by evapotranspiration, or if there is no major topographic valley, the discharge area may be quite widespread. In such cases, the regional flow pattern will not have converging flowlines. Figure 6.3A shows a flow net for a cross section of a homogeneous drainage basin with a gently sloping water table and a horizontal, impermeable lower boundary (2). The entire upper half of the drainage basin is in the recharge area, while the lower portion forms a discharge area. There is no vertical exaggeration in the

---

*For a discussion of the problem of drawing streamlines, see R. O. Von Everdingen, *Groundwater Flow Diagrams in Sections with Exaggerated Vertical Scale*, Geological Survey of Canada Paper 63-27, 1963.

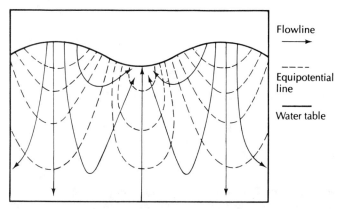

Flowline

→

- - - -
Equipotential
line

———
Water table

FIGURE 6.2. Flowlines based on the equipotential field of Figure 6.1. The aquifer is assumed to be very deep. SOURCE: M. K. Hubbert, *Journal of Geology*, 48, no. 8, (1940):795-944. Used with permission of University of Chicago Press.

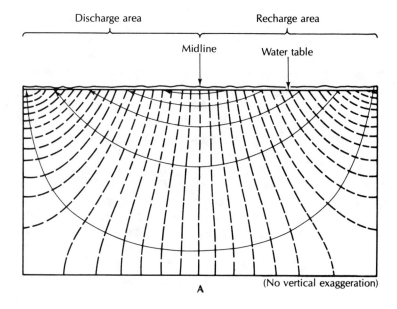

Discharge area   Recharge area

Midline   Water table

A   (No vertical exaggeration)

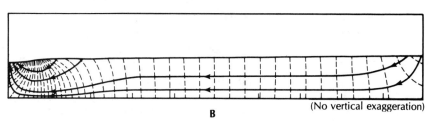

B   (No vertical exaggeration)

FIGURE 6.3. **A.** Regional flow pattern in an area of sloping linear topography and water table. The flow pattern is symmetrical about the midline; **B.** Regional flow pattern in an area with a break in the regional topographic slope forming a major valley. The discharge area is controlled by the valley. SOURCES: Part A. J. A. Tóth, *Journal of Geophysical Research*, 67 (1962):4375-87. Part B. R. A. Freeze and P. A. Witherspoon, *Water Resources Research*, 3 (1967):623-34.

155

drawing, so that streamlines cross equipotential lines at right angles. In recharge areas, the angle between the water table and equipotential lines is oblique and points upstream; at the hinge or midline, it is a right angle; and in groundwater discharge areas, it is oblique and points downstream (5). If a major valley is present in the discharge area, groundwater discharge is concentrated in the valley (5). This changes the potential field, and the entire area above the valley is a recharge area. Figure 6.3B shows a flow net for this situation. There is no vertical exaggeration in the figure.

In Parts A and B of Figure 6.3, there is only one flow system; i.e., a single recharge area and a single discharge area. This occurs because the surface topography—hence, the water table—is linear with a regional slope. However, if the surface topography has well-defined local relief, a series of local groundwater flow systems can form in humid regions (2, 5). This is due to the fact the topographic relief causes undulations in the water table. A local groundwater flow system has the recharge area at a topographic high spot and its discharge area at an adjacent topographic low. Figure 6.4A shows a flow net of a groundwater drainage basin with a series of local flow systems (3, 4, 5). The basin depth is one-twentieth of the basin length from the regional groundwater divide to the lowest part of the basin, and there is no vertical exaggeration.

If the basin depth-to-width ratio increases, other flow systems may also develop. Intermediate flow systems have at least one local flow system between their recharge and discharge areas. Regional flow systems have the recharge area at the basin divide and the discharge area at the valley bottom (Figure 6.4B). Depending upon the drainage basin topography and the basin-shape geometry, flow systems may have regional; local; local and intermediate; or local, intermediate, and regional components.

In addition to the influence of the drainage basin depth/length ratio, it has been shown that the more pronounced the relief of the undulating water table, the deeper the local flow systems extend. In some basins, both local and regional flow systems may exist, while in other basins with a similar depth/length ratio but with a more pronounced water-table relief, only deep local flow systems develop. This is illustrated in Figure 6.5A, which has local and intermediate flow systems, and in Part B of the figure, where the more pronounced relief of the water table has resulted in the exclusive formation of local flow systems.

One of the features of complex flow systems is the presence of **stagnation points** in the flow field (3). At a stagnation point, the magnitude of the vectors in the flow field are equal but opposite in direction and cancel each other. The value of the hydraulic potential is higher at the stagnation point than at any part of the surrounding region. Groundwater flow paths diverge around stagnation points, which are found at the juncture of local and regional flow systems. Figure 6.6 illustrates the potential field and flow paths at a stagnation point. Stagnation points can exist in materials that are completely isotropic and homogeneous.

It has been suggested that "dead cells" or stagnation points might be appropriate areas in which to inject waste fluids for permanent disposal (6). Diffusion would then be the only physical mechanism to disperse the fluid.

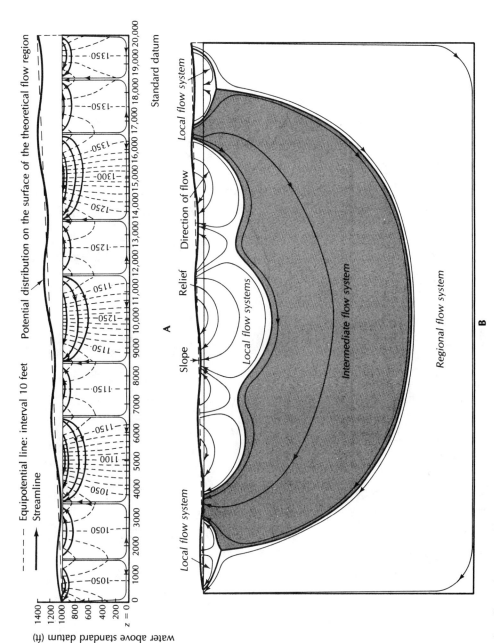

FIGURE 6.4. The effect of increased basin depth is shown on these two figures. In Part A, the basin depth/length ratio is 1:20; in Part B, it is 1:2. The shallow basin has only local flow systems, while the deep basin has local, intermediate, and regional flow systems. The water-table configuration is the same for both basins. SOURCE: J. A. Tóth, *Journal of Geophysical Research,* 68 (1963):4795-4811.

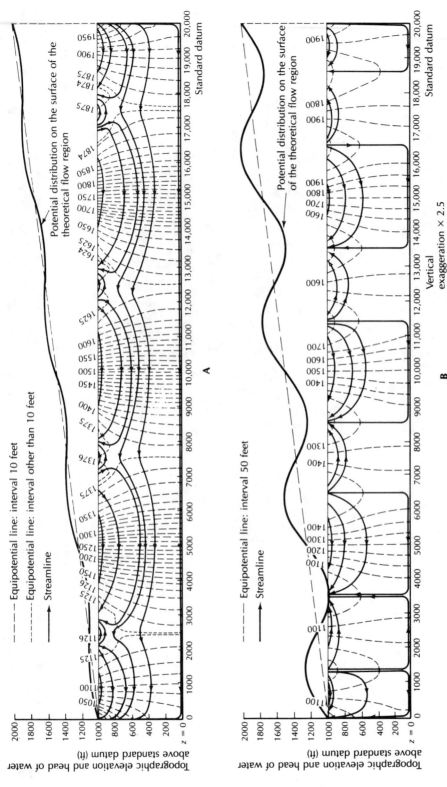

FIGURE 6.5. The amplitude of the undulations of the water table controls the depth of local flow systems. For shallow basins, this can determine whether both local and regional flow systems will develop (Part A), or, with deeper undulations, only local flow systems will form (Part B). SOURCE: J. A. Tóth, *Journal of Geophysical Research*, 68 (1963):4795-4811.

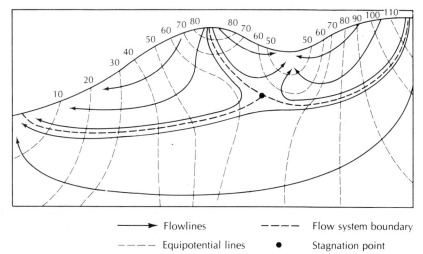

| Flowlines | Flow system boundary |
| Equipotential lines | Stagnation point |

FIGURE 6.6.   The potential field and flowlines in the vicinity of a stagnation point which will develop at the intersection of three flow systems.

However, if the waste fluid were not of the exact density and temperature as the native groundwater, the original potential field might be disrupted causing flow in the area of the stagnation point, with resultant movement of the waste fluid. Likewise, groundwater pumpage could change the potential field, shifting or eliminating the location of stagnation points.

If regional flow systems develop, the flow paths are long compared with those of local flow systems (3). In aquifers composed of soluble rock material, the degree of mineralization is a function of both the initial chemistry of the water and the length of time it is in contact with the aquifer (7). Referring back to Figure 6.4B, we see the boundaries of local, intermediate, and regional flow systems for a deep aquifer with an undulating water table. The surface area where recharge to the regional flow system takes place is quite small in relation to the volume of water stored in that region of the aquifer. The water moves slowly and circulates deeply within the aquifer, as the flow paths are long. At the point of discharge, the water from the regional flow system is likely to have relatively high mineralization and an elevated temperature due to the geothermal gradient. (The temperature of the earth increases with depth at a more or less constant rate of 1° C per 100 meters of depth.)

Local flow systems are shallower with short flow paths. The size of the recharge area is much greater with respect to the volume of water in the aquifer. Thus, water has a shorter contact time with the rocks and is potentially mineralized to a lesser degree than that of the regional system. The temperature of water discharging from the local flow systems is close to the mean annual air temperature. Local flow systems are areas of rapid circulation of groundwater; therefore, groundwater in these systems is much more active in the hydrologic cycle than groundwater in regional flow systems (3). Spring discharge of local flow systems is closely related to recharge of precipitation and shows wide fluctuations. This is illustrated in Figure 6.7. Intermediate flow systems have properties falling between those of local and regional flow systems.

159

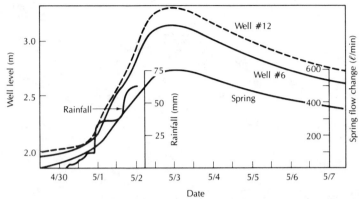

FIGURE 6.7. The water table and spring flow of a local flow system will fluctuate with recharge from rainfall. SOURCE: R. S. Sartz et al., *Water Resources Research*, 13 (1977):524-30.

If a flow system has extended areas with a flat water table, the potential is the same in all parts of the field. Neither local nor regional flow systems can develop, and the groundwater is stagnant. Evapotranspiration is the only method of groundwater discharge. Groundwater under such conditions is likely to be highly mineralized due to a long contact time with the aquifer rocks.

6.2.3. HETEROGENEOUS AQUIFERS

Piezometers may sometimes yield water levels that are apparently anomalous with respect to the expected regional flow pattern (2). A set of piezometers at various depths may show a water elevation equal to the water table for a shallow well, a lower water elevation for a piezometer of intermediate depth, and then a water elevation equal to the water-table elevation for a deep well. Geologic logs of these piezometers might show the shallow one to end in a fine, silty sand; the one of intermediate depth to end in coarse sand; and the deepest one to end in the fine, silty sand. Figure 6.8A shows a cross section of the potential distribution where a body of material of high hydraulic conductivity is surrounded by material with a lower conductivity. The high-conductivity zone acts as a conduit for flow, attracting water from much of the aquifer. The result is that the potential field bends away from the high-conductivity zone on either end. Flow will thus converge toward the high-conductivity zone on the upstream end and diverge away from it on the downstream side.

A line of piezometers of equal depth, extending to line A-A', would have a potentiometric profile that would differ from the water-table profile (Figure 6.8B). Upstream of the midpoint of the high-conductivity layer, the piezometric profile would be lower than the water table. It would cross the water table at the midpoint and be higher than the water table below the midpoint. Such a profile would not occur in a homogeneous aquifer.

If a lens of low-permeability material is buried in an aquifer, it acts as a partial barrier to groundwater flow. The groundwater streamlines diverge around the lens. While a modicum of the flow is carried in the low-conductivity layer, the majority of the flow tends to be in the aquifer.

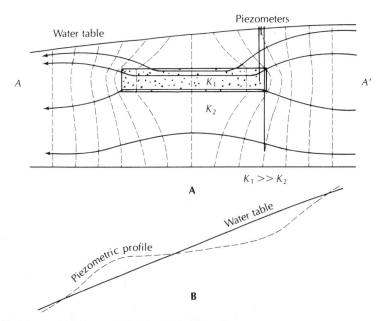

FIGURE 6.8.   A. Equipotential field and flowlines in a region where a high-conductivity body is buried in a lower conductivity aquifer; **B**. The water table and the piezometer profile of a line of piezometers, each ending at the same elevation along line A-A' of Part A.

Layered aquifers are especially prevalent in sedimentary basins, with individual geohydrologic units having different hydraulic conductivities. If a lower formation has a substantially higher conductivity than the surface layer, it acts as the major conduit of flow (5). Figure 6.9A shows the potential distribution in a layered aquifer when the lower unit has a conductivity ten times that

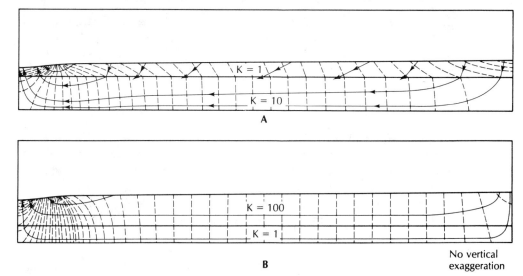

FIGURE 6.9.   Regional groundwater flow in layered aquifers. The greater proportion of the flow occurs in the layer with higher hydraulic conductivity. SOURCE: R. A. Freeze and P. A. Witherspoon, *Water Resources Research,* 3 (1967):623-34.

of the upper. Flow in the lower unit is horizontal, while the upper unit has vertical flow components in the recharge and discharge areas. As the difference in conductivity between the upper and lower layer increases, the components of vertical flow in the upper unit increase as more of the flow is carried in the lower unit.

   If a high-conductivity layer overlies a unit of substantially lower conductivity, the potential field is very similar to that of an isotropic aquifer. Most of the flow is carried in the upper, more conductive, layer, as Figure 6.9B illustrates. The potential field of Figure 6.9B is quite similar to that of Figure 6.3B (page 155), which is homogeneous.

6.2.4 ANISOTROPIC AQUIFERS

There is considerable evidence that many aquifers are anisotropic. For deposits as uniform as glacial outwash, the horizontal permeability may be two to twenty times as great as the vertical (8). Figure 6.10 shows the potential distribution in an aquifer in which horizontal conductivity is ten times as great as the vertical. In anisotropic aquifers, the flowlines do not cross the equipotential lines at right angles; the correct angles can be obtained graphically (9). Figure 6.3B shows the same section in an isotropic medium. The vertical components of flow are more pronounced in the anisotropic aquifer. The greatest variation in the potential field occurs at the extreme ends of the groundwater basin.

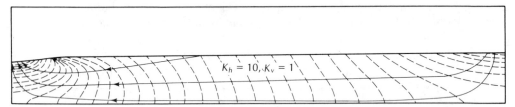

$K_h = 10, K_v = 1$

FIGURE 6.10.   Effect of anisotropy on regional groundwater flow. This figure is the same as Figure 6.3B, except that, here, the horizontal hydraulic conductivity is ten times the vertical. SOURCE: R. A. Freeze and P. A. Witherspoon, *Water Resources Research,* 3 (1967):623-34.

6.3 CONFINED AQUIFERS

Aquifers that are overlain by a layer of substantially lower hydraulic conductivity are confined. The hydraulic gradient is generally greater in the confining bed than in the aquifer. Since the frictional resistance to flow is so much greater in the confining layer, most of the available energy of the potential field is dissipated there.

   Confined aquifers may be either sloping or flat. If the aquifer crops out near a topographic high, substantial recharge takes place in the outcrop area. In the sloping aquifer shown in Figure 6.11A, the confining layer is a barrier to flow. Wells drilled though it to the underlying aquifer would yield artesian flow. Streamlines refract as they cross the confining layer. Discharge of the regional flow system is concentrated in the valley bottom.

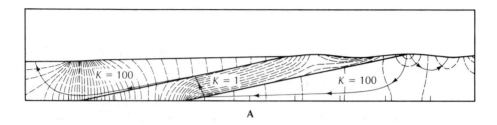

A

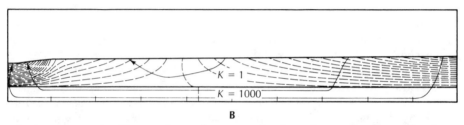

B

FIGURE 6.11. Regional groundwater flow in confined aquifers: **A.** Aquifer confined by a sloping confined layer; **B.** aquifer confined by a flat-lying confining layer. SOURCE: R. A. Freeze and P. A. Witherspoon, *Water Resources Research,* 3 (1967):623-34.

The flat-lying confined aquifer in Figure 6.11B does not have an outcrop area. Recharge to the aquifer occurs by downward flow through the confining layer. Almost all of the energy of the potential field is consumed as flow moves through the confining layer in recharge and discharge areas. Only one equipotential line crosses the aquifer. The volume of water flowing through the buried aquifer is less than it would be if it cropped out in the recharge area. This is due to the use of much of the available potential energy in forcing recharge through the low-conductivity layer. If a well were drilled in the discharge area, artesian conditions would occur.

Unless the confined aquifer is capped by a completely impermeable layer, there will be some discharge from the aquifer in the form of upward leakage in the area of upward hydraulic gradient. Many confined aquifers have heads above the land surface when the first wells are drilled (10). The amount of upward flow occurring through the confining beds is typically small, and the water does not circulate rapidly. Groundwater withdrawals in many confined regional aquifers have lowered the potentiometric head. This actually increases the rate of lateral groundwater flow in the aquifer, as the hydraulic gradient between the recharge area and the well-field area is increased.

163

6.4 TRANSIENT FLOW IN REGIONAL GROUNDWATER SYSTEMS

The systems we have considered have been in a state of **dynamic equilibrium**. The amount of water recharging the aquifer is balanced by an equal amount of natural discharge, and the potential field is more or less constant. If a well field is established in the groundwater basin, the withdrawal of well water increases the discharge from the system, disrupting the equilibrium. Thus, a new equilibrium must be established (11).

In the case of an unconfined aquifer, the water table around the well field will be drawn down. As the discharge exceeds the recharge, the difference comes from gravity drainage of groundwater stored in the aquifer. The cone of depression around the well field will slowly expand until it affects the flow system enough to create a new equilibrium condition. This will occur when the area of the cone of depression is large enough to intercept sufficient aquifer recharge to supply the well discharge. This will reduce natural discharge somewhere else, and a new condition of dynamic equilibrium will be reached. Should the rate of withdrawal be so great that the cone of depression reaches the boundaries of the aquifer without intercepting sufficient recharge, the aquifer will not reach equilibrium and eventually could be drained.

In confined and leaky confined aquifers, pumping will reduce the heads near the wells. As a result, the potentiometric surface will decline. The cone of depression will expand rapidly due to the small value of storativity of confined aquifers. Initially, the water being pumped comes from storage in the aquifer. The cone of depression in a leaky confined aquifer will stabilize when enough downward leakage is induced to balance pumpage. This, of course, will upset the natural equilibrium in the overlying aquifer that is furnishing the water.

In a confined aquifer, the cone of depression will grow until it reaches either the recharge area of the aquifer, or the discharge area, or both. The resulting change in the potential field will induce either increased recharge, or decreased natural discharge, or both. If this is sufficient to balance recharge and discharge, the aquifer will again be in dynamic equilibrium. If not, the water levels will continue to decline.

6.5 NONCYCLICAL GROUNDWATER

There is a certain amount of water in the ground that is not encompassed by the hydrologic cycle. When sediments are deposited, water is present in the pores. The same may be true for undersea volcanic rocks. Later geologic events may bury the sediment or rock and its contained pore water. Water buried with the rock is termed **fossil water** (12). Interstitial water which was not buried with the rock but which has been out of contact with the atmosphere for an appreciable part of a geologic period is called **connate water** (12).

Magmatic water is associated with a magma. It may be in part **juvenile water**, having never before circulated in the hydrologic cycle (13). However, most magmatic water comes from the recycling of connate or fossil water. Magmatic water can re-enter the hydrologic cycle through volcanic eruptions or thermal springs.

SPRINGS

Springs have played a role in the settlement pattern of many lands where they have served as a local water supply. Mineralized and thermal springs have been thought to have a therapeutic value. The importance of springs is evident from the many localities named for the springs found there (e.g., Tarpon Springs, Florida; Palm Springs, California; Hot Springs, Arkansas; Steamboat Springs, Colorado).

A spring may have a discharge that is fairly constant, or the discharge may vary. Springs can be permanent or ephemeral. The water may contain dissolved minerals of many different types, or certain dissolved gases or petroleum. The temperature of the water may be close to the mean annual air temperature, or be lower or higher—even boiling. Flow may range from a barely perceptible seepage to 90 or more cubic meters (1000 cubic feet) per second.

Topographic low spots provide the simplest mechanism for the formation of springs. **Depression springs** are formed when the water table reaches the surface (14). The change in topography creates a corresponding undulation in the water-table configuration. A local flow system is thus created, with a spring formed at the local discharge zone (Figure 6.12A).

Where permeable rock units overlie rocks of much lower permeability, a **contact spring** may result (14). A lithologic contact is often marked by a line of springs, which may be either in the main water table, or in a perched water table. It is not necessary for the underlying layer to be impermeable, merely that the difference in hydraulic conductivity be great enough to preclude transmission of all of the water that is moving through the upper horizon (Figure 6.12B).

A classic occurrence of contact springs is found along the eastern side of Chuska Mountain, New Mexico. A sandstone cliff rises 60 to 150 meters above a terrace composed of shale, which also underlies the sandstone. More than thirty springs are found at the foot of the cliff at the contact of the sandstone and shale (15). One of the most spectacular series of springs in the world is in the Snake River Canyon below Shoshone Falls in Idaho. Along a 64-kilometer reach of the canyon, there are eleven springs with a discharge of more than 9.3 cubic meters (100 cubic feet) per second. The springs issue from permeable basalt flows (16).

Faulting may also create a geologic control favoring spring formation. A faulted rock unit that is impermeable may be implaced adjacent to

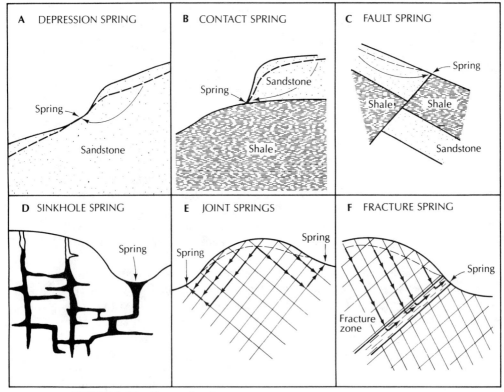

FIGURE 6.12. Types of springs.

an aquifer. This can form a regional boundary to groundwater movement and force water in the aquifer to discharge as a **fault spring** (Figure 6.12C).

Some of the largest springs are found in areas of limestone bedrock. In such areas, the runoff may be carried in part or totally as subterranean flow. It may be either diffused flow in pores and fractures in the rock, or channelized flow in caverns. Springs may be found where a cavern is connected to a shaft that rises to the surface. Many of the famous springs of Florida cover an area of several hectares in which water rises to the surface through sinkholes (Figure 6.12D). The water in these **sinkhole springs** is under artesian pressure and comes from the principal artesian aquifer, or Floridan aquifer, which underlies Florida (17). This aquifer is in Tertiary-age limestones.

Joint springs or **fracture springs** may occur from the existence of jointed or permeable fault zones in low-permeability rock. Water movement through such rock is principally through fractures, and springs can form where these fractures intersect the land surface at low elevations (Figure 6.12E,F).

Springs in limestone terrane can be interconnected to topographic depressions caused by collapsed caverns (sinkholes) at higher elevations. Water level in the sinkholes may rise and fall due to variations in runoff (18). Discharge of these springs, known as **karst springs**, may correspond with the elevation of water in the sinkholes (Figure 6.13).

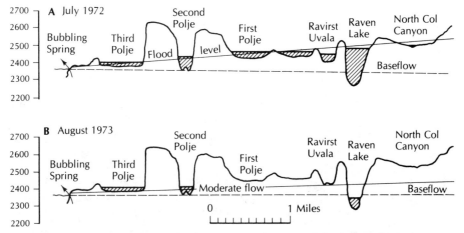

FIGURE 6.13. Hydrogeology of a karst spring, the rate of discharge of which is a function of the water level in upstream sinkholes. SOURCE: G. A. Brook, in *Karst Hydrogeology,* ed. J. S. Tolson and F. L. Doyle (Huntsville, Ala.: UAH Press, 1977), pp. 99-108. Used with permission.

GEOLOGY OF REGIONAL FLOW SYSTEMS 6.7

CASE STUDY: REGIONAL FLOW SYSTEMS IN THE GREAT BASIN

The Basin and Range Province contains a number of topographically closed basins. These intermontaine basins are characterized by an accumulation of relatively permeable clastic sediments. The mountains surrounding the basins are composed of bedrock, which also underlies the basins at depth. The hydraulic conductivity of the bedrock types is extremely variable (20).

Annual precipitation is greatest in the mountains and least in the valleys (21). Below an elevation of 1800 meters (6000 feet), annual precipitation is less than 20 centimeters (8 inches). This is almost all evaporated, with virtually no recharge of groundwater. Above 2750 meters (9000 feet), there may be more than 50 centimeters (20 inches) of precipitation, with up to 13 centimeters (5 inches) of groundwater recharge. The areas of greatest precipitation and recharge are in topographic highs, which are good recharge zones.

Those mountain areas formed by crystalline rocks or low-permeability sedimentary rocks have near-surface permeability due to fracturing. Such mountain areas have many small springs and perennial streams, as the groundwater is discharged by local flow systems (6). Mountains underlain by highly permeable carbonate rocks are generally dry. The groundwater appears as the discharge of relatively large springs at the foot of the mountains or in the intermontaine valleys. In the areas of

167

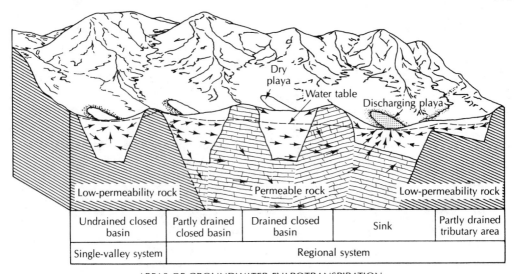

AREAS OF GROUNDWATER EVAPOTRANSPIRATION

Phreatophytes   Discharging playa

FIGURE 6.14. Flow systems of the Great Basin Region. SOURCE: T. E. Eakin et al., U.S. Geological Survey Professional Paper 813-G, 1976.

carbonate aquifers, the water table is relatively flat, and may extend with a regional slope beneath topographic divides (6, 20, 21). A block diagram of single-valley hydrologic systems and regional flow systems is shown in Figure 6.14.

In the White River area of southeastern Nevada, a regional interbasin groundwater flow system has been identified (21). There are thirteen topographic basins: seven of them are closed; the other six were drained by the White River during the Pleistocene. The mountains are 2450 to 3050 meters (8000 to 10,000 feet) high, with the valley bottoms 600 to 1220 meters (2000 to 4000 feet) lower. The principal water-bearing units are Paleozoic limestone and dolomites, up to 9150 meters (30,000 feet) thick. There are some volcanic rocks (tuffs and welded tuffs) which can form locally perched aquifers. The valleys are filled with Tertiary-age clastic sedimentary rocks and evaporites.

Groundwater is discharged by means of several large springs. The flow of Muddy River Springs, which is the largest, is highly uniform, which suggests a regional flow system as the source (22). A longitudinal profile of the area, which shows both the topography and the potentiometric surface, reveals that the regional hydraulic gradient is unaffected by crossing topographic divides (Figure 6.15).

The amount of groundwater recharge is much greater than discharge in the topographically higher basins. The water balance is reversed for the topographically lower basins where the large springs are located. However, the regional water budget is balanced when all thirteen of the basins are included (21).

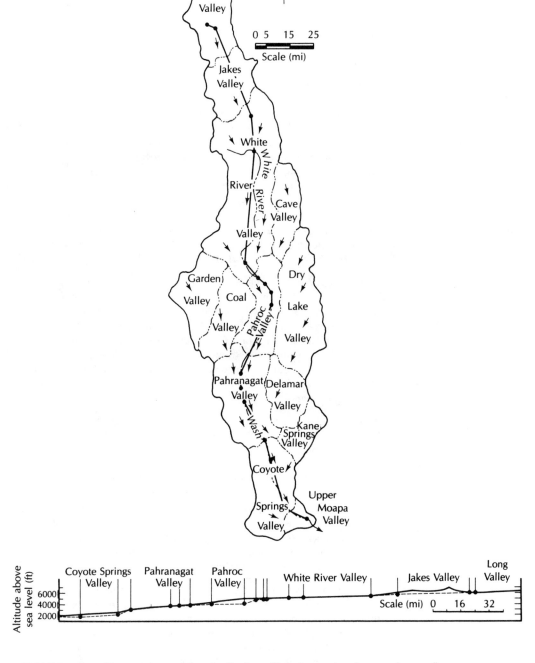

FIGURE 6.15. Flow paths and longitudinal profile of a regional groundwater flow system in Nevada. SOURCE: T. E. Eakin, *Water Resources Research,* 2 (1966):251-71.

In the Great Basin area, local flow systems have small drainage areas and short flow paths within the same topographic basin. Springs have a wide fluctuation in discharge. The temperature of the spring water is about the same as mean annual air temperature, and the dissolved ion content is relatively low. Regional flow systems have long flow paths, which often cross basin divides. The drainage basin area is large. Springs are large with fairly constant discharge, elevated temperatures, and higher concentrations of dissolved salts (6).

Another large groundwater flow system has been delineated in southern Nevada at the Nevada Test Site (24). The area has typical basin and range topography. Rocks range in age from Recent sediments in the valleys to Precambrian. A basement unit of Cambrian and Precambrian siltstones and quartzites is a regional confining layer. Above this layer is a major regional aquifer—the lower carbonate aquifer of Lower Paleozoic age. There are solution openings in the aquifer, but the hydraulic conductivity is due primarily to fractures. The saturated thickness ranges from a hundred to a thousand meters or more. There are several caves in the outcrop area. One of them, Devils Hole, is partially filled with water and extends vertically more than 100 meters (300 feet) below the water table. The lower carbonate aquifer feeds many large springs.

No other aquifer has a regional extent, although several are important locally. There is a regional confining layer which overlies the lower carbonate aquifer. This is the welded tuff confining layer of Tertiary age. The clay-rich beds of this layer restrict groundwater movement. The three regional units control groundwater flow. The welded tuff confining layer permits slow downward movement of water to the lower carbonate aquifer in recharge zones and upward movement in discharge zones. The lower carbonate aquifer underlies most of the valleys and ridges.

In some of the upland valleys, such as Yucca Flats and Frenchman Flats, the water table lies from 210 to 610 meters (700 to 2000 feet) below the valley floor (Figure 6.16). Water drains from these valleys downward to the lower carbonate aquifer. Water flows laterally in this unit until it reaches the Ash Meadows area of the Amargosa Desert Basin. Running across Ash Meadows is a normal fault with a displacement of at least 160 meters (500 feet). This has formed a hydraulic barrier which impedes the groundwater flow. The result is a line of springs discharging at the outcrop of the lower carbonate aquifer. The average discharge of all the springs is 640 liters per second (10,000 gallons per minute).

The inferred groundwater drainage basin contains ten intermontaine valleys and is on the order of 11,500 square kilometers (4500 square miles) in area. The Ash Meadows Basin may be hydrologically interconnected with other intermontaine valleys from which groundwater flow may be obtained. There may also be underflow past the spring line at Ash Meadows which feeds into the central Amargosa Desert. The lack of wells and the extremely complex structure of the area make exact delineation of flow systems difficult.

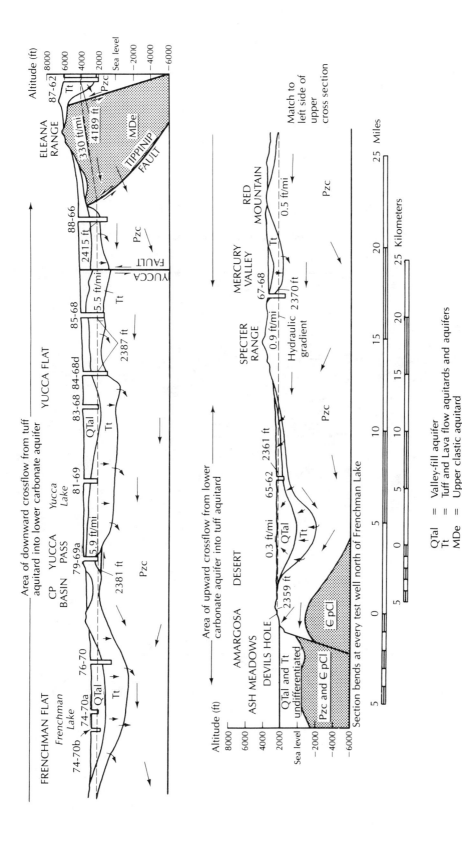

FIGURE 6.16. Regional groundwater flow in the vicinity of the Nevada Test Site, Southern Nevada. SOURCE: I. J. Winograd and W. Thordarson, U.S. Geological Survey Professional Paper 712-C, 1975.

171

CASE STUDY: REGIONAL FLOW SYSTEMS IN THE COASTAL ZONE OF THE SOUTHEASTERN UNITED STATES

Groundwater flow regimes in humid regions characteristically have a much thinner unsaturated zone than those in arid regions. The saturated zone is close to the surface in recharge areas as well as in discharge areas. Because of the greater amounts of precipitation, the volume of recharge is much higher, and proportionally more water circulates through groundwater flow systems.

The coastal zone of the southeastern United States is underlain by extensive aquifers contained in Tertiary- and Quaternary-age limestones, forming some of the most productive aquifers in the

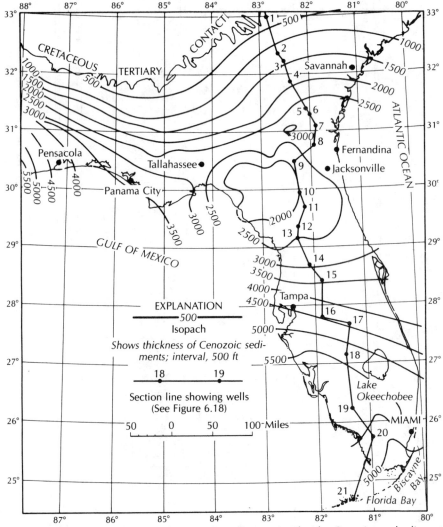

FIGURE 6.17. Isopach map of Cenozoic sediments in Florida, Georgia, and adjacent parts of Alabama and South Carolina. SOURCE: V. T. Stringfield, U.S. Geological Survey Professional Paper 517, 1966.

United States (25, 26). Sediments on the coastal plain dip seaward, and the units thicken as the coast is approached. Figure 6.17 shows the isopach map of the Cenozoic sediments and sedimentary rocks of the region. Starting at a feather edge near the interior border of the Atlantic Coastal Plain, they thicken to more than 1650 meters (5400 feet) in southwestern Florida. The structure is disrupted by the Ocala Uplift north of Tampa, which results in a thinning of the sediments.

There are four principal hydrostratigraphic units* in the region (25) (Table 6.1). The uppermost unit, Number 1, consists of limestone ranging in age from Late Miocene to Recent. In southeastern Florida, this forms the very productive Biscayne aquifer with an estimated maximum thickness of 75 meters (250 feet). Elsewhere it is rather thin and nonproductive.

TABLE 6.1. Major stratigraphic and hydrologic units in Florida

Age	Formation	Hydrologic Unit	Remarks
Recent	Sands, clays, and phosphate rock	Surficial aquifer	Domestic wells in sands.
Pleistocene	Anastania Fm. Miami Oolite Ft. Thompson Fm. Key Largo Ls.	Biscayne aquifer	Biscayne aquifer is in Miami Area.
Pliocene	Caloosahatchee Marl		
Miocene	Tamiami Fm. Hawthorn Fm. upper units	Upper unit confining bed	Upper beds of the Hawthorn Fm. act as a confining unit.
	Hawthorn Fm. lower units		
	Tampa Ls.		
Oligocene	Suwannee Ls. Byram Fm. Marianna Ls.	Floridan aquifer	Floridan aquifer is known to under- lie Florida and southeast Georgia. The upper part is the permeable sand and limestone of the Hawthorn Fm.
Upper Eocene	Ocala Ls.		
Middle Eocene	Avon Park Ls. Talahassee Ls. Lake City Ls.		
Lower Eocene	Oldsmar Ls.		These units do not contain any fresh water. They are used for deep-well disposal of wastes.
Paleocene	Cedar Keys Ls.		

*See Section 4.14 on hydrostratigraphic units.

173

A regional confining layer consisting primarily of the Hawthorn Formation of Miocene age forms Hydrostratigraphic Unit 2. It contains up to 150 meters (500 feet) of interbedded, low-permeability clays, marls, sands, and limestones. The sands and limestones may be locally productive. In Florida it is missing in the area of the Ocala Uplift, where it has been removed by erosion.

The Floridan aquifer (principal artesian aquifer) forms Hydrostratigraphic Unit 3. This is the source of water for the large springs of Florida. It is as thick as 460 meters (1500 feet), and consists of limestone ranging from middle Eocene to middle Miocene in age. It underlies all of Florida, southeastern Georgia, and southeastern Alabama, cropping out in western Florida, southern Alabama, central Georgia, and southwestern South Carolina. The top of the aquifer is more than 150 meters (500 feet) above sea level at the outcrop area in Georgia and 35 meters (100 feet) above sea level where it crops out on the Ocala Uplift. As it dips beneath southern Florida, it is more than 300 meters (1000 feet) below sea level. From bottom to top, the Floridan aquifer contains the Lake City, Tallahassee, Avon Park, Ocala, Marianna, and Suwannee limestones; Flint River Formation; Tampa Limestone; and the lower part of the Hawthorn Formation.

Underlying the principal artesian aquifer is a unit consisting primarily of the Oldsmar and Cedar Keys formations. These are limestones, dolomites, and evaporites of varying conductivity. Beneath Florida, the chloride content of the water is high, and is not used for water supply. It is extensively used for disposal of wastewater, which is injected into the rocks through deep wells. A stratigraphic cross section from central Georgia to the Florida Keys shows the hydrostratigraphic units (Figure 6.18). The vertical scale is greatly exaggerated.

Precipitation in Florida averages about 140 centimeters (55 inches) annually with about 100 centimeters (39 inches) of evaporation (27). The 40-centimeter (16-inch) difference represents the recharge to aquifers and overland flow into streams. The vast majority of the available water, 35 centimeters (14 inches), flows into surface water, with only 3 to 5 centimeters (1 to 2 inches) recharging the Floridan aquifer. The major recharge areas are discernible as the areas of high elevation on the potentiometric map (Figure 6.19). The Floridan aquifer has two major hydraulic regions separated by a hydrologic divide (27). This is shown as a dashed line in Figure 6.19.

The northern portion of the Floridan aquifer is recharged in the outcrop area of central Georgia where the elevations are 30 to 70 meters (100 to 230 feet) above sea level. Recharge to the aquifer also occurs in a region extending from north of Valdosta, Georgia, to east of Gainesville, Florida. This is an area of high topography with many sinkhole lakes. The sinkholes breach the Hawthorn Formation, a confining layer, so recharge occurs through the sinkhole lakes into the underlying Floridan aquifer. Water flows from the recharge areas,

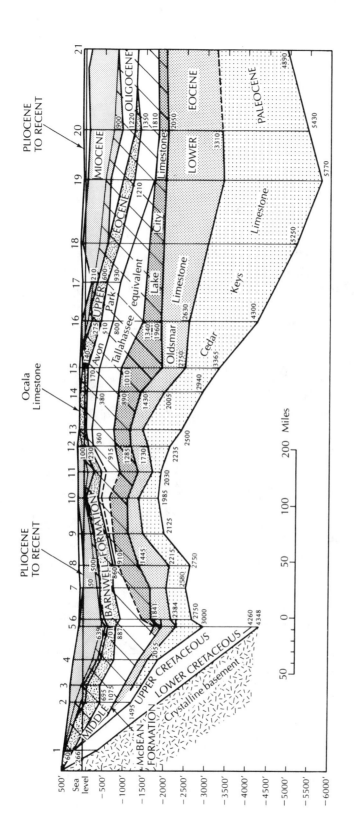

Floridan aquifer
(Upper boundary approximate)

FIGURE 6.18. Cross section of the Floridan aquifer from the inner margin of the coastal plain in Georgia to southern Florida. SOURCE: V. T. Stringfield, U.S. Geological Survey Professional Paper 517, 1960.

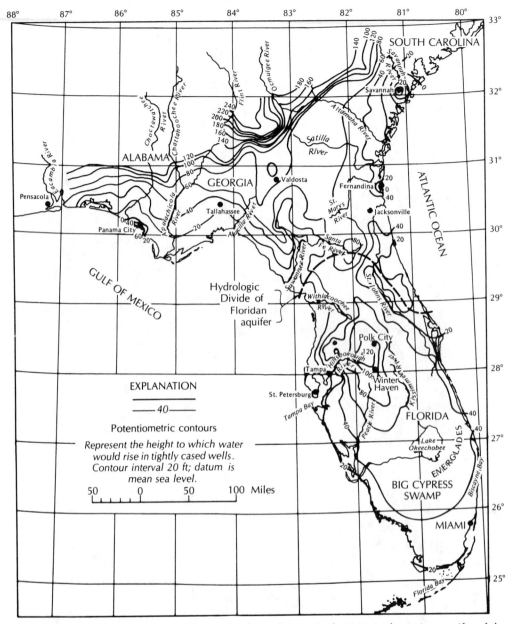

FIGURE 6.19. Potentiometric surface of water in the principal artesian aquifer of the southeastern United States. SOURCE: V. T. Stringfield, U.S. Geological Survey Professional Paper 517, 1966.

following the regional hydraulic gradient, and discharges along the coastlines. There are also lesser amounts of recharge from downward leakage through the Hawthorn Formation over much of Florida.

The southern portion of the Floridan aquifer is hydraulically separated from the north by the potentiometric divide. The Withlacoochee River and the St. Johns River act as groundwater drains to lower the

176

potentiometric surface and isolate the southern aquifer. In the Polk City area, north of Winter Haven, the potentiometric surface is as much as 35 meters (115 feet) above sea level (28). This is another area in which sinkhole lakes as deep as 60 meters (200 feet) are present and can serve as recharge conduits.

The general circulation of groundwater in the Floridan aquifer is downward in the recharge areas and then laterally toward the coasts. To the west, discharge is by upward leakage through the Hawthorn Formation into the Gulf of Mexico or into coastal springs located in sinkholes. Flow eastward can reach the Atlantic Ocean floor, where some units of the Floridan aquifer crop out. There is also discharge into the rivers that drain the coastal area. Major rivers, such as the Suwannee and St. Johns in northern Florida, have a profound effect on the potentiometric surface, as they are regions of major discharge from the aquifer.

Zones of high-salinity groundwater are found in the lower parts of the Floridan aquifer. These are naturally occurring, but heavy pumping of overlying fresh water is causing an upconing of the saline water. In a number of areas in the Southeastern Coastal Plain, the principal artesian aquifer has been heavily pumped, with a subsequent lowering of the potentiometric surface. In an area east of Tampa, the potentiometric surface declined by as much as 18 meters (60 feet) from 1949 to 1969 (28). In the Savannah area, the cone of depression resulting from groundwater pumping has reversed the regional hydraulic gradient and has caused saltwater encroachment (29). This is becoming common throughout much of the Southeastern Coastal Plain.

GROUNDWATER-LAKE INTERACTIONS

One of the important aspects of lake hydrology is the interaction between lakes and groundwater. This interaction plays a critical role in determining the water budget for a lake (30). Lakes may be classified hydrogeologically on the basis of domination of the annual hydrologic budget by surface water or groundwater. Surface-water–dominated lakes typically have inflow and outflow streams, while seepage lakes are groundwater-dominated (31).

In a field study of the groundwater regime of permanent lakes in hummocky moraine of western Canada, both local and intermediate flow systems were identified (32). The lakes were seepage-type (no surface outlet) and received inflow from either local, or local plus regional flow systems. Figure 6.20 shows the water table and two lakes. Interlake areas are recharge areas, with the higher elevation lake receiving water from local flow systems; the lower lake is fed by both local and regional flow systems.

In early spring, the water table was high, and most lakes were like those of Figure 6.20A. However, as the water table fell during the dry season, the groundwater divide disappeared between some lakes. These lakes became flow-through lakes, with groundwater flowing in on one side and out on the other. The upper lake in Figure 6.20B illustrates this condition.

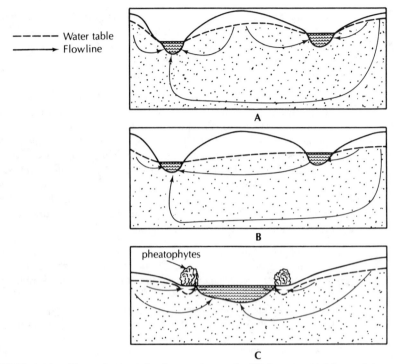

----- Water table
——→ Flowline

FIGURE 6.20. Groundwater-lake interactions: **A**. High water table and interlake groundwater divide; **B**. Low water table and no interlake divide; **C**. depressed water table due to fringe of phreatophytes. SOURCE: Redrawn from P. Meyboom, *Journal of Hydrology*, 5 (1967):117-42. Used with permission of Elsevier Scientific Publishing Company, Amsterdam.

The flow conditions close to each lake were complicated by a growth of willow trees around the lake. Willows are **phreatophytes**—plants that use large amounts of groundwater. During the growing season, the water table may be locally depressed below lake level beneath the phreatophyte fringe (Figure 6.20C). This can cause outflow at the margins of the lake, still leaving inflow beneath the bottom. Phreatophyte water usage is diurnal, with resulting fluctuations in the seepage balance of the lake (32).

A very interesting aspect of lake hydrology has been revealed through the numerical modeling of lake-groundwater systems. If the water table is higher than the lake level on all sides of a seepage lake, groundwater will seep into the lake from all sides. Figure 6.21 shows a cross section through a seepage lake, with the water table higher than the lake surface. Potential distribution is based on a two-dimensional steady-state numerical model (33). There is a stagnation zone beneath the lake, indicating both local and regional groundwater flow. Upward seepage takes place throughout the lake bottom. As long as the stagnation zone is present, the lake will not lose water through the bottom. However, should an aquifer or high-conductivity layer underlie the lake, the stagnation zone could be eliminated (33). Without the stagnation zone, the lake will lose water through part or all of the bottom, even with the presence of a water-table mound downslope (Figure 6.22). Flow patterns of considerable

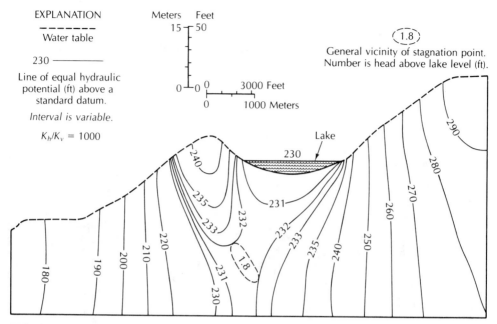

FIGURE 6.21. Hydrogeologic cross sections showing head distribution in a one-lake system with a homogeneous, anisotropic aquifer system. Results are based on a two-dimensional steady-state numerical-simulation model. SOURCE: T. C. Winter, U.S. Geological Survey Professional Paper 1001, 1976.

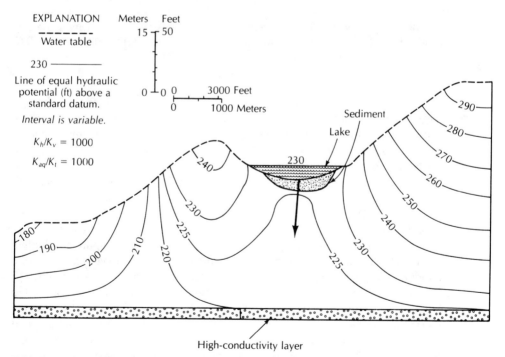

FIGURE 6.22. Hydrogeologic cross section showing head distribution in a one-lake system with a layered aquifer system. The high-conductivity layer has a conductivity 1000 times as great as the low-conductivity layer. The lake loses water to the aquifer. SOURCE: T. C. Winter, U.S. Geological Survey Professional Paper 1001, 1976.

complexity can form in multiple-lake systems, with some lakes having inflow through the bottom and some outseepage (Figure 6.23).

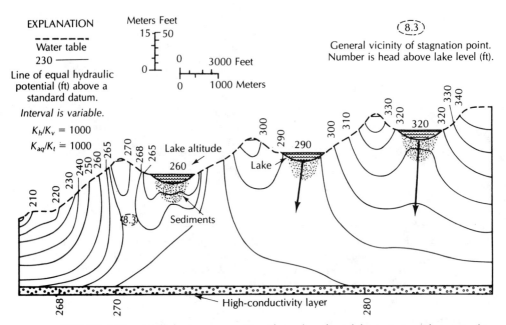

FIGURE 6.23. Hydrologic cross section through a three-lake system with a complex aquifer. Local and regional groundwater flow systems are present. SOURCE: T. C. Winter, U.S. Geological Survey Professional Paper 1001, 1976.

Three-dimensional numerical analysis of groundwater lake systems has shown that the stagnation zone, if present, is beneath the downgradient side of the lake, and follows the lakeshore (34). The area of outseepage, if present, can be estimated from the three-dimensional model. The one-lake system in Figure 6.24 is shown in top view and two cross sections. Outseepage occurs through the center of the lake, while seepage into the lake takes place all around the edges.

Whether or not a stagnation zone will develop is largely determined by the following factors: (a) the height of adjacent water-table mounds relative to the lake level, (b) the position and conductivity of highly conductive layers in the groundwater system, (c) the ratio of horizontal to vertical hydraulic conductivity, (d) the regional slope of the water table, and (e) the lake depth (33).

As the downgradient groundwater mound height increases, the less the likelihood that the lake will leak. If the ratio of horizontal to vertical hydraulic conductivity is low (less than 100), the model studies have indicated that stagnation points typically are present. As the ratio increases, so does the likelihood of the stagnation point disappearing and outseepage developing. Also, the presence of high-conductivity layers increases the possibility of out-

180

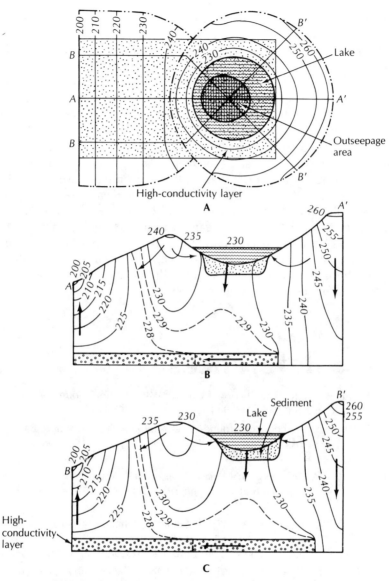

FIGURE 6.24. Three-dimensional analysis showing the area of outseepage in the center of the lake. Seepage into the lake occurs over the rest of the bottom. Areal extent of the high-conductivity layer is indicated by stipples. SOURCE: T. C. Winter, *Water Resources Research,* 14, no. 2 (1978):245-54.

seepage, as does deeper lakes. Flow systems with a low regional water-table gradient, as well as thin aquifers, are more likely to have only local flow systems —hence, lakes that do not leak if there is a downgradient water-table divide.

In field studies of groundwater-lake interactions, there is no substitute for piezometers in defining the flow field. There must be several shallow piezometers all around the lake in order to define the water-table configuration (33, 34). It is necessary to determine whether a groundwater divide exists on the downgradient side of a lake or between two lakes. If a divide exists, then the maximum water-table elevation must be found. The existence of a stagnation point can be determined by placing a nest of closely spaced piezometers of different depths below the shoreline on the downslope side of the lake. If the head at all depths is greater than the elevation of the lake surface, then a stagnation point is indicated.

In order to calculate the seepage rate into a lake, the seepage per unit area can be measured at the sediment-lake interface (35). This is done with a device made of half an oil drum, with the open part pushed into the sediments. The rate of flux across the sediments is measured. Inflow generally occurs in the littoral zone of a lake (34, 35). Outseepage in lakes with no downgradient water-table divide can occur in both the downgradient littoral zone, as well as in the deeper part of the lake. If the lake has a downgradient water-table divide, and there still is outseepage, it will be through the deeper part of the lake bottom. This area typically contains low-conductivity sediments, limiting the rate of water loss (34).

References

1. HUBBERT, M. K. "The Theory of Ground-Water Motion." *Journal of Geology,* 48, no. 8 (1940):785–944.

2. TÓTH, J. A. "A Theory of Ground-Water Motion in Small Drainage Basins in Central Alberta, Canada." *Journal of Geophysical Research,* 67, no. 11 (1962):4375–87.

3. TÓTH, J. A. "A Theoretical Analysis of Ground-Water Flow in Small Drainage Basins." *Journal of Geophysical Research,* 68, no. 16 (1963):4795–4811.

4. FREEZE, R. A. and P. A. WITHERSPOON. "Theoretical Analysis of Regional Groundwater Flow: 1. Analytical and Numerical Solutions to the Mathematical Model." *Water Resources Research,* 2, no. 4, (1966):641–56.

5. FREEZE, R. A. and P. A. WITHERSPOON. "Theoretical Analysis of Regional Groundwater Flow: 2. Effect of Water-Table Configuration and Subsurface Permeability Variation." *Water Resources Research,* 3, no. 2 (1967):623–34.

6. MAXEY, G. B. "Hydrogeology of Desert Basins." *Ground Water,* 6, no. 5 (1968):10–22.

7. BACK, W. and B. B. HANSHAW. "Comparison of the Chemical Hydrogeology of the Carbonate Peninsulas of Florida and Yucatan." *Journal of Hydrology,* 10 (1970):330–68.

8. WEEKS, E. P. "Determining the Ratio of Horizontal to Vertical Permeability by Aquifer-Test Analyses." *Water Resources Research,* 5, no. 1 (1969):196–214.

9. LIAKOPOULOS, A. C. "Variation of the Permeability Tensor Ellipsoid in Homogeneous, Anisotropic Soils." *Water Resources Research,* 1, no. 1 (1965):135–41.

10. WEIDMAN, S. and A. R. SCHULTZ. *The Underground and Surface Water Supplies of Wisconsin.* Wisconsin Geological and Natural History Survey Bulletin 35, 1915, 664 pp.

11. THEIS, C. V. "The Significance and Nature of the Cone of Depression in Ground-Water Bodies." *Economic Geology,* 38 (1938):889–902.

12. WHITE, D. E. "Magmatic, Connate and Metamorphic Water." *Bulletin, Geological Society of America,* 68, no. 12 (1957):1659–82.

13. WHITE, D. E. "Thermal Waters of Volcanic Origin." *Bulletin, Geological Society of America,* 68, no. 12 (1957):1637–58.

14. BRYAN, K. "Classification of Springs." *Journal of Geology,* 27 (1919):522–61.

15. GREGORY, H. E. *The Navajo Country.* U.S. Geological Survey Water Supply Paper 380, 1916, 219 pp.

16. MEINZER, O. E. *Large Springs of the United States.* U.S. Geological Survey Water Supply Paper 557, 1927, 94 pp.

17. STRINGFIELD, V. T. *Artesian Water in Tertiary Limestone in the Southeastern States.* U.S. Geological Survey Professional Paper 517, 1966, 226 pp.

18. BROOK, G. A. "Surface and Groundwater Hydrogeology of a Highly Karsted Sub-Arctic Carbonate Terrain in Northern Canada." In *Karst Hydrogeology,* ed. J. S. Tolson and F. L. Doyle. Huntsville, Ala.:UAH Press, 1977, pp. 99–108.

19. SARTZ, R. S. et al. "Hydrology of Small Watersheds in Wisconsin's Driftless Area." *Water Resources Research,* 13, no. 3 (1977):524–30.

20. MIFFLIN, M. D. *Delineation of Ground-Water Flow Systems in Nevada* (Technical Report Ser. H-W, Pub. No. 4.) Reno:University of Nevada, Desert Research Institute, 1968, 54 pp.

21. EAKIN, T. E. "A Regional Interbasin Groundwater System in the White River Area, Southeastern Nevada." *Water Resources Research,* 2, no. 2 (1966):251–71.

22. EAKIN T. E. and D. O. MOORE. *Uniformity of Discharge of Muddy River Springs.* U.S. Geological Survey Professional Paper 501-D, 1964, pp. 171–76.

23. EAKIN, T. E. et al. *Summary Appraisals of the Nation's Groundwater Resources—Great Basin Region.* U.S. Geological Survey Professional Paper 813-G, 1976, 37 pp.

24. WINOGRAD, I. J. and W. THORDARSON. *Hydrogeologic and Hydrochemical Framework, South Central Great Basin, with Special Reference to the Nevada Test Site.* U.S. Geological Survey Professional Paper 712-C, 1975, 126 pp.

25. STRINGFIELD, V. T. and H. E. LE GRAND. *Hydrology of Limestone Terranes in the Coastal Plain of the Southeastern United States.* Geological Society of America Special Paper 93, 1960, 46 pp.

26. STRINGFIELD, V. T. *Artesian Water in Tertiary Limestone in the Southeastern States.* U.S. Geological Survey Professional Paper 517, 1966, 226 pp.

27. PARKER, G. G. "Water and Water Problems in the Southwest Florida Water Management District and Some Possible Solutions." *Water Resources Bulletin,* 11 (1975):1–20.

28. STEWART, J. W. et al. *Potentiometric Surface of Floridan Aquifer, Southwest Florida Water Management District.* U.S. Geological Survey Hydrologic Investigations Atlas HA-440, 1971.

29. COUNTS, H. B. and E. DONSKY. *Salt-Water Encroachment, Geology and Ground Water Resources of the Savannah Area, Georgia and South Carolina.* U.S. Geological Survey Water Supply Paper 1611, 1963, 100 pp.

30. WINTER, T. C. "Classification of the Hydrogeologic Settings of Lakes in the North Central United States." *Water Resources Research,* 13 (1977):753–67.

31. BORN, S. M., S. A. SMITH, and D. A. STEPHENSON. "The Hydrogeologic Regime of Glacial-Terrain Lakes, with Management and Planning Applications." *Journal of Hydrology,* 43, no. 1/4 (1979):7–44. (Special issue: *Contemporary Hydrogeology—The George Maxey Memorial Volume,* ed. William Back and D. A. Stephenson.)

32. MEYBOOM, P. "Mass-transfer Studies to Determine the Ground-Water Regime of Permanent Lakes in Hummocky Moraine of Western Canada." *Journal of Hydrology,* 5 (1967):117–42.

33. WINTER, T. C. *Numerical Simulation Analysis of the Interaction of Lakes and Groundwaters.* U.S. Geological Survey Professional Paper 1001, 1976, 45 pp.

34. WINTER, T. C. "Numerical Simulation of Steady-State Three-Dimensional Groundwater Flow Near Lakes." *Water Resources Research,* 14 (1978):245–54.

35. LEE, D. R. "A Device for Measuring Seepage Flux in Lakes and Estuaries." *Limnology and Oceanography,* 22 (1976):140–47.

184

Geology of Groundwater Occurrence

7.1 INTRODUCTION

Groundwater inevitably occurs in geological formations. Knowledge of how these earth materials formed and the changes they have undergone is vital to the hydrogeologist. The earth is basically heterogeneous, and geological training is prerequisite to understanding the distribution of geologic materials of varying hydraulic conductivity and porosity.

A groundwater study of an area necessarily includes a review of previous geologic studies. In an area of limited geologic knowledge, a detailed geologic study must be made along with the groundwater study. The hydrogeologist is concerned with the distribution of earth materials as they affect the porosity and hydraulic conductivity of the earth. The results of a geologic study conducted by a groundwater geologist may well differ substantively from a study in the same area by a paleontologist or a petrologist. The methods employed by each would be very similar, however. Rock outcrops would be examined, certain properties noted, and the results mapped. Test borings and geophysical methods would be used to examine the subsurface. Earth materials would be grouped according to similar properties. The hydrogeologist most often begins with the proper formation names that have been assigned to an area. But these formations may then be grouped or split according to hydraulic characteristics rather than by lithology or fossil species present. For example, some bedding planes in a limestone may have been enlarged by solution, so as to transmit significant amounts of water. A hydrogeologic study might identify these on the basis of outcrop and subsurface data and map their location. Other geological studies might not even note their occurrence.

A locality almost always has a variety of geologic materials resulting from different processes. For example, the Keweenaw Peninsula of Michigan has Precambrian bedrock famous for copper mineralization in the Portage Lakes lava series and Copper Harbor conglomerate. Some water wells obtain a small supply of water from fractures in the dense bedrock, which has a very low primary permeability. One community obtains a supply from an old mine adit originally constructed to intercept seepage into the lower working levels of the mine. Sediments deposited as fluvio-glacial deposits and as deltaic or long-shore deposits in Lake Superior (especially during the Pleistocene when lake levels were higher) also serve as aquifers for community water supplies (1). A hydrogeological study of groundwater exploration in this area must examine a number of different geologic settings. A study focused on bedrock aquifers might locate wells yielding 0.06 to 1 liter per second (1 to 15 gallons per minute) from fractured bedrock. Sand deposited by long-shore currents along Lake Superior can yield up to 6.4 liters per second (100 gallons per minute) to wells. If present, this is a preferred alternative to bedrock wells.

   The hydrogeologist should consider all alternatives before selecting one or more sites for detailed exploration. The preliminary survey usually consists of examination of air photos, topographic and geologic maps, logs and reports of existing wells, and then a "walking survey" of the area. This is followed by detailed study using methods such as test borings, fracture-trace analyses, and geophysical surveys. If the results of the detailed study are favorable, one or more test wells are usually installed and pumping tests made.

   Hydrogeologic studies are also made to evaluate the suitability of a site for such uses as sanitary landfills, land treatment of wastewater, seepage ponds for wastewater disposal and spray irrigation of wastewater, or nuclear power plant siting. The basic methods of study and evaluation are similar. However, the target geologic terrain will vary according to the application. Landfills are usually placed in areas of low-permeability material in order to minimize the possibility of groundwater pollution. On the other hand, seepage ponds for artificial aquifer recharge would be located in as coarse a material as possible. While this chapter will focus on groundwater as a resource, the basic principles of occurrence are the same, regardless of the application.

UNCONSOLIDATED AQUIFERS 7.2

Materials ranging in texture from fine sand to coarse gravel are capable of being developed into a water-supply well. Material that is well sorted and free from silt and clay is best. The hydraulic conductivities of some deposits of unconsolidated sands and gravels are among the highest of any earth materials. The filterlike nature of many fine- to medium-grain sediments removes such particulate matter as bacteria and viruses, so that epidemiological water quality is usually good. Unconsolidated materials are also often close to a source of recharge, such as a stream or lake. Shallow unconsolidated aquifers are in the region of rapid circulation of water—usually in local flow systems.

   Unconsolidated deposits range in thickness up to tens of thousands of meters in sedimentary basins such as the Gulf of Mexico. Local well-construction regulations often call for about 10 meters (30 feet) of surface casing to prevent surface water or very shallow groundwater from entering a well. An unconsolidated deposit must extend more than 10 meters (30 feet) deep to be useful for a water supply. Even for a domestic well, at least 1 or 2 meters (3 to 6 feet) of slotted casing, well screen, or a well point is needed below the minimum casing depth.

7.2.1 GLACIATED TERRAIN

The midcontinental area of North America has been covered a number of times in the past by great sheets of glacier ice. The moving ice eroded and deposited as it waxed and waned across the land surface. As a result, deposits of glacial drift from less than a meter to scores of meters thick mantle the bedrock. This covering of glacial drift is a potential source of groundwater, although the hydrogeology of glaciated areas can be very complex.

Glacially related sediments have a wide range of hydraulic conductivity values. Materials carried by the glaciers ranged in size from clay to large boulders. The mode of deposition determined how well the sediment was size-sorted. Glacial till was deposited directly from the glacial ice without significant sorting by running water. As a result, it can contain particles of any size. If the ice held a wide range of particle sizes, so will the till it deposited. Generally the till will have a low hydraulic conductivity, especially if it is clay-rich. Mountain glaciers, in particular, have deposited some very coarse till materials. The hydraulic conductivity of clay-rich glacial tills from the mid-western United States varies over four orders of magnitude, from 7.9×10^{-8} centimeter per second (3.0×10^{-4} gallon per day per square foot) to 2.4×10^{-3} centimeter per second (0.9 gallon per day per square foot) (2) (Figure 7.1). Little, if any, work has been published on the hydraulic conductivity of lacustrine clays and silts associated with glacial lakes; unpublished data for Wisconsin show values of 10^{-6} to 10^{-8} centimeter per second (4×10^{-4} to 4×10^{-6} gallon per day per square foot) (3). Laboratory permeability studies and field tests of glacial silts and clays in Wisconsin have shown values of hydraulic conductivity in the range of 10^{-5} to 10^{-8} centimeter per second (4×10^{-3} to 4×10^{-6} gallon per day per square foot) (3). Brittle glacial till and clay can be fractured, so that significant secondary permeability is developed.

Glacial drift may also have been well sorted by running water—usually the meltwater from the glacier. Such sediments can be seen forming in the braided stream deposits of meltwater streams draining modern mountain glaciers. The coarse gravel is close to the terminus, sand is deposited further downstream, and the silt and clay are carried away to be deposited far downstream in lakes or the ocean. Coarser, water-sorted glacial materials can have very high hydraulic conductivities. The general range for these sediments is 10^{-3} to 15 centimeters per second (0.4 to 6000 gallons per day per square foot). The materials with lower hydraulic conductivities are not as well sorted. Some silty outwash materials have hydraulic conductivities as low as 10^{-5} centimeter per second (4×10^{-3} gallon per day per square foot). Well-sorted glacial deposits may be found associated with recognizable geomorphic features such as moulin kames, kame terraces, and eskers. However, these features are topographically high and, except for the base, may be unsaturated (Figure 7.2). If they are buried, then they can be excellent sources of water. Such buried gravel deposits can sometimes be located by electrical resistivity methods. Outwash plains and valley-train deposits are not as obvious geomorphic features, although they are often found near terminal moraines. Outwash deposits

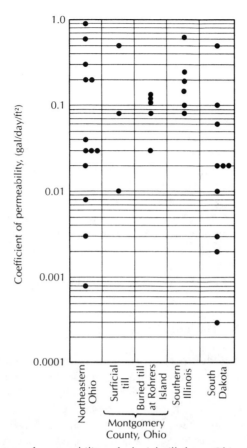

FIGURE 7.1. Range of permeability of glacial till from Ohio, Illinois, and South Dakota. Results from northeastern Ohio, South Dakota, and surficial till in Montgomery County, Ohio, are based on laboratory determination of permeability. Field tests using aquifer pumping tests were used in southern Illinois and buried till in Montgomery County. SOURCE: S. E. Norris, U.S. Geological Survey Professional Paper 450-E, 1963.

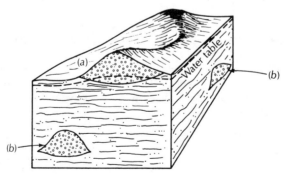

FIGURE 7.2. Eskers in glacial till. The esker at the surface (a) is composed of permeable sand and gravel, but only the base is saturated; hence, it has little potential as an aquifer. The buried esker (b) is highly permeable and fully saturated.

are usually well-sorted sand or gravel, or mixtures of both. Some glacial out-wash materials are tens of meters thick and make prolific aquifers. The city of Tacoma, Washington, obtains groundwater from outwash deposits in the valley of the North Fork of the Green River. Six production wells were tested at rates of 480 to 550 liters per second (7500 to 8600 gallons per minute) with total drawdowns of 0.58 to 2.22 meters (1.9 to 7.3 feet). **Specific capacities*** as high as 0.9 square meter per second were obtained (4).

A key mark of glaciated terrain is the variability of hydraulic conductivity. A single test boring might reveal sequences of glacial deposits with a variation of hydraulic conductivity of 10^{-1} centimeter per second for well-sorted glacial sands to 10^{-7} centimeter per second for interbedded tills or clays. Figure 7.3 shows glacial stratigraphic cross sections for three areas in the Mesabi Iron Range of Minnesota with a high potential for groundwater development (5).

During the Pleistocene, meltwaters from continental glaciers flowed across much of the North American landscape. This was true even in areas south of the limit of glaciation. These rivers carried large volumes of sediment. Well-sorted glacio-fluvial sediments can provide excellent supplies of groundwater.

The pre-Pleistocene landscape of the midcontinent was a bed-rock erosional surface. Deeply incised rivers drained the land with a well-developed drainage network. The glacial sculpting of the North American landscape resulted in many changes of the preglacial drainage patterns. Many of the deep bedrock valleys were filled with sediment. Layers of till, lacustrine silts, and clays alternate with glacio-fluvial deposits. Modern rivers follow the course of some of the buried channels, while others lie inconspicuously beneath the farmland of the Midwest. The course of buried bedrock valleys must be determined from compilation of well-log data and geophysical methods. Seismic refraction has proven to be a very reliable method of delineation.

CASE STUDY: HYDROGEOLOGY OF A BURIED VALLEY AQUIFER AT DAYTON, OHIO

There is a classic buried valley running beneath the city of Dayton, Ohio. One of the factors promoting the growth of Dayton has been the ready availability of a source of high-quality groundwater. Permeable layers of glacial drift in the bedrock valley furnish water in large quantities to wells. In turn, these are recharged by infiltration of precipitation, as well as by water from the Miami River and its tributaries.

The bedrock in the area is the Richmond Shale of Ordovician age. During the Tertiary, an erosional surface developed, which was cleft

---

*The specific capacity of a well is the yield divided by the drawdown of the water level from the nonpumping level. The units are thus flow/distance or $m^3/s/m$ (m^2/s) or gpm/ft. One gpm/ft = 2.06×10^{-4} m^2/s.

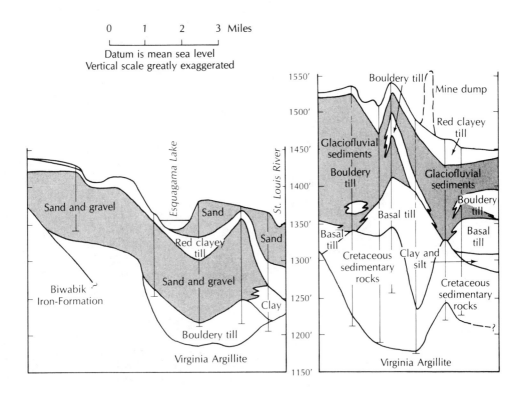

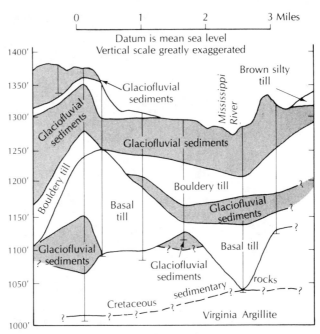

FIGURE 7.3. Complex glacial stratigraphy in the Mesabi Iron Range, Minnesota. Sand and gravel and glacio-fluvial sediments are potential aquifers. SOURCE: T. C. Winter, U.S. Geological Survey Water Supply Paper 2029-A, 1973.

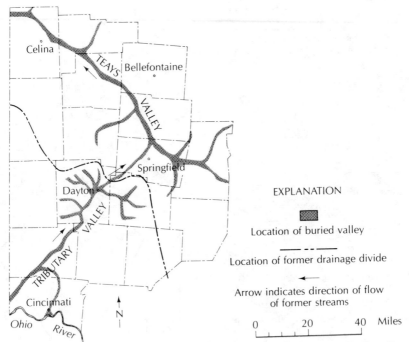

FIGURE 7.4. Approximate location of Teays Stage (late Tertiary) bedrock valleys in western Ohio. SOURCE: S. E. Norris and A. M..Spieker, U.S. Geological Survey Water Supply Paper 1808, 1966.

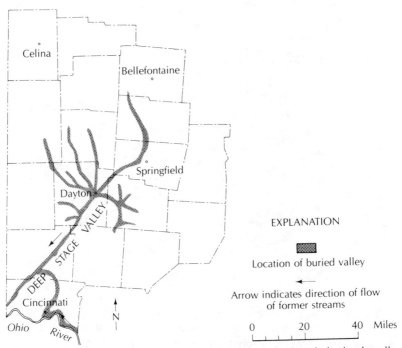

FIGURE 7.5. Approximate location of Deep-Stage (interglacial) bedrock valleys in western Ohio. SOURCE: S. E. Norris and A. M. Spieker, U.S. Geological Survey Water Supply Paper 1808, 1966.

by deeply incised rivers. During the late Tertiary, the main river draining the area was the Teays (Figure 7.4). This drainage system cut a number of valleys in the bedrock. Early Pleistocene glaciation dammed the rivers, so that lacustrine silts are found filling many parts of the Teays system. During the Kansas-Illinois interglacial period, a radically different drainage system, the Deep Stage, prevailed in southwestern Ohio (Figure 7.5). Bedrock valleys were deeply entrenched during this time. The Deep Stage Valley passed through the present site of Dayton. Glacial processes of Illinoian and Wisconsin age filled the Deep Stage Valley with layers of till and outwash. The character of the sediment is quite heterogeneous, as revealed by a test hole in the bedrock valley south of Dayton (Figure 7.6). The valley-train, gravel aquifers alternate with confining till

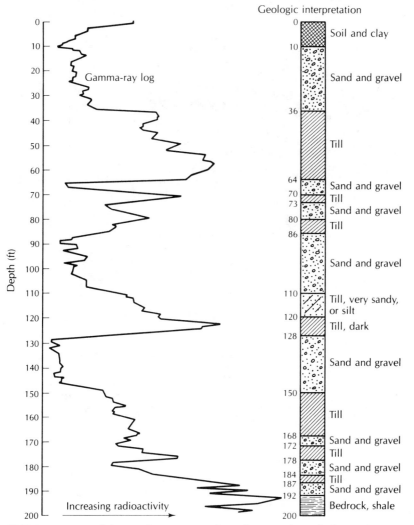

FIGURE 7.6. Well log and gamma-ray log of uncased test hole in glacial deposits filling a buried bedrock valley south of Dayton, Ohio. SOURCE: S. E. Norris and A. M. Spieker, U.S. Geological Survey Water Supply Paper 1808, 1966.

193

sheets. The aquifer layers beneath the till sheets are recharged where the till is missing, either due to nondeposition or to river-channel erosion. A cross section of the valley just south of Dayton shows the upper and lower aquifers separated by a discontinuous till sheet (Figure 7.7). The potentiometric surface in the lower aquifer lies

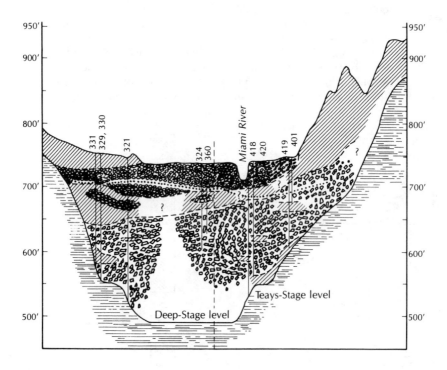

EXPLANATION

Lower aquifer

 Sand and gravel deposits generally occurring between the till-rich zone and bedrock; contains interbedded lenses and masses of till and clay, expecially near the bedrock surface

Till-rich zone

 Fairly widespread sheets, lenses, and masses of till; contains pockets and lenses of sand and gravel; occurs as a layer of low permeability and generally separates the sand and gravel deposits into an upper and lower aquifer

Upper aquifer

 Sand and gravel deposits occurring at or near the surface; generally overlies the till-rich zone

Shale of Ordovician age with thin interbedded limestone layers

————— – – – – – – –
Geologic contact
Dashed where approximate

· ·
Potentiometric surface in lower aquifer

FIGURE 7.7. Cross section of buried bedrock valley at Dayton, Ohio, showing upper (water-table) aquifer and lower (confined) aquifer. SOURCE: S. E. Norris and A. M. Spieker, U.S. Geological Survey Water Supply Paper 1808, 1966.

below the bed of the Miami River. The potentiometric surface rises sharply near the river when the discharge and stage of the Miami River rise. This indicates a good hydraulic connection between the river and the aquifer (Figure 7.8).

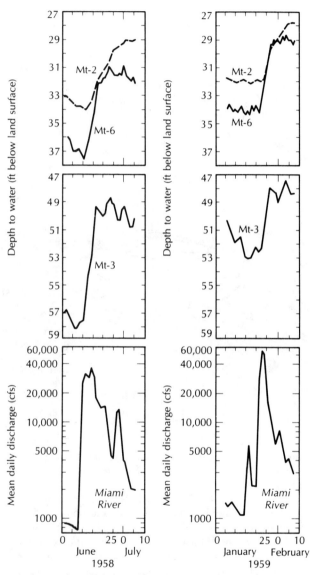

FIGURE 7.8. Hydrographs of water-table elevations in observation wells near the Miami River and daily stage of the Miami River at Dayton during two periods of peak river discharge. A major increase in discharge is followed a few days later by a rise in the water table. SOURCE: S. E. Norris and A. M. Spieker, U.S. Geological Survey Water Supply Paper 1808, 1966.

The Rohrers Island Well Field is located on an island in the Mad River, a tributary of the Miami River. The upper sand and gravel aquifer yields up to 3940 liters per second (90 million gallons per day) with an average pumpage of 1750 liters per second (40 million gallons per day). The upper aquifer, which is up to 20 meters (65 feet) thick, is artificially recharged with river water that floods onto about 8 hectares (20 acres) of infiltration lagoons during periods when the river turbidity is low. The lagoon bottoms are annually cleaned of silt and clay (Figure 7.9). This is a classic example of induced stream infiltration used as a water resource management technique. Virtually all of the water pumped from this aquifer comes indirectly from the river. The filtration through the sediments reduces turbidity and removes pathogens.

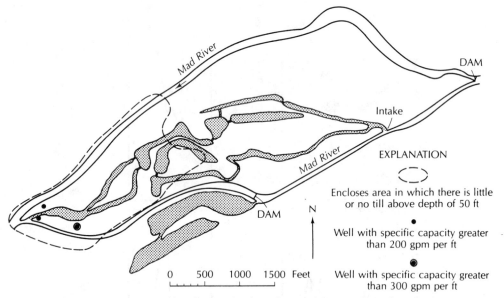

FIGURE 7.9. Map of Rohrers Island in the Mad River. River flow is diverted into artificial recharge ponds during periods of low turbidity. Only a few of the producing wells are shown. SOURCE: S. E. Norris and A. M. Spieker, U.S. Geological Survey Water Supply Paper 1808, 1966.

The river water is of low quality due to point- and diffused-source waste discharges. This may lead to a lowering of groundwater quality. Past practices of locating landfill sites and other waste-disposal facilities in old gravel pits in the buried valley aquifer have also resulted in localized reduction in groundwater quality (7).

The Miami Valley Conservancy District plays an important role in the management of the water resources of the area. This agency was originally formed to construct dams for flood protection. Over the years it has assumed many responsibilities for area-wide planning and management of surface-water and groundwater resources.

7.2.2 ALLUVIAL VALLEYS

Flowing rivers deposit sediment, generally termed **alluvium.** During periods of flooding, alluvium is deposited in the channel as well as in the floodplain. As the flood peak passes, flow velocities start to drop, the energy available to transport sediment decreases, and deposition begins. Coarse gravel is deposited in the stream channel, sand and fine gravel forms natural levees along the banks, and silt and clay come to rest on the floodplain. Point bars are formed by deposition on the inside of a bend in the river. Point-bar formation is not limited to floods.

The alluvium may be reworked by a meandering stream, even during quiescent periods for the river. As the channel swings back and forth across the floodplain, point-bar deposits of sand and coarse gravel are left behind. If the stream is aggrading, the general land level subsiding, or both, the alluvial deposits will thicken with time.

Rivers draining glaciers, either modern or Pleistocene, have very heavy sediment loads. Braided streams and gravel bars in the river are typical, and connote the aggrading nature of such rivers. Thick deposits of sand, gravel, or mixtures of both are formed. Downcutting by a river through previously deposited sediment can form terraces on the sides of the lower stage floodplain. Many modern rivers, even some outside of glaciated regions, have terraces formed of Pleistocene-age gravel deposits.

Alluvial valleys can be excellent sources of water. There are zones of gravel in old channel and point-bar deposits with a very high conductivity. The task of the hydrogeologist is to find these gravel zones, since silt and clay floodplain deposits may also be present, obscuring their location. A knowledge of fluvial processes is helpful in this regard. Surface surveys of electrical resistivity* can often be used to locate sand and gravel deposits where they are surrounded by cohesive sediments. Test drilling is usually necessary to confirm the preliminary conclusions of the hydrogeologist and geophysicist.

In evaluating stream-terrace deposits, it should be kept in mind that erosion, as well as deposition, can result in terrace formation. The stream in Figure 7.10 has a terrace on each bank. The bedrock terrace represents a former erosional surface which was buried in gravel but has since been

---

*See Chapter 12 for a discussion of electrical resistivity.

exhumed by erosion. It would not be a suitable well-field site, since the alluvium is missing.

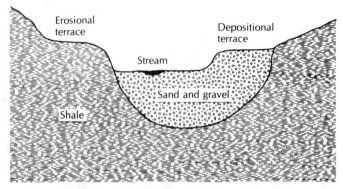

FIGURE 7.10. Cross section of a stream with two terrace deposits. Only the gravel terrace is a potential source of groundwater.

7.2.3 ALLUVIUM IN TECTONIC VALLEYS

Many major valley systems are products of tectonic activity, rather than fluvial or glacial erosion. During mountain-building episodes, the uplift of mountain masses will result in intermontaine basins being formed. Fault-block valleys can also be created by down-dropping of large crustal pieces along faults. Erosion of the mountains creates sediment which is carried into the valleys, forming talus slopes, alluvial fans, and alluvial and lacustrine deposits. These sediments can be very coarse with high hydraulic conductivities—alluvial fan gravels and channel deposits, for example. Lacustrine clays, on the other hand, can be fine with low conductivity. Gravel aquifers confined by lacustrine sediment are quite typical of such basins.

Intermontaine valleys of the interior basins of southeastern California and the Great Basin area of Nevada and Utah are typically tectonic. The unconsolidated sediments in the basins may be part of either local or regional flow systems (Figure 7.11). In regions that are semiarid to arid, groundwater is recharged by precipitation in the mountains. Bedrock beneath the mountains may receive direct recharge and then feed the valley-fill sediments. Streams originating in the mountains may also lose water to the alluvium when the flow goes across the valley bottoms (Figure 7.12).

The water table is generally closer to the surface in the lower elevations of intermontaine basins than it is at the edges next to the mountains, so wells in the former location would have a smaller pumping lift—hence, lower energy use. The intermontaine deposits typically slope downward from the mountain flanks. Beneath the upper parts of alluvial fans, the depth to groundwater may be hundreds of meters. Surface streams flowing onto high alluvial fans may disappear as they lose water to the coarse sediment.

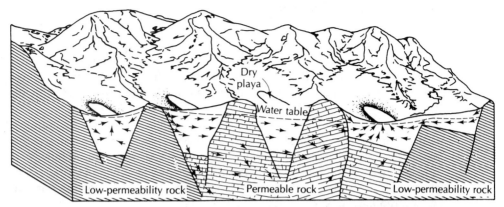

FIGURE 7.11. Common groundwater flow systems in tectonic valleys filled with sediment. Basins bounded by impermeable rock may form local or single-valley flow systems. If the interbasin rock is permeable, regional flow systems may form. In closed basins, groundwater discharges into playas where it is discharged by evaporation and transpiration by phreatophytes. SOURCE: Modified from T. E. Eakin, D. Price, and J. R. Harrill, U.S. Geological Survey Professional Paper 813-G, 1976.

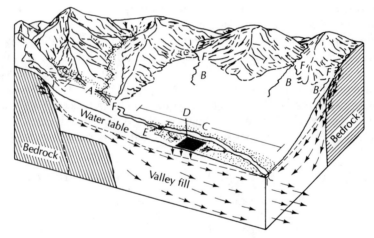

A, Gaining reach, net gain from groundwater inflow although in localized areas stream may recharge wet meadows along flood plain. Hydraulic continuity is maintained between stream and groundwater reservoir. Pumping can affect streamflow by inducing stream recharge or by diverting groundwater inflow which would have contributed to streamflow.

B, Minor tributary streams, may be perennial in the mountains but become losing ephemeral streams on the alluvial fans. Pumping will not affect the flow of these streams because hydraulic continuity is not maintained between streams and the principal groundwater reservoir. These streams are the only ones present in arid basins.

C, Losing reach, net loss in flow due to surface-water diversions and seepage to groundwater. Local sections may lose or gain depending on hydraulic gradient between stream and groundwater reservoir. Gradient may reverse during certain times of the year. Hydraulic continuity is maintained between stream and groundwater reservoir. Pumping can affect streamflow by inducing recharge or by diverting irrigation return flows.

D, Irrigated area, some return flow from irrigation water recharges groundwater.

E, Flood plain, hydrologic regimen of this area dominated by the river. Water table fluctuates in response to changes in river stage and diversions. Area commonly covered by phreatophytes (shown by random dot patterns).

F, Approximate point of maximum stream flow.

FIGURE 7.12. Groundwater-surface water relationships in valley-fill aquifers located in arid and semiarid climates. SOURCE: T. E. Eakin, D. Price, and J. R. Harrill, U.S. Geological Survey Professional Paper 813-G, 1976.

199

In the Great Basin region of the western United States, saturated valley-fill deposits over 300 meters (1000 feet) thick are common in the large valleys (8). Artesian conditions are often found in the lower elevations of these basins. Pleistocene lacustrine clays near the surface overlying coarse alluvium creates this situation.

In tectonic valleys, groundwater outflow can occur by transpiration, evaporation from surface water or saturated soils, spring discharge, and/or underflow into adjacent basins. Water pumped and used consumptively for irrigation is also an outflow. If the basin is topographically closed, there is no stream discharge. In such a case, evapotranspiration and underflow are the natural drainage methods.

Wells in tectonic valleys should be located where the aquifer material is coarse, the depth to water is not great, and a source of recharge water is available. These criteria are often met in well fields located near modern rivers. However, there may not be any surface streams near areas where wells are needed. Artesian aquifers overlain by thick lacustrine deposits may be too deep to find by surficial geophysical methods, such as electrical resistivity. Test drilling may be the only recourse in such cases, albeit an expensive one. Groundwater quality in tectonic valleys can have significant spatial variation. In general, groundwater that is not actively circulating may be high in dissolved solids. Shallow groundwater, with a high rate of direct evaporation, may also have high salinity. In the Great Basin area, individual wells have been developed with yields of up to 540 liters per second (8600 gallons per minute) with specific capacities of up to 0.6 square meter per second (3000 gallons per minute per foot). These are prodigious wells. The average yield is lower, but still substantial at 65 liters per second (1000 gallons per minute) (8).

CASE STUDY: TECTONIC VALLEYS—SAN BERNARDINO AREA

The San Bernardino area is in the upper Santa Ana Valley of the coastal area of southern California (9). The valley is tectonic in origin and bounded on the north, east, and south by mountains of consolidated rocks. The alluvial valley is subdivided into separate groundwater basins by faults in the alluvial materials, which are barriers to groundwater movement. The Bunker Hill Basin has mountains on three sides and the San Jacinto Fault on the fourth side (Figure 7.13).

The alluvium in the basin consists of Pleistocene-age deposits overlain by Recent alluvium, river-channel deposits, and dune sands. The alluvium is from 210 to 425 meters (690 to 1400 feet) thick at the center of the basin. The older deposits consist of alluvial fans, terrace deposits, and stream channels. The deposits can be highly permeable, but facies of low permeability exist and form confining layers. Faulting cuts the older alluvium, and the faults are generally barriers to groundwater movement. They may impede groundwater flow by (a) offsetting of gravel beds against clay layers, (b) folding of impermeable beds upward along the fault, (c) cementation by carbonate formation, and (d) formation of clayey gouge.

200

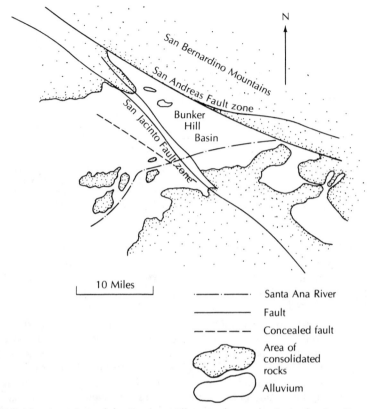

FIGURE 7.13. Location of the Bunker Hill groundwater basin in the San Bernardino area of southern California. SOURCE: Modified from L. C. Dutcher and A. A. Garrett, U.S. Geological Survey Water Supply Paper 1419, 1963.

Recent alluvium includes highly permeable river-channel deposits through which much of the groundwater recharge occurs. They underlie active and abandoned stream channels and are above the water table, for the most part. The Recent floodplain material is less permeable than the stream channels. It is 15 to 30 meters (50 to 100 feet) thick, and is not known to be faulted.

The lower parts of the valley floor receive 30 to 43 centimeters (12 to 17 inches) of precipitation per year. Due to orographic effects, precipitation in the San Bernardino Mountains to the east is as much as 71 centimeters (28 inches). Runoff from the mountains flows into the Santa Ana River and its tributaries. In the upper reaches of the basin, the river is a losing stream. Groundwater recharge by floodwater takes place through the permeable river-channel deposits.

A cross section of the Bunker Hill area is shown in Figure 7.14. Groundwater recharge takes place in the upper parts of the basin and flows toward lower elevations. The hydrographs of two nearby wells in this area are shown in Figure 7.15A. The elevation in the deeper

201

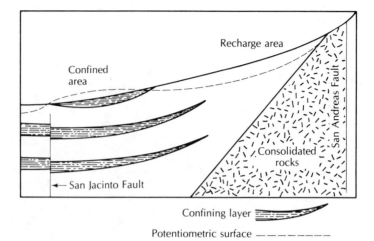

Confining layer
Potentiometric surface

FIGURE 7.14. Cross section through the unconsolidated deposits of the Bunker Hill groundwater basin. The section is along the course of the Santa Ana River which provides recharge to the aquifers. SOURCE: Modified from L. C. Dutcher and A. A. Garrett, U.S. Geological Survey Water Supply Paper 1419, 1963.

well is lower than the shallower one, indicating downward flow. Both wells show seasonal fluctuations due to pumping of irrigation water, but the deeper well is affected to a greater extent. There are several confining layers in the lower basin; hence, there are several identifiable aquifer zones. In Figure 7.15B, well hydrographs on the upstream side of the San Jacinto Fault show that the head in a shallow well is lower than in deeper wells, indicating upward flow. In the San Bernardino area, the younger alluvium is confining, and when wells were first drilled in the area they were flowing.

At one time, considerable groundwater outflow occurred from the Bunker Hill Basin in the area where the Santa Ana River crosses the San Jacinto Fault Zone (Colton Narrows). Although the fault zone does not offset the Recent alluvium, it forces deeper water to move upward. Except for Recent river-channel deposits beneath the Santa Ana River, the surficial deposits are, for the most part, impermeable. At the turn of the century, large springs made this area marshy, and ponds were present. Heavy groundwater withdrawals for irrigation have lowered the water table and reduced the amount of outflow. Figure 7.16 shows the water-table contours where the Santa Ana River crosses the San Jacinto Fault as they existed in 1939. Groundwater contours northeast (upstream) of the fault indicate that groundwater is moving toward the surface. Well hydrographs for three deep wells in the area also indicate upward groundwater movement—the deeper the well, the greater the hydraulic head (Figure 7.15 B). The water-table contours on either side of the fault indicate a very steep gradient across the fault zone.

202

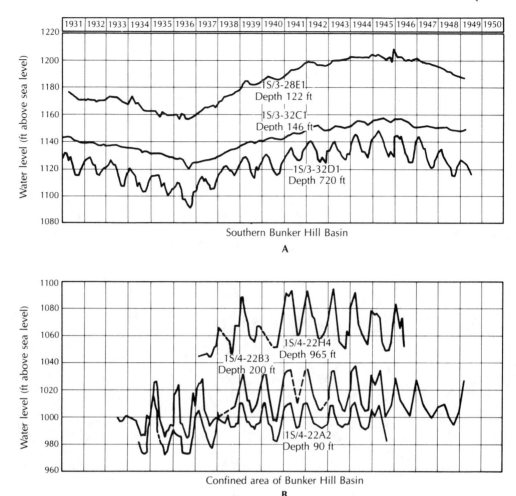

FIGURE 7.15. Well hydrographs of observation wells located in the Bunker Hill groundwater basin: **A.** Recharge area—head decreases with depth; **B.** Discharge area —head increases with depth. SOURCE: L. C. Dutcher and A. A. Garrett, U.S. Geological Survey Water Supply Paper 1419, 1963.

By the late 1970s the basin had undergone extensive groundwater overdrafts. The water level in the swampy area along the Santa Ana River dropped to 15 to 45 meters (50 to 150 feet) below the land surface. This drained the wetland, so that today the land is dry. Urban development in the area has been intense—even in this former wetland.

Should groundwater conditions return to their former state, serious consequences for the residents of the San Bernardino area would

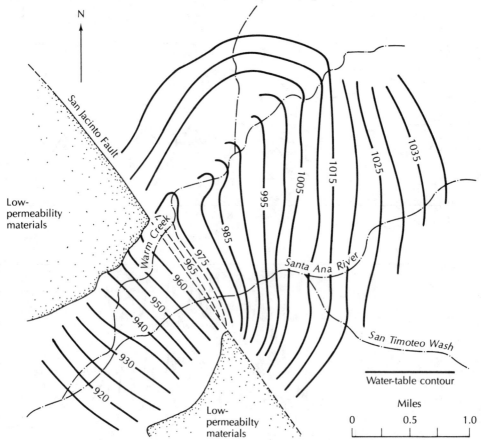

FIGURE 7.16. Water-table contours in March 1939 of shallow deposits at the Colton Narrows area where the Santa Ana River crosses the San Jacinto Fault. The contours for 970 and 965 are shown, but the exact position was not known. SOURCE: L. C. Dutcher and A. A. Garrett, U.S. Geological Survey Water Supply Paper 1419, 1963.

result. Abandoned and forgotten wells could begin to flow again. Basements would be flooded and soils waterlogged. This might occur if there were a climatic change to increase natural recharge, if pumpage were reduced, if imported water were used to artificially recharge the aquifer, or some combination of these events.

A computer model of the groundwater basin has been developed (10). This is being used as a water management tool, so that local officials can predict the impact of various water resource management schemes to prevent the rise of groundwater levels in the critical area. This is an especially sensitive problem, as the critical area is located at the point of natural groundwater discharge from the Bunker Hill Basin. Planning must ensure that the drafts equal or exceed recharge to the basin; otherwise, water levels will naturally rise.

LITHIFIED SEDIMENTARY ROCKS 7.3

Clastic sedimentary rocks are typically composed of silicate, carbonate, or clay minerals. Chemically precipitated sedimentary rocks are primarily limestone, dolomite, salt, or gypsum. Coal and lignite can also be considered to be sedimentary deposits, with the original sediments being organic matter.

Sedimentary rock sequences were formed with younger beds laid down upon older ones. The original sediments may have been subaqueous or terrestrial. Rarely do sedimentary rocks occur as a single unit; there is typically a sequence of many beds. The original layered sequence may be undisturbed, or it may be extensively folded and faulted.

A typical sedimentary rock aquifer is found in southeastern Wisconsin and northeastern Illinois (11). There are beds of sandstone, siltstone, shale, and dolomite which strike north-south and dip gently to the east. A number of the sandstone and dolomite formations have been grouped into the sandstone aquifer, a hydrostratigraphic unit, while the Maquoketa Shale forms a confining bed (Figure 7.17).

System	Geologic Unit	Dominant Lithology	Saturated Thickness (ft)	Hydrologic Unit	Areal Extent	Yield
Quaternary	Holocene and Pleistocene deposits	Unsorted mixture of clay, silt, sand, gravel, and boulders	0–300	Sand and gravel aquifer	Entire report area, but aquifer is localized as outwash, alluvium, and buried deposits	Small to large yields; not extensively developed for large yields
Devonian	Undifferentiated	Shale and dolomite	0–155	Niagara aquifer	Near Lake Michigan north from Milwaukee	Some small yields where creviced
Silurian	Undifferentiated	Dolomite	0–560		Eastern two-thirds of report area	Small to large yields depending upon number and size of solution channels and crevices

Figure 7.17 continued on pages 206–7

205

System	Geologic Unit	Dominant Lithology	Saturated Thickness (ft)	Hydrologic Unit	Areal Extent	Yield
Ordovician	Maquoketa Shale	Shale	0–270	Confining bed	Eastern three-fourths of report area	Generally cased out in deep wells; very small yields locally from minor amounts of interbedded dolomite
	Galena Dolomite, Decorah and Platteville formations	Dolomite	0–340	Leaky confining bed in recharge area	Entire report area except southeastern corner of Jefferson County	Small to moderate yields from crevices; developed as sole unit only where Maquoketa Shale is absent
	St. Peter Sandstone	Sandstone	0–260		Entire report area except Hartford area and southeastern corner of Jefferson County	Moderate yields; generally not used as sole unit; tends to cave
	Prairie du Chien Group	Dolomite	0–150		Missing or very thin in much of report area	Small yields
Cambrian	Trempealeau Formation	Dolomite	0–10(?)	Sandstone aquifer	Entire report area except Hartford area	Small yields generally, but some large yields in areas of well-developed solution channels
	Franconia and Galesville sandstones	Sandstone	0–225		Entire report area except Hartford and part of Milwaukee area	Moderate to large yields, especially from lower part

System	Geologic Unit	Dominant Lithology	Saturated Thickness (ft)	Hydrologic Unit	Areal Extent	Yield
Cambrian	Eau Clarie Sandstone	Sandstone, siltsone, and shale	0–160		Entire report area except Hartford area	Small yields, decreasing to south
	Mount Simon Sandstone	Sandstone	0–1500+	Sandstone aquifer	Entire report area except Hartford area	Moderate to large yields; not fully penetrated east and south of Waukesha
Precambrian	Undifferentiated	Crystalline rock	Unknown	Confining bed	Entire report area	Very small yields locally from crevices

FIGURE 7.17.   Paleozoic sedimentary rock sequence and aquifer properties in southeastern Wisconsin, where there are two major aquifers and a regional confining layer.

CASE STUDY: SANDSTONE AQUIFER OF NORTHEASTERN ILLINOIS-SOUTHEASTERN WISCONSIN

Tilted sedimentary rock units can form the classical artesian aquifer system. A formation of lower hydraulic conductivity acts as a confining layer for underlying aquifers. The sandstone aquifer of southeastern Wisconsin is confined by the Maquoketa Shale (Figure 7.18). The Maquoketa Shale thins to the west where it has been eroded. Recharge to the sandstone aquifer occurs wherever the hydraulic gradient is downward. The rate is greater where the shale is thin or absent. In the "recharge area" that is west of the shale outcrop, recharge does not have to pass through the shale. In northeastern Illinois, the downward leakage to the sandstone aquifer in the recharge area was estimated for 1958 conditions to be 26.3 cubic meters per day per square kilometer (18,000 gallons per day per square mile). Where the Maquoketa Shale is present, with a vertical hydraulic conductivity of 10^{-9} centimeter per second (4×10^{-7} gallon per day per square foot), the recharge was 3.07 cubic meters per day per square kilometer (2100 gallons per day per square mile) (12).

When wells were first drilled in eastern Wisconsin, the artesian head in the sandstone aquifer was 40 to 60 meters (130 to 200 feet) above the level of Lake Michigan. There was a hydraulic gradient of 0.0009 from the recharge zone toward the Milwaukee area. By 1973, the potentiometric surface in the Milwaukee area had dropped more than 100 meters (325 feet) due to pumping, and was many meters below

207

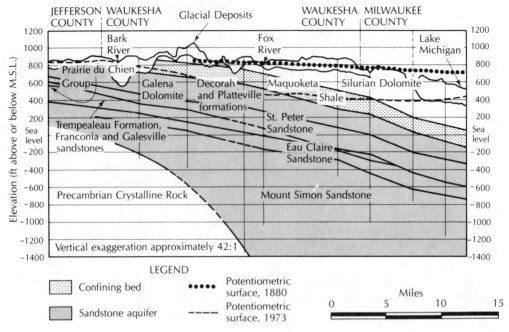

FIGURE 7.18. Artesian flow system of southeastern Wisconsin.
SOURCE: U.S. Geological Survey.

land surface. The hydraulic gradient from the recharge area had increased to 0.003.

In the Milwaukee area, the original vertical hydraulic gradient across the Maquoketa Shale was upward, indicating upward flow. Because of the decline in the potentiometric surface in the sandstone aquifer, it is now downward. Recharge to the sandstone aquifer is thus occurring not only in the "recharge zone," but also over much of the area underlain by the confining shale. The flow in the aquifer from the direct-recharge area toward the Milwaukee pumping center has increased by 300 to 400 percent due to the increase in the horizontal hydraulic gradient.

The fact that a formation may change in lithology from one locality to another accounts for some of the difficulties associated with studies of sedimentary rock units. For example, a sandstone may grade into a shaly sandstone or siltstone, yet still have the same fossil fauna assemblage and the same stratigraphic nomenclature. The Eau Claire Formation of Cambrian age is a sandstone in east-central Wisconsin, but grades into a shale and siltstone in northern Illinois. In Wisconsin, it has high conductivity and is an aquifer; in Illinois it becomes a confining layer (12, 13).

Complex stratigraphy can be a very real hindrance to groundwater exploration. On the Hualapai Plateau of northwestern Arizona, the best potential aquifer is the stratigraphic sequence of sedimentary rocks in the Rampart Cave Member of the Mauv Limestone Formation (14). Springs discharge from this member into the Grand Canyon along the flanks of the plateau. The

Bright Angel Shale stratigraphically underlies the Rampart Cave Member of the Mauv Limestone, causing the groundwater to be perched (Figure 7.19). However, because these units were being deposited in a transgressing sea, the various beds are interfingering. As a result, at some localities the Bright Angel Shale can be both above and below the Rampart Cave Member (Figure 7.20). Drillers in the area have had unsuccessful wells because they stopped drilling when the Bright Angel Shale was first reached (15). Unfortunately, the target formation had not yet been penetrated. A well drilled at Location *A* of Figure

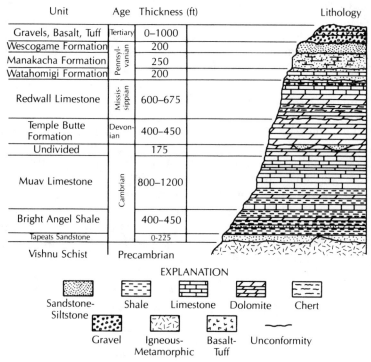

FIGURE 7.19. Stratigraphy of the Grand Canyon area. SOURCE: P. W. Huntoon, *Ground Water,* 15 (1977):426-33.

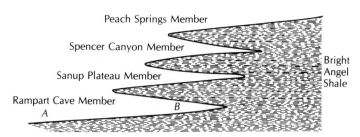

FIGURE 7.20. Interfingering of sedimentary rock units of the Hualapai Plateau area. SOURCE: Adapted from P. W. Huntoon, *Ground Water,* 15 (1977):426-33.

7.20 would not hit Bright Angel Shale until the Rampart Cave Member was penetrated. However, at location B, three shale members alternating with limestone would have to be penetrated before the Rampart Cave Member was reached. The area of these interfingering members extends for tens of kilometers. The hydrogeologist working in such areas of complex stratigraphy must be aware of the possible hydrogeologic consequences.

Folding and faulting of sedimentary rocks can create very complex hydrogeologic systems, in which determination of the location of recharge and discharge zones and flow systems is confounded. Not only must the hydrogeologist determine the hydraulic characteristics of rock units and measure groundwater levels in wells to determine flow systems, but detailed geology must also be evaluated. In most cases, the basic geologic structure will have already been determined; however, logs of test wells and borings must be reconciled with the pre-existing geologic knowledge.

Fault zones can act as either barriers to groundwater flow or as groundwater conduits, depending upon the nature of the material in the fault zone. If the fault zone consists of finely ground rock and clay (gouge), the material may have a very low hydraulic conductivity. Significant differences in groundwater levels can occur across such faults (see Figure 7.16, page 204). Inpounding faults can occur in unconsolidated materials with clay present, as well as in sedimentary rocks where interbedded shales, which normally would not hinder lateral groundwater flow, can be smeared along the fault by drag folds (Figure 7.21) **Clastic dikes** are intrusions of sediment that are forced into rock fractures. If they are clay-rich, they can act as groundwater barriers in either sediments or in a lithified sedimentary rock. Clastic dikes are known to occur in alluvial sediments, glacial outwash, and lithified sedimentary rock.

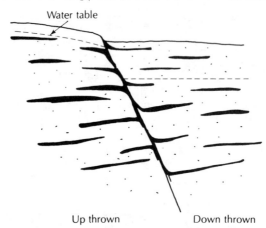

FIGURE 7.21. A low-permeability fault zone formed in sandstone with interbedded shale that is smeared along the fault zone by drag folding.

In consolidated rocks, faults are more often groundwater conduits. Broken and brecciated rock in the fault zone may have a high porosity and hydraulic conductivity. Water may move along a fault to discharge as a spring (Figure 7.22A). A fault may also contain water under great pressure at depth. One of the dangers of digging deep tunnels in low-permeability rock is the possibility of broaching an unexpected fault zone (Figure 7.22B). Damaging and dangerous flooding can occur if the fault contains groundwater with a high hydraulic head.

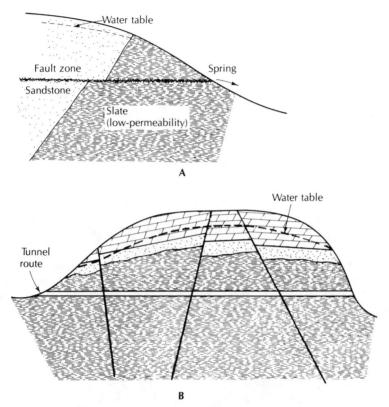

FIGURE 7.22. **A.** Fault zone acting as a groundwater conduit forming a spring; **B.** Faults cutting through impermeable shale unconformably overlain by permeable sandstone and limestone. These faults can act as groundwater conduits, creating dangerous flooding conditions during tunnel construction.

Overthrust faulting can create conditions in which a rock, normally found as an impermeable basement unit, is overlying the sedimentary rock units, typically a groundwater source. In such a case, the hydrogeologist might recommend drilling through the "basement" rocks to attempt to obtain a groundwater supply from younger sedimentary units, provided there is an opportunity for recharge to occur and the water is not known to have a high mineral content (Figure 7.23).

The presence of faulting may have no significant effect on hydrogeology. Faults with a very small movement will usually have the same hydraulic characteristics as the parent rock. Brecciated fault zones are potential conduits for movement of hydrothermal solutions, and faults can become mineralized. The crystallization of minerals in a fault zone can seal it, so that its original high conductivity is lost.

211

Folding can affect the hydrogeology of sedimentary rocks in several ways. The most obvious is the creation of confined aquifers at the center of synclines. The nature of the fold will affect the availability of water. A tight,

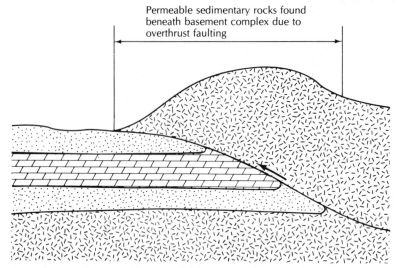

Permeable sedimentary rocks found
beneath basement complex due to
overthrust faulting

FIGURE 7.23. Overthrust faulting creating a situation in which a sedimentary rock aquifer is found beneath the basement complex. In the area shown on the cross section, drilling through the basement complex material would yield a well in the hidden aquifer.

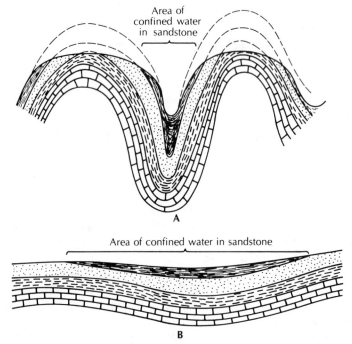

Area of
confined water
in sandstone

A

Area of confined water in sandstone

B

FIGURE 7.24. Effect of folding on the occurrence of confined aquifers and the depth of wells necessary to reach them.

deeply plunging fold might carry the aquifer too deep beneath the surface to be economically developed (Figure 7.24A). Deeply circulating groundwater is also typically warmed by the geothermal gradient, and may be highly mineralized. A broad, gentle fold can create a relatively shallow confined aquifer, which extends over a large area (Figure 7.24B). This might be a good source of water if sufficient recharge can occur through the confining layer, or if the aquifer can transmit enough water from areas where the confining layer is absent.

Another effect of folding is to create a series of outcrops of soluble rock, such as limestone, alternating with rock units that are not as permeable. Smaller streams flowing across the limestone might sink at the upper end, only to reappear at the lower outcrop. The type of trellis drainage that can develop on folded rocks is shown in Figure 7.25. Surface streams follow the strike of rock outcrops usually along fault or fracture traces. When streams traverse the strike of formations, they typically follow a fault or fracture trace.

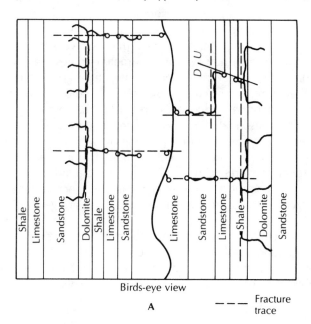

Birds-eye view

A

— — — Fracture trace

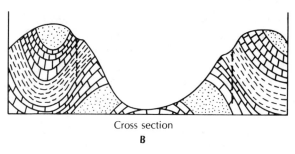

Cross section

B

FIGURE 7.25. Drainage pattern developed in an area of longitudinally folded rock strata: **A.** Top view; **B.** Cross section.

213

Complex folded and faulted sedimentary rock units are a challenge to the hydrogeologist. Competency in geology is necessary in order to construct geologic cross sections based on well logs, drill-core samples, and outcrops. Regional flow systems can be controlled by the structural and stratigraphic relations of the confining beds and aquifer units. In addition, distribution of hydraulic potential must be determined, very often from limited data.

7.3.1   CLASTIC SEDIMENTARY ROCKS

Hydraulic conductivity of clastic sedimentary rocks, based on primary permeability, is a function of the grain size, shape, and sorting of the original sediment. The same factors that affect the permeability and porosity of loose sediments also are important in sedimentary rocks. **Cementation**, in which part of the voids are filled with precipitated material, such as silica, calcite, or iron oxide, can reduce the original porosity. Solution of the original material may occur during and after diagenesis, resulting in an increase in porosity.

Consolidated rocks also contain secondary porosity and permeability due to fracturing. Microfractures may add very little to the original hydraulic characteristics; however, major fracture zones may have localized hydraulic conductivities several orders of magnitude greater than the unfractured rock. Fracturing can occur through several geologic processes. Rock at depth is under great pressure due to the weight of the overburden. As uplift and erosion bring the consolidated rock to the surface, it expands as the pressure is reduced. The expansion can cause fracturing of the rock, with the majority of the expansion fractures occurring within about 100 meters (300 feet) of the surface. Vertical fractures carry recharging precipitation downward, and provide a very important function in bypassing low-permeability layers near the surface. Wells located in surface fracture zones are generally highly successful.

Fracturing may also be associated with tectonic activity. Rock deformed by faulting or folding may fracture when it is subjected to tension or compression. Such activity can take place at substantial depths; thus, secondary permeability is not strictly a near-surface phenomenon. However, greater pressure on the deeper fractures does not permit them to be as open (have as great a porosity) as shallow fractures.

The hydrologic properties of clastic sedimentary rocks are summarized in Table 7.1. The data reflect primary porosity and permeability, as they are laboratory values. Rock samples small enough to fit a permeameter would rarely exhibit secondary effects of fracturing. The table is based on limited data and should not be used in lieu of field and laboratory studies. The hydraulic conductivity values are for vertical water movement. For individual samples, the hydraulic conductivity in the horizontal direction is generally higher than the vertical. This is due to the layering effect of sedimentary processes.

The yield to wells is proportional to the transmissivity of the aquifer. This, in turn, is proportional to the aquifer thickness, if the hydraulic

TABLE 7.1. Hydraulic properties of clastic sedimentary rocks

	Sandstone		Siltstone	Claystone	Shale
	Fine-grained	Medium-grained			
Vertical Hydraulic Conductivity (gpd/ft^2)					
Range	0.008–37	0.05–220	0.00002–0.03	0.002	
Mean	5	77	0.004	0.002	
Porosity— Undisturbed					
Range	13.7–49.3	29.7–43.6	21.2–41.0	41.2–45.2	1.4–9.7
Mean	33	37	35	43	6
Specific Yield					
Range	2.1–39.6	11.9–41.1	0.9–32.7		
Mean	21	27	12		

SOURCE: D. A. Morris and A. I. Johnson, U.S. Geological Survey Water Supply Paper 1839-D, 1967.

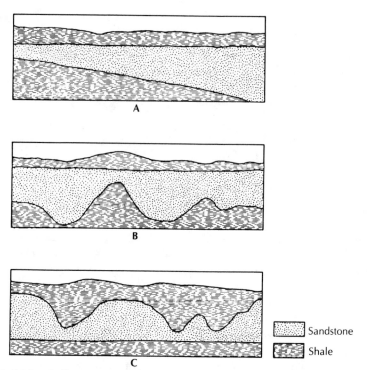

▨ Sandstone

▨ Shale

FIGURE 7.26. Sedimentary conditions producing a sandstone aquifer of variable thickness: **A.** Sandstone deposited in a sedimentary basin; **B.** Sandstone deposited unconformably over an erosional surface; **C.** Surface of sandstone dissected by erosion prior to deposition of overlying beds.

conductivity is uniform throughout the aquifer. Sedimentary aquifers were deposited in sedimentary basins in which units gradually thicken. Variable thickness of a sedimentary aquifer may also be due to the deposition of the aquifer material over an eroded surface with high relief, or a dissection of the top of the aquifer after deposition (Figure 7.26). Higher well yields will be obtained from thicker sections of the aquifer. The relationship between the specific capacity of wells and the uncased thickness of two sandstone aquifers in northern Illinois is shown in Figure 7.27.

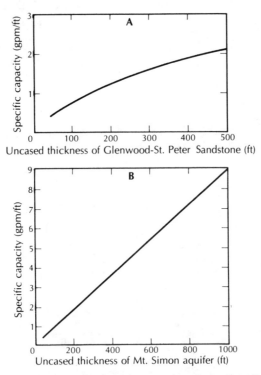

FIGURE 7.27. Relation between the specific capacity of a well (gallons per minute of yield per foot of drawdown) and the uncased thickness of the sandstone aquifer: **A.** Glenwood-St. Peter Sandstone; **B.** Mt. Simon Sandstone. Both of northern Illinois. SOURCE: W. C. Walton and S. Csallany, Illinois State Water Survey Report of Investigation 43, 1962.

The thickness of the Glenwood-St. Peter Sandstone in northern Illinois is shown as Figure 7.28. The lower St. Peter Sandstone was deposited unconformably on an erosional surface of the Prairie du Chien Dolomite, and it is thicker in areas that were once Ordovician valleys on that surface. In such areas, the Glenwood-St. Peter Sandstone is more than 100 meters (300 feet) thick.

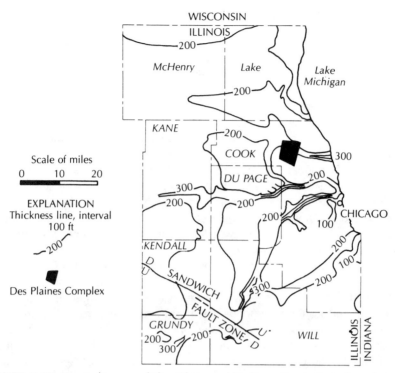

FIGURE 7.28. Isopach map of the Glenwood-St. Peter Sandstone of northeastern Illinois. SOURCE: M. Suter et al., Illinois State Geological Survey and Water Survey Cooperative Ground-Water Report 1, 1959.

Wells in sandstone aquifers should be located in such a manner as to penetrate the maximum saturated thickness of the aquifer. If one area of the aquifer is known to have a higher hydraulic conductivity than other areas, the combination of hydraulic conductivity and thickness should be considered in order to locate the well in the area of greatest aquifer transmissivity.

The yield of sandstone wells can sometimes be increased by the detonation of explosives in the uncased hole. The shots are generally located opposite the most permeable zones of the sandstone. The loosened rock and sand is bailed from the well prior to the installation of the pump. The shooting process has two effects on the borehole: it enlarges the diameter of the well in the permeable zones and also breaks off the surface of the sandstone, which may have been clogged by fine material during drilling. Fractures near the well may be opened all the way to the borehole by shooting. Old wells may be rehabilitated by shooting if the yield has decreased due to mineral deposition on the well face. Shooting has increased the specific capacities of sandstone wells in northern Illinois by an average of 22 to 38 percent, depending upon the formation (17).

217

7.3.2 CARBONATE ROCKS

The primary porosity of limestone and dolomite is variable. If the rock is clastic, the primary porosity can be high. Chemically precipitated rocks can have a very low porosity and permeability if they are crystalline. Bedding planes can be zones of high primary porosity and permeability.

Limestone and (to a much lesser extent) dolomite are soluble in water that is mildly acidic. In general, if the water is unsaturated with respect to calcite or dolomite, it will dissolve the mineral until it becomes saturated. (Carbonate equilibrium is discussed in detail in Section 9.8).

Secondary permeability in carbonate aquifers is due to the solution enlargement of fractures and bedding planes. The rate of solution is a function of the amount of groundwater moving through the carbonate rock and the degree of saturation (with respect to the particular carbonate minerals present). Initially, more groundwater flows through the larger fractures and bedding planes, which have a greater hydraulic conductivity. These become enlarged with respect to lesser fractures; hence, even more water flows through them. Solution mechanisms of carbonate rocks favor the development of larger openings at the expense of smaller ones. Carbonate aquifers can be highly anisotropic and nonhomogeneous if water moves only through fractures and bedding planes that have been preferentially enlarged. Water entering the carbonate rock is typically unsaturated. As it flows through the aquifer, it approaches saturation, and dissolution slows and finally ceases. It has been shown experimentally that solution passages form from the recharge area to the discharge area, and that as they follow fracture patterns, many smaller solution openings join to form fewer but larger ones (18) (Figure 7.29). Eventually, many passages join to form one outlet. Greater groundwater movement; hence, solution, takes place along the intersection of two joints or a joint and a bedding plane. Groundwater moving along a bedding plane tends to follow the strike of intersecting joints.

A second mode for the entry of unsaturated water into a carbonate aquifer occurs near valley bottoms. In **karst*** regions, flow in valleys with permanent streams is usually discharged from carbonate aquifers recharged beneath highlands. Water tables in many karst areas are almost flat due to the high hydraulic conductivity. Floodwaters from surface streams can enter the carbonate aquifers and reverse the normal flow. If the floodwaters are unsaturated with respect to the mineral in the aquifer, solution will occur (19).

Swallow holes, or shafts leading from surface streams, can carry surface water underground into caverns. Swallow holes can drain an entire stream, or only a small portion of one.

Geochemical studies have shown that there are two types of groundwater found in complex carbonate aquifer systems (20). The joints and bedding planes that are not enlarged by solution contain water that is satu-

---

*Karst is a term applied to topography formed over limestone, dolomite, or gypsum where there are sinkholes, caverns, and lack of surface streams.

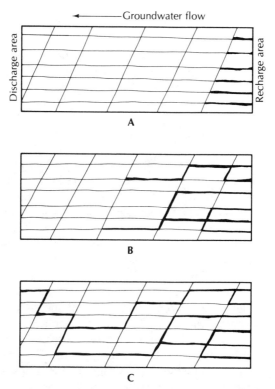

FIGURE 7.29. Growth of a carbonate aquifer drainage system starting in the recharge area and growing toward the discharge area. **A**. At first, most joints in the recharge area undergo solution enlargement; **B**. As the solution passages grow, they join and become fewer; **C**. Eventually, one outlet appears at the discharge zone.

rated with respect to calcite (or dolomite). Because of the low hydraulic conductivity of these openings, this water mass moves slowly. Another mass of water, generally undersaturated, moves more rapidly through well-defined solution channels close to the water table. It is this second body of water that forms the passageways. Most carbonate solution; hence, cave formation, takes place below the water table. This is true of all horizontal passages. Vertical shafts can form in the vadose zone by undersaturated infiltrating water trickling down the rock surface (21). Caves that are presently dry were formed below the water table when the regional water table was higher. The regional base level of a karst region is typically a large river. If the river is downcutting, the regional water table will be lowering. The result will be a series of dry caves at different elevations, each formed when the regional water table was at a different level.

Carbonate aquifers show a very wide range of hydrologic characteristics. There are, to be sure, a number of "underground rivers" where a surface stream disappears and flows through caves as open channel flow. At the other extreme, some carbonate aquifers behave almost like a homogeneous, isotropic porous medium. Most lie between these extremes.

219

Three conceptual models for carbonate aquifers have been proposed (19). **Diffuse-flow carbonate aquifers** have had little solutional activity directed toward opening large channels; these are to some extent homogeneous. **Free-flow carbonate aquifers** receive diffused recharge, but have well-developed solution channels along which most flow occurs. There are no structural or stratigraphic controls on flow or free-flow aquifers. **Confined-flow carbonate aquifers** have solution openings in the carbonate units, but low-permeability noncarbonate beds exert control over the direction of groundwater movement.

Diffuse-flow aquifers are typically found in dolomitic rocks or shaly limestones, neither of which is easily soluble. Water movement is along joints and bedding planes that have been only modestly affected by solution. Moving groundwater is not concentrated in certain zones in the aquifer and, if caves are present, they are small and not interconnected. Discharge is likely to be through a number of small springs and seeps. The Silurian-age dolomite aquifer of the Door Peninsula of Wisconsin is an example. Well tests have shown that the horizontal flow of water is along seven different bedding planes in the dolomite. Vertical recharge is through fractures. The bedding-plane zones can be identified by borehole geophysical means (caliper logs) and correlated across several miles (22). Because water movement takes place along broad bedding planes, the yield of wells is fairly constant from place to place. Wells in vertical fractures have a higher yield, as they possess both vertical and horizontal conductivity. The water table in diffuse-flow aquifers is well defined and can rise to a substantial elevation above the regional base level.

Free-flow carbonate aquifers have substantial development of solution passages. Not only are many joints and bedding planes enlarged, but some have formed large conduits. While all of the openings are saturated, the vast majority of flow occurs in the large channels, which behave hydraulically as pipe flow. Velocities are similar to surface streams. Flow is turbulent, and the stream may carry a sediment load—both as suspended material and bedload. Water quality is similar to surface water, and the regional discharge may occur through a few large springs. Because of the rapid drainage, the water table is nearly flat, having only a small elevation above the regional base level. The very low hydraulic gradient indicates that diffused flow through the unenlarged joints and fractures is exceedingly slow. Recharge to the subterranean drainage system is rapid, as water drains quickly through the vadose zone. The water level in the open pipe network may rise rapidly in a recharge event (23). The spring discharge will also increase in response to the amount of recharge, so that the spring hydrograph may resemble the flood peak of a surface stream. The water levels in the open-pipe network will also fall rapidly as the water drains. Caving expeditions have been known to end tragically when a "dry" cave passage became filled with surcharged water during a rapid recharge event.

The depth of major solution openings below base level is probably less than 60 meters (200 feet), unless artesian flow conditions are present (19). However, in areas where the regional base level was formerly at a lower level (for example, where a buried bedrock valley is present), cavern development may have taken place graded to that base level. In the coastal aquifers of the southeastern United States, the drilling fluid may suddenly drain from a well being drilled when it is 100 or more meters deep. Cavernous zones found at these depths are well below the present water table. They formed when mean sea level, and the regional water table, were lower during the Pleistocene. The development of a sinkhole in Hernando County, Florida, was initiated by drilling in the Suwannee Limestone. The drilling fluid was lost several times, and the drill-bit would drop through small caverns. At 62 meters (202 feet), the drill broke into a cavern and, within ten minutes, a large depression had formed, with the drill rig sinking into the ground and the drillers narrowly escaping the same fate. The present-day water level is close to the land surface (24).

Sinking surface streams may also feed the pipe-flow network of a karst region. Lost River of southern Indiana is a typical headwater surface stream flowing across a thick clay layer formed as a weathering residuum on the St. Louis and St. Genevieve limestones. Where the clay thins, a karst landscape is present, and Lost River sinks beneath an abandoned surface channel. It appears as a large spring some miles away. There is no surface drainage in the karst region other than some ephemeral streams flowing into sinkholes (25).

If the carbonate rock beneath the uplands between regional drainage systems is capped by a clastic rock, karst landforms will not form. Dry caves in the uplands capped with clastics are less likely to have collapsed than caves in areas that are not capped. Recharge to the phreatic zone occurs through vertical shafts located at the edge of the caprock outcrop (22). These shafts, which may be as large as 10 meters (30 feet) in diameter and more than 100 meters (300 feet) deep, only extend to the water table. Water flows from them through horizontally oriented drains.

The central Kentucky karst region, including the Mammoth Cave area, is a capped carbonate aquifer system in some areas (26, 27). A cross section through the Mammoth Cave Plateau is shown in Figure 7.30. The plateau is capped by the Big Clifty Sandstone Member of the Galconda Formation, with cavern development in the underlying limestone formations, including the Girkin, St. Genevieve and St. Louis limestones. Contact springs are found at the margins of the top of the plateau, as there is a thin shale layer at the top of the Girkin Formation. Recharge to the main carbonate rock aquifer takes place through vertical shafts formed in the plateau where the shale layer is absent. Karst drainage in the Pennyroyal Plain also contributes to the regional water table. Drainage is to the Green River through large springs, such as the River Styx outlet and Echo River outlet. These streams may also be seen underground where they flow in cave passages. Wells in the area draw water from the Big Clifty Sandstone, which yields enough for domestic supplies. There are some perched water bodies in the limestone, but the amount of water in storage is limited. A more permanent supply is reached if the well penetrates the regional water table, below the level of the Green River. The water level in these deep wells can rise 8 meters (25 feet) in a few hours during heavy rains, and then can fall almost as fast (26).

If a carbonate rock is confined by strata of low hydraulic conductivity, they may control the rate and direction of groundwater flow (20). Such

221

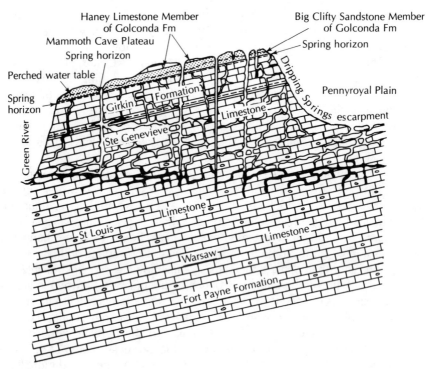

FIGURE 7.30. Diagrammatic cross section through the Mammoth Cave Plateau. Groundwater flow in the carbonate aquifer is from south to north. SOURCE: R. F. Brown, U.S. Geological Survey Water Supply Paper 1837, 1966.

confined systems may have groundwater flowing to great depths, with solution openings that are much deeper than those found in free-flow aquifers. Flow is not localized, and a greater density of joint solution takes place. The cavern formation of the Pahasapa Limestone of the Black Hills is apparently of this type, as water flows through the equivalent Madison Limestone aquifer eastward from a recharge area at the eastern side of the Black Hills. It has been suggested that because of the low gradient of the potentiometric surface (0.00022), the Madison Limestone is highly permeable, due to solution openings, for at least 200 kilometers (130 miles) east of the Black Hills (28).

In the preceding discussion of karst hydrology, it was assumed that the various types of carbonate rock aquifers were isolated; however this may not actually be the case. Highly soluble carbonate rock may be adjacent to a shaly carbonate unit with only slight solubility. The slope and position of the water table is a reliable indicator of the relative hydraulic conductivity of different carbonate rock units. In general, the water table will have a steeper gradient in rocks of lower hydraulic conductivity (29). This may be due to either a change in lithology, or in the degree of solution enlargement of joints.

Figure 7.31 illustrates some conditions that might result in a change in the water-table gradient. In Part A, the hilltop is capped by sandstone, with a low-permeability shale between the sandstone and the underlying limestone. Only a limited amount of recharge occurs through the shale. A spring horizon exists at the sandstone-shale contact, with small streams flowing across the shale outcrop area only to sink into the limestone terrane. The much greater amount of water circulating through the limestone in the area where the shale is absent has created a highly permeable, cavernous unit. Beneath the caprock, the limestone is less dissolved, due to lower groundwater recharge. Because of the lower hydraulic conductivity, the water-table gradient is steeper. This type of situation has been reported in the central Kentucky karst, with the hydraulic gradient beneath the caprock near the drainage divide being much steeper (0.01) than that near the Green River (0.0005) (30). If an area has two rock units, one of which is more soluble, the more soluble rock may develop large solution passages and, hence, have greater conductivity. Parts B and C of Figure 7.31 illustrate two situations in which the upland is underlain by shaly dolomite and the lowland by cavernous limestone. The difference in rock solubility creates a change in hydraulic gradient.

The general concept of a water table in free-flowing karstic regions may be different from the model water table found in sandstones or sand and gravel aquifers. Because of the extremely high conductivity of some lime-

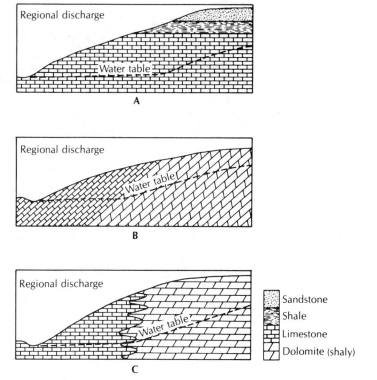

FIGURE 7.31. Geologic conditions resulting in a difference in hydraulic conductivity; hence, a difference in the water-table gradient.

stones, the water table can occur far beneath the land surface in mountains. Water can "perch" in solution depressions above the main water table. The level of free-flowing streams in caves is controlled by the regional water table, and they can have losing and gaining reaches, just as surface streams. Finally, because the solution of carbonate rock can be isolated along such features as fracture zones, the water table may be discontinuous. Wells drilled between fractures may not have any water in them, whereas nearby wells of the same depth may be in a fracture and therefore measure a water table (Figure 7.32).

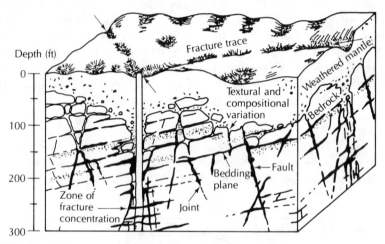

FIGURE 7.32. Concentration of groundwater along zones of fracture concentrations in carbonate rock. Wells that do not intercept an enlarged fracture or a bedding plane may be dry, thus indicating a discontinuous water table. SOURCE: L. H. Lattman and R. R. Parizek, *Journal of Hydrology*, 2 (1964):73-91. Used with permission of Elsevier Scientific Publishing Company, Amsterdam.

The selection of well locations in carbonate terrane is one of the great challenges for the hydrogeologist. As the porosity and permeability may be localized, it is necessary to find the zones of high hydraulic conductivity. One of the most productive approaches to the task is the use of **fracture traces** (31, 32). Fracture traces (up to 1.5 kilometers) and lineaments (1.5 to 150 kilometers) are found in all types of geologic terrane. As they represent the surface expression of nearly vertical zones of fracture concentrations, they are often areas with hydraulic conductivity 10 to 10^3 times that of adjacent rock. The fracture zones are from 2 to 20 meters (6 to 65 feet) wide and have surface expressions such as swales and sags in the land surface; vegetation differences, due to variations in soil moisture and depth to the water table; alignment of vegetation type; straight stream and valley segments; and alignment of sinkholes in karst. The surface features can reveal fracture traces covered by up to 95 meters (300 feet) of residual or transported soils. Fracture traces are found over carbonate rocks, siltstones, sandstones, and crystalline rocks.

Because of differential solution in carbonate rocks, if a fracture zone has somewhat higher conductivity than the unfractured rock, flowing groundwater will eventually create a much larger conductivity difference. Wells located in a fracture trace, or especially at the intersection of two fracture traces, have a statistically significant greater yield than wells not located on a fracture trace (33). The same relationship is apparently true for wells located on lineaments, as opposed to those not on lineaments (32).

In areas where topography is influenced by structure, valleys may form along fracture traces. In central Pennsylvania 60 to 90 percent of a valley may be underlain by fracture traces (33). Under such conditions, valley bottoms are good places to prospect for groundwater. In the Valley and Ridge Province of Pennsylvania, the valleys are structural, and wells in valley bottoms are statistically more productive. In Illinois, just the reverse is true. For the shallow dolomite aquifer, the yields of wells in bedrock uplands are greater than those in bedrock valleys (34). Figure 7.33 shows specific-capacity data for wells in bedrock uplands and bedrock valleys in northern Illinois.

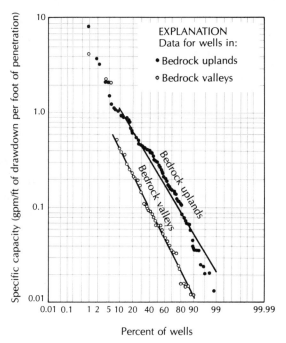

FIGURE 7.33.   Relation between specific capacity and bedrock topography for shallow dolomite wells in northern Illinois. SOURCE: S. Csallany and W. C. Walton, Illinois State Water Survey Report of Investigation 46, 1963.

In central Pennsylvania, the yield of wells drilled into anticlines is greater than the yield of wells drilled into synclines. However, the proximity to a fault trace and rock type are more significant than structure in determining yield (33). In other karst areas, synclines have been noted as major water producers (35). The relation between structure and well yield is not clear; thus, local experience must be used as a guide.

225

One of the integral parts of carbonate terrane hydrogeology is the **regolith**—the layer of soil and weathered rock above bedrock. The regolith can be composed of weathering residuum of insoluble minerals remaining after solution of the carbonate minerals. This is typically reddish in color, due to iron oxides, and contains a high proportion of clay minerals. The regolith may also include transported materials, such as glacial drift. If recharge must first pass through a low-permeability regolith, the rapid response of a carbonate aquifer will be reduced or eliminated. As the regolith slowly releases water from storage, spring discharge from areas overlain by a thick regolith will be more constant than if the regolith were absent. The regolith may also be a local aquifer. The weathering residuum of the Highland Rim area of Tennessee contains localized zones of chert, which can yield water in small amounts (36, 37).

7.3.3   COAL AND LIGNITE

Organic plant material that has accumulated in aqueous environments and has been partially decomposed is known as **peat**. When buried and subject to heat and pressure, chemical changes occur which result in the formation, first of lignite, and then, sequentially, of coal. The porosity of the peat is reduced during the coalification process.

Coal contains bedding planes cut by fractures that are termed **cleat**. Cleat is similar to joint sets in other rock. It is formed as a response to local or regional folding of the coal. There are typically two trends of cleat which are normal to the bedding planes and also cut each other at about a 90-degree angle (38). Coal is often an aquifer and yields water from the cleat and bedding. The quality of water from coal aquifers is variable, and sometimes can be poor. Such coals are typically anisotropic with the maximum hydraulic conductivity oriented along the face cleat, which develops perpendicular to the axis of folding (Figure 7.34).

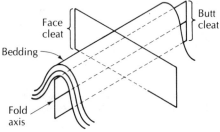

FIGURE 7.34.   Relationship of the face-cleat and butt-cleat systems to structure in coal, or regional structure if coal is unfolded. The dominant cleat is the face cleat, to which maximum hydraulic conductivity is parallel. SOURCE: R. Stone and D. F. Snoeberger, *Ground Water*, 15 (1977):434-38.

There is not a great deal of information on the hydraulic characteristics of coals. The maximum hydraulic conductivity of the Felix No. 2 Coal,

Wasatch Formation, was determined to be 3.1×10^{-4} centimeter per second (0.12 gallon per day per square foot). The calculated storativity is 1.2×10^{-3} (38). Mean hydraulic conductivity of lignite from four sites in western North Dakota is 4×10^{-4} centimeter per second (0.15 gallon per day per square foot) (39).

Thick coal beds, such as those of the Powder River Basin of Wyoming, may be important regional aquifers. Well yields of 0.6 to 6 liters per second (10 to 100 gallons per minute) are possible from these coals (40). Conflicts arise between water supply from coal aquifers and energy development. Strip mining will take place at the outcrop areas of the coal. If these are recharge areas for the coal aquifer, mine dewatering will adversely affect the potentiometric level in the coal aquifer. If the infiltration capacity of the area is reduced by mining, spoil disposal, or the like, this can also reduce the available recharge to the aquifer.

IGNEOUS AND METAMORPHIC ROCK 7.4

7.4.1   INTRUSIVE IGNEOUS AND METAMORPHIC ROCK

Water in unweathered crystalline rocks occurs in joints, faults, fractures, and in other cracks and crevices. The porosity is small, perhaps on the order of 1 percent for unweathered and unfractured rock (41). A number of studies have suggested that jointing is a near-surface phenomenon (41, 42, 43). Fractures associated with faulting can occur at any depth at which rock is brittle.

Yield tests from wells in crystalline rocks have shown that the yield, expressed in gallons per minute divided by the depth of the saturated zone penetrated by the well, decreases rapidly with depth (41). Figure 7.35 shows this relation for the eastern United States. There was no significant difference between wells in granite and wells in schist. As a general rule, for randomly sited wells drilled in crystalline rock, if sufficient water is not encountered in the first 100 meters (300 feet), a second hole should be tried rather than deepening the dry hole. Wells in crystalline rock might be drilled deeper if the intention is to reach a specific fault zone that is known to be fractured. Likewise, in some areas, such as the Piedmont of the eastern United States, jointed crystalline rock is known to occur to depths as great as 150 meters (500 feet) (44). Drilling wells to this depth is likely to be economical only for persons needing a yield greater than a few liters per second (100 to 150 gallons per minute).

Chemical weathering of igneous and metamorphic rocks yields a product called **saprolite**. This material contains alteration products of the parent rock. If the unweathered rock exhibits schistosity, then so will the saprolite.

227

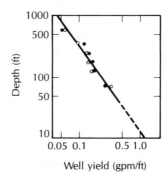

Well yield (gpm/ft)

FIGURE 7.35. Yields of wells in crystalline rock in the eastern United States. Open circles represent grouped mean yields of granite rock wells and black dots represent grouped mean yield for schist wells. SOURCE: S. N. Davis and L. J. Turk, *Ground Water*, 2 (1964):6-11.

Mica crystals, especially, will cause the saprolite to be anisotropic, with the greater hydraulic conductivity parallel to the orientation of the mica. Laboratory tests made on saprolite developed on schist and gneiss in the Piedmont (45) show porosity and specific yield to be relatively constant from 0 to 12 meters (40 feet), but to decrease rapidly between 12 and 21 meters (40 and 70 feet) where it grades into bedrock (Figure 7.36). This suggests that shallow wells in saprolite are more likely to be successful than deep ones.

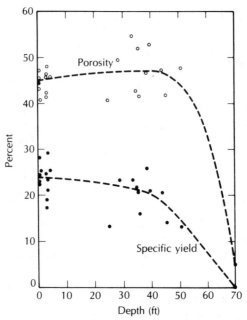

FIGURE 7.36. Relationship of porosity and specific yield and depth in saprolite at the Georgia Nuclear Laboratory. SOURCE: J. W. Stewart, U.S. Geological Survey Professional Paper 450-B, 1962.

The probability of obtaining a high-yield well in crystalline rock areas can be maximized if drilling takes place in an area where fractures are localized. It has been observed that zones of high conductivity in crystalline rock areas underlie linear sags in the surface topography (45). Such sags are the surface features that overlie major zones of fracture concentration. These show as fracture traces and lineaments on aerial and satellite photographs (31).

Fracture traces in granite have been found to have linear surface depressions 1 to 2.7 meters (3 to 9 feet) deep and 5 to 66.6 meters (16 to 218 feet) wide (32). Most work in relating fracture traces to well yield has been done in carbonate rocks, although the technique has been successfully employed in folded siltstones, sandstones, and shales (32). The technique should be equally successful in crystalline rock, where the hydraulic conductivity is fracture-controlled. Well yields in some areas of igneous rock are greater when the wells are located in valley bottoms (46). Many of these valley bottoms probably developed along fracture zones.

7.4.2 VOLCANIC ROCKS

Because volcanic rocks crystallize at the surface, they can retain porosity associated with lava-flow features and pyroclastic deposition. Hydraulic conductivity of volcanic rocks such as lava flows and cinder beds is typically quite high. However, ash beds, intrusive dikes, and sills may have a much lower hydraulic conductivity. Younger basalt flows tend to have greater conductivity than older ones. Flow features such as clinker zones and gas vesicles in aa (a type of lava), and lava tubes and gas vesicles in pahoehoe (another lava type), as well as vertical contraction joints and surface irregularities and stream gravels buried between successive flows contribute to overall conductivity (47). Some of the most productive aquifers are located in basalt flows, as is described in the following case studies.

CASE STUDY: VOLCANIC PLATEAUS—COLUMBIA RIVER BASALTS

The Columbia Plateau area of Washington, Oregon, and Idaho consists of a very thick sequence of Miocene-age basaltic lava flows that erupted from fissures. The lava was very fluid: individual flows are 50 to 150 meters (150 to 500 feet) thick, and some can be traced for 200 kilometers (650 miles). For the most part, the basalt flows are either flat-lying or gently tilted. In some places, the basalts have been folded by later tectonic activity. The basalt flows of the Columbia River Group of east central Washington are as thick as 3000 meters (10,000 feet), although in most areas a thickness of 1400 meters (4600 feet) is more typical (48). The basalt flows dip at angles of 1 to 2 degrees from northwest to southeast. Water occurs in distinct zones, probably related to interflow boundaries. Figure 7.37 shows a diagrammatic cross section of the basalt aquifers overlying basement crystalline rocks. The regional potentiometric surface is shown in Figure 7.38.

A number of irrigation wells have been drilled in the upper 300 meters (1000 feet) of the basalt flows. Typical well yields of 60 to 120

229

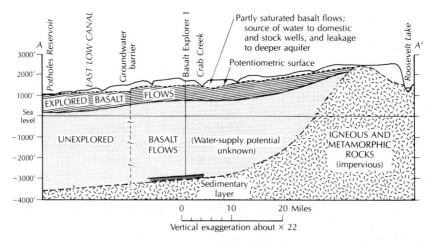

FIGURE 7.37. Cross section through basalt aquifers of east-central Washington. The groundwater barrier is of unknown origin. Line of section is shown in Figure 7.38. SOURCE: J. E. Luzier and J. A. Skrvian, U.S. Geological Survey Water Supply Paper 2036, 1975.

liters per second (1000 to 2000 gallons per minute) are obtained. In the area around Odessa and Lind, Washington, confined aquifers located at a depth of between 150 and 300 meters (500 and 1000 feet) supply these large yields. Shallower aquifers are not as productive, but have a hydraulic head up to 30 meters (100 feet) greater than the deeper aquifers. Uncased well bores are draining water from the shallower aquifers into the deeper ones (49). Recharge to the aquifer systems comes from precipitation and loss from ephemeral streams. Precipitation; hence, recharge, increases to the north and east. As these are also topographic high areas, regional groundwater flow is away from them, to the southwest. Age determinations based on carbon-14 dating of groundwater of the basalts range from modern to as old as 32,000 years B.P. Age appears to increase with depth, but relationships are obscured by mixing of water of different ages coming from different aquifer zones (49).

The classic conceptual model of a series of confined aquifers corresponding to individual basalt flows might not be valid in all parts of the Columbia Plateau. Vertical joints form in basalt as it shrinks during cooling. These can provide substantial vertical conductivity, producing unconfined aquifer conditions in thick sequences of lava flows. In Horse Heaven Plateau, Washington, there is reported to be an unconfined aquifer consisting of the Saddle Mountains Basalt and the upper part of the underlying Wanapum Basalt. A deeper basalt, the Grande Ronde, is confined by a saprolite layer, which formed by weathering of a basalt flow (50).

CASE STUDY: VOLCANIC DOMES—HAWAIIAN ISLANDS

In contrast to the very large area of similar geology of the Columbia Plateau Basalts, lava flows associated with some volcanic eruptions can have very heterogeneous aquifer systems. The Hawaiian Islands provide the classic example (47, 51, 52, 53).

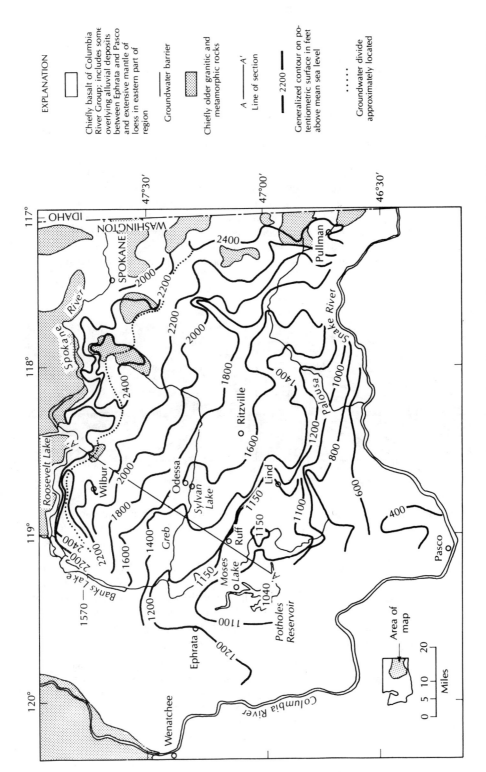

FIGURE 7.38. The extent of the basalt aquifer system of east-central Washington and the regional potentiometric surface. SOURCE: J. E. Luzier and J. A. Skrvian, U.S. Geological Survey Water Supply Paper 2036, 1975.

Each of the Hawaiian Islands consists of shield volcanoes forming from one to five volcanic domes. Each dome is composed of thousands of individual basaltic lava flows coming from either craters or fissures. Lava flows cooling above sea level are thin-bedded, highly fractured, or composed of vesicular and very permeable basalt. Those cooled under water are more massive and less permeable. However, due to lowered sea levels during the Pleistocene and isostatic sinking of the islands, highly permeable, air-cooled basalt is found below present-day sea level. Interbedded between the lava flows are ash beds, having a lower permeability. In the zones in which lava flows originated, igneous dikes with low porosity and permeability cut across the lava beds. The original dome structure may be partially eroded. Sediments have accumulated along some of the coastal areas. These coastal plain sediments consist of both terrestrial and marine deposits. The cross section of Figure 7.39 is through an idealized Hawaiian volcanic dome, with both the original and eroded states illustrated.

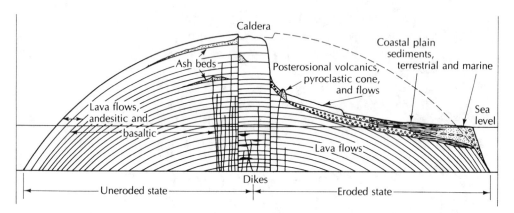

FIGURE 7.39. Geologic structure of an idealized Hawaiian volcanic dome. SOURCE: D. C. Cox, *Hawaiian Planters Record*, 54 (1954). Used with permission.

The Hawaiian Islands are typical examples of oceanic islands where fresh groundwater is underlain by salty groundwater. Because fresh water is less dense than salty water, the fresh groundwater beneath an oceanic island can be thought of as "floating" as a thin lens in the salty groundwater. The fresh groundwater grades into salty groundwater in a zone of mixing.

Groundwater in the Hawaiian Islands is contained in the highly permeable basalt flows. It is recharged by rainfall, which can average as much as 635 centimeters (20.8 feet) per year on Oahu. In the interior of the islands, the basalt flows are isolated by cross-cutting igneous dikes, which are groundwater dams. The groundwater is trapped at a high elevation behind these dams. High-level groundwater is also found as perched water in lava beds overlying low-permeability ash beds. If the infiltration is not trapped on an ash bed or behind a dike dam, it moves downward to the basal groundwater body, which is a freshwater lens in dynamic balance with salty groundwater. The basal groundwater may be unconfined or, in areas of coastal plain sediments, it may be confined by the low-permeability sediments locally termed **caprock**. A cross section of the occurrence of groundwater is shown in Figure 7.40. A birds-eye view of the island of Oahu indicates areas of high-level water bodies impounded by dikes (Figure 7.41).

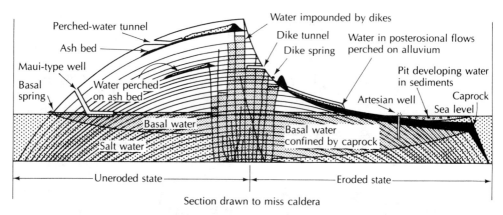

FIGURE 7.40. Occurrence and development of groundwater in an idealized Hawaiian volcanic dome. SOURCE: D. C. Cox, *Hawaiian Planters Record* 54 (1954). Used with permission.

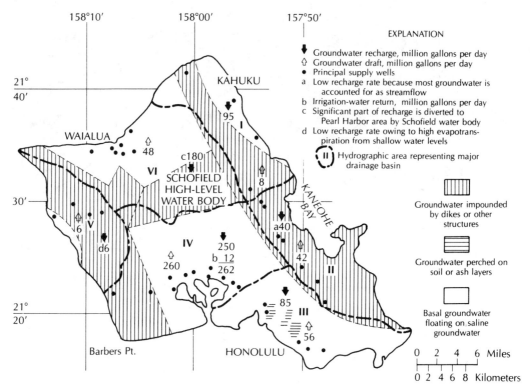

FIGURE 7.41. Map of the island of Oahu showing the approximate outline of groundwater reservoirs, recharge, 1975 draft, and principal supply wells by hydrographic areas representing major drainage basins. SOURCE: K. J. Takasaki, U.S. Geological Survey Professional Paper 813-M, 1978.

233

Springs issue from ash-bed perched aquifers and also from dike-dammed water bodies. Some of these are a hundred meters (300 feet) or more above sea level. High-level water is developed by tunnels into the tops of ash beds or penetrating dike dams. Most groundwater is developed from the basal groundwater body. Unconfined basal water is collected in horizontal skimming tunnels, called **Maui tunnels**, which are at sea level and slightly below. These skimming tunnels can develop water where the freshwater lens is very thin, and conventional wells would pump brackish or salt water. Maui tunnels are capable of producing up to 2×10^3 liters per second (3.1 $\times 10^4$ gallons per minute), although in most cases the yield is much less.

Where the basal water is confined by coastal plain sediments, conventional wells are used for groundwater development. On southern Oahu, the most extensive development of the basal aquifer

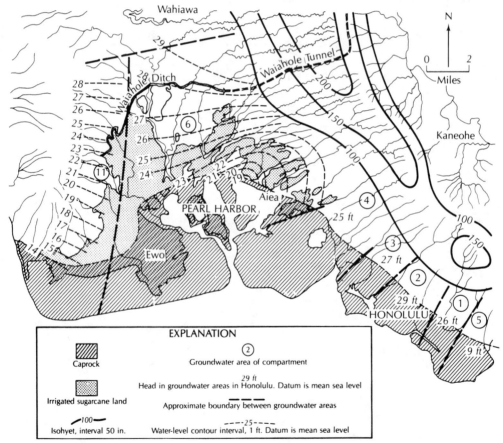

FIGURE 7.42. Map of southern Oahu showing groundwater areas, the distribution of caprock, and the distribution of irrigated sugarcane land. The figures for heads in the Honolulu area and for water-level contours in the Pearl Harbor area indicate conditions on May 31, 1958, after a four-month period during which there was no pumping for irrigation. SOURCE: F. N. Visher and J. F. Mink, U.S. Geological Survey Water Supply Paper 1778, 1964.

has occurred. The lava flows are continuous; however, the aquifer is divided into compartments of permeable lava separated by low-permeability sediments which fill ancient valleys in the lava surface to a depth far below sea level. These groundwater areas are recognizable by the distinctive artesian head. In Figure 7.42, the various groundwater compartments in Southern Oahu are locally identified by number. Recharge takes place by rainfall in the interior mountains as well as by excess irrigation water applied to areas not underlain by caprock. The caprock extends more than 300 meters (1000 feet) below sea level at the coast.

One of the problems with water supply development on Oahu is the intrusion of salty groundwater into the freshwater zone of the basal aquifer (54). An increase in chloride content is the usual measure of saltwater intrusion. The position of the boundary between fresh and saline groundwater is a function of the hydraulic characteristics of the aquifer and the volume of freshwater discharge. If the freshwater discharge is reduced, the zone of saline water will move upward and landward. This has happened in the Pearl Harbor area. Figure 7.43 shows the positions of the 200-ppm chloride isocon* and the

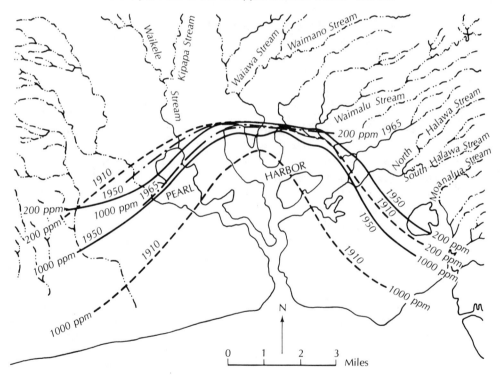

FIGURE 7.43. Map of Pearl Harbor area, Oahu, Hawaii, showing the position of 200-parts per million and 1000-parts per million chloride isocons in 1910, 1950, and 1965. SOURCES: Based on F. N. Visher and J. F. Mink, U.S. Geological Survey Water Supply Paper 1778, 1964 and R. H. Dale, Hydrologic Atlas 267, 1967.

---

*An isocon connects points of equal geochemical concentration.

235

1000-ppm chloride isocon in 1910, 1950, and 1965. Although the 200-ppm isocon has not shown much movement, the 1000-ppm isocon has. Fresh water is discharged naturally by mixing with the saline water and flowing upward through the confining caprock. The increase in chloride content with depth of a well drilled at the edge of Pearl Harbor is shown in Figure 7.44.

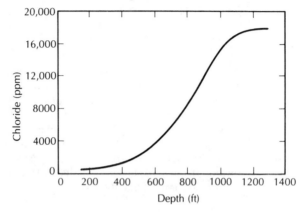

FIGURE 7.44. Increase in chloride content of groundwater with depth in a well in the confined groundwater area near Pearl Harbor. SOURCE: R. H. Dale, U.S. Geological Survey Hydrologic Investigations Atlas HA-267, 1967.

7.5 GROUNDWATER IN PERMAFROST REGIONS

In polar latitudes and high mountains, the mean annual temperatures may be sufficiently low for the ground to be at a temperature below 0° C. If this temperature persists for two or more years, the condition is known as **permafrost** (55, 56, 57). During the summer, warm temperatures may cause the upper meter or two (3 to 6 feet) of the soil or rock to thaw. This is called the **active layer**, but underneath it the ground may be frozen to depths up to 400 meters (1300 feet).

The magnitude of the annual temperature fluctuation of the soil is greatest at the surface, and diminishes with depth until, at some depth, there is zero annual temperature amplitude. The depth at which the maximum annual soil temperature is 0° C is the **permafrost table**. In some areas, the permafrost may occur in layers, with zones of unfrozen ground between them. This condition is usually the result of past climatic events, and the permafrost distribution is not congruent with the present climatic and thermal regime. The local depth of permafrost is a function of the geothermal gradient and the mean annual air temperature (58).

The insulating cover of glacier ice and large lakes may prevent the formation of permafrost. Permafrost is likewise not present beneath the ocean. At higher elevations and on north-facing slopes, where the mean temperature is lower, permafrost may be thicker (Figure 7.45). Lakes and streams also affect the permafrost. Shallow lakes—2 meters (6 feet) deep or less—freeze to the bottom in winter and have little effect on the permafrost table, but the permafrost may be warmer; hence, not as thick. Deeper lakes are unfrozen at the bottom and have an insulating effect. Small deep lakes create a saucer-shaped depression in the permafrost table, which, in turn, creates an upward indentation in the bottom of the permafrost (59). Permafrost is absent beneath large deep lakes—even in the continuous permafrost zone (Figure 7.45B). The effects of large rivers on distribution of permafrost are similar to those of large deep lakes.

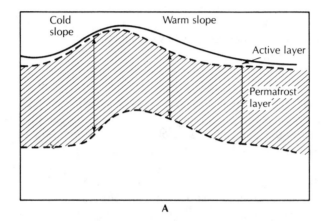

A

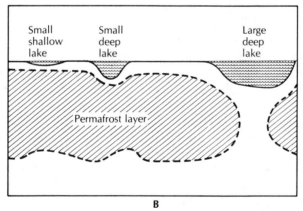

B

FIGURE 7.45. **A.** Effect of topography on the permafrost layer with the thinnest layer found beneath a warm, south-facing slope; **B.** Effect of lakes on the permafrost layer.

The permafrost table creates perched water in the active layer. This results in poorly drained soils and the typical muskeg and marsh vegetation of tundra regions. Deeper aquifers are recharged only in the absense of permafrost, as the permafrost layer acts as a confining layer.

237

Water below the permafrost layer is confined. The potentiometric surface may be in the permafrost layer or even above land surface (57). Sub-permafrost water may discharge into large rivers and lakes, beneath which unfrozen conduits are open. Saline groundwater may exist beneath permafrost, but it is typically not actively circulating (57). Discharge of groundwater at the surface, especially in winter, may result in the development of sheets of surface ice or large conical hills called **pingos**. Pingos have an ice-core and are formed by the upward arch of the ground surface due to hydraulic pressure of groundwater confined by permafrost.

Alluvial river valleys are good sources of groundwater in permafrost regions. The permafrost beneath and along the river may be thin, or absent in places, and alluvial gravel deposits are good aquifers. Large rivers will have more unfrozen ground; hence, more available water, than smaller tributaries and headwater streams.

In the far northern parts of Alaska, Canada, and the Soviet Union, permafrost is present nearly everywhere. It is in this continuous permafrost region that maximum permafrost depths are found. To the south, the permafrost layer is discontinuous. It may be up to 180 meters (600 feet) thick locally, but elsewhere there could be unfrozen ground. In the continuous permafrost regions of Alaska, alluvial valleys of large rivers may be the only water available. As the active layer freezes in winter, baseflow to even the largest rivers may be reduced to zero. Subpermafrost groundwater is typically saline or brackish. In the discontinuous permafrost region, the permafrost is thinner, and the areas free of permafrost are much more extensive in alluvial river valleys.

Alluvial fans are found in Alaska at the margins of many of the mountain ranges. The fans are composed of glacial-fluvial deposits, which tend to be coarse sand and gravel. Groundwater can be obtained from these deposits, either below the permafrost or near rivers or lakes where the permafrost may be thin or absent. In the more southerly alluvial fans, the water table may be below the permafrost layer. In this case, the permafrost has little impact on the hydrogeology, other than restricting recharge to areas where it is absent.

The distribution of permafrost in consolidated rocks is similar to that in unconsolidated deposits. If a rock unit has significant hydraulic conductivity in both the horizontal and vertical directions, groundwater hydrology should be similar to that of unconsolidated deposits. In highly anisotropic aquifers, even discontinuous permafrost bodies could act as groundwater dams, preventing horizontal flow and significantly reducing or eliminating vertical recharge. For example, groundwater in fracture zones is replenished by downward recharge. If the fracture zone were covered by a patch of permafrost, recharge would be difficult, even if the fracture zone extended below the permafrost layer.

CASE STUDY: ALLUVIAL AQUIFERS—FAIRBANKS, ALASKA

In the area of Fairbanks, Alaska, the Chena and Tanaua rivers have alluvial deposits up to 250 meters (800 feet) thick. The alluvium consists of interbedded gravel, sand, and silt. The floodplains are interspersed with terraces from 1 to 8 meters (3 to 25 feet) in height. The distribution of permafrost in the area is irregular. Permafrost is absent or nearly so in the alluvium beneath the river channels. Thin permafrost may be found under islands. The youngest, low-terrace deposits are underlain by unfrozen alluvium, with some isolated permafrost bodies up to 24 meters (80 feet) thick. Higher terraces have more continuous permafrost which can be 60 meters (200 feet) deep. The older terraces have nearly continuous permafrost up to 85 meters (280 feet) deep (57). Permafrost thickness increases beneath progressively older terraces. Wells in the alluvial valley obtain water from unfrozen areas or, if there is permafrost, from either above or below it. Transmissivity of the alluvium is as much as several thousand square meters per day (several hundred thousand gallons per day per foot) (60).

COASTAL PLAIN AQUIFERS 7.6

Coastal plains are found on all of the continents, with the exception of Antarctica. They are regional features, bounded on the continental side by highlands and seaward by a coastline. Coastal plains exist in areas of both stable basement rock as well as in those areas where the basement is sinking. The coastal plain may include large areas of former sea floor, and the geology of the coastal zone may be very similar to that of the adjacent continental shelf.

Sediment and sedimentary rocks of coastal plains were formed either as terrestrial or marine deposits. The terrestrial deposits tend to be landward and the marine deposits seaward, although fluctuating sea levels have resulted in alternating continental and marine strata—units deposited at a given time grade from continental to marine. Individual units have a variety of shapes, although the overall sequence usually thickens seaward. Coastal plains almost always contain Quaternary sediments; many also contain Tertiary- and Mesozoic-age deposits as well. The older rock units tend to crop out on the landward side, whereas younger, Quaternary and Recent, rock and sediments are found at the surface near the coast.

The coastal plain sediments may be unconsolidated or lithified, with no relationship to age. The Magothy aquifer of the northeastern United States is an unconsolidated sand of Cretaceous age, while the Biscayne aquifer of southeastern Florida consists of Pleistocene-age marine limestone. Aquifers typically are continental sands, gravels, and sandstones, or marine sands or limestones. Confining beds consist of marine and continental silts and clays.

Typically, alternating and interfingering facies of different lithology are encountered. Some units are thick and extend for hundreds of kilometers; others are often not traceable for more than a few kilometers. One or more aquifers may be found at most locations. Baton Rouge, Louisiana, has ten aquifers in the first 900 meters (3000 feet) of sediments (61). Some of the world's most prolific aquifers are found in coastal plain deposits. A notable example is the Atlantic and Gulf Coastal Plain of North America.

239

The sediments of coastal plain areas were deposited either adjacent to, or in shallow marine waters. Thus, pore water was originally saline. Fluctuating sea levels during the Pleistocene inundated many areas that are now land. As a result, saline waters occupied many contemporary freshwater coastal plain aquifers in the not-too-distant geologic past. Relative sea level was up to 100 meters (300 feet) lower during the last Wisconsin glaciation than present sea level, uncovering more of the continental shelf than is now exposed. During this period, saline water was flushed from inland aquifers to considerable depths.

Because of the seaward-sloping nature of coastal plain strata, deeper aquifers are recharged in inland areas. Fresh water flows downgradient, and then discharges by several mechanisms to the coastal waters. The amount of aquifer flushing that has occurred is a function of the volume of aquifer recharge and the amount of fresh water that can escape downgradient via the available mechanisms. Saline water has been flushed to a depth of 1800 meters (5900 feet) in Karnes County, Texas (62), and 1060 meters (3495 feet) below sea level in St. Tammany Parish, Louisiana (63), although this is unusual. Fresh water is found at depths of 300 meters (1000 feet) or more below sea level in many areas of the coastal plain aquifer (62). Fresh water has been found to occur in deep, confined coastal plain sediments many miles at sea. In a deep test well on Nantucket Island, located 64 kilometers off the New England coast, fresh water was found in confined aquifers at depths of 222 to 250 meters (730 to 820 feet) and 274 to 283 meters (900 to 930 feet). These are Cretaceous-age sands and are confined by clays (64). They represent aquifers that were recharged during the Pleistocene, when sea level was lower (64, 65).

Fresh water can discharge from a coastal aquifer via several natural mechanisms: (a) evapotranspiration; (b) direct seepage into springs, streams, tidal water, and the ocean floor; (c) mixing with saline groundwater in a zone of diffusion; (d) flow across a semipermeable layer under the influence of a hydraulic gradient; and (e) flow across a semipermeable layer due to osmotic pressure caused by a salinity gradient. Examples of the various discharge mechanisms in a coastal area are shown in Figure 7.46. Mechanisms (a) and (b) are very efficient in discharging water from unconfined aquifers. Freshwater springs in the sea bottom occur in unconfined aquifers or confined aquifers where the confining layer is breached. Deep confined aquifers utilize only the last three methods—(c), (d), and (e)—which are not as efficient. Assuming that all aquifers have equal transmissivities, the deeper the aquifer, the less fresh water it can discharge because of the reduction in the number and efficiency of discharge methods. Confined coastal aquifers can contain fresh water, even though overlying aquifers are salty.

On Long Island, New York, there is an unconfined aquifer and two deep confined aquifers, all recharged by precipitation at the center of the island. Most of the fresh groundwater discharges through the unconfined aquifer, with much lower volumes flowing in the confined aquifers. Wells drilled in a barrier island 8500 meters (28,600 feet) from the coastline show fresh water the full depth of both confined aquifers, with 1.5 to 3 meters (5 to 10 feet) of

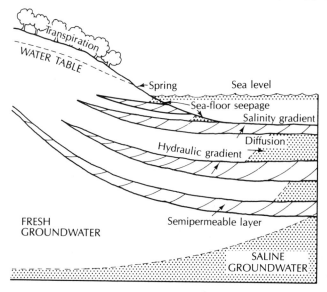

FIGURE 7.46.   Typical freshwater-saltwater relationship in a layered coastal aquifer.

artesian head in the upper confined aquifer and 6 meters (20 feet) in the lower. This is a distance of at least 18 to 20 kilometers (11 to 12 miles) from the closest part of the recharge area for the confined aquifers (66).

The most characteristic type of water quality degradation occurring in coastal plain aquifers is saline water intrusion. Sources of saline water are found as connate water below inland freshwater aquifers, as subsurface sea water below island aquifers, on the seaward edge of coastal aquifers, and as surface tidal waters in natural estuaries and artificial canals.

The shape and position of the boundary between saline groundwater and fresh groundwater is a function of the volume of fresh water discharging from the aquifer (excluding intrinsic aquifer characteristics). Any action that changes the volume of freshwater discharge results in a consequent change in the saltwater-freshwater boundary. It should be noted that minor fluctuations in the boundary position occur with tidal actions and seasonal and annual changes in the amount of freshwater discharge. For this reason, the boundary is in a state of quasi-equilibrium. Natural changes in the equilibrium position can result from long-term changes in climatic patterns or the position of relative sea level.

Human action that results in saline groundwater entering a freshwater aquifer is termed **saline water encroachment**. It occurs as a result of a diversion of fresh water that previously had discharged from a coastal aquifer. Saltwater encroachment can be either **active** or **passive** (67).

Passive saline water encroachment occurs when some fresh water has been diverted from the aquifer—yet the hydraulic gradient in the aquifer is still sloping toward the saltwater-freshwater boundary. In this case, the boundary will slowly shift landward until it reaches an equilibrium position based on the new discharge conditions. Passive saline water encroachment is

241

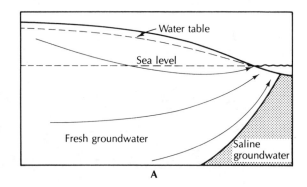

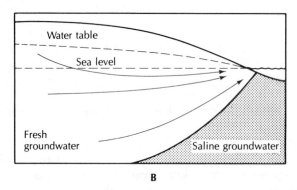

FIGURE 7.47. **A**. Unconfined coastal aquifer under natural groundwater discharge conditions; **B**. Passive saline water encroachment due to a general lowering of the water table. Flow in the freshwater zone is still seaward.

taking place today in many coastal plain aquifers where groundwater resources are being developed. It acts slowly and, in some areas, it may take hundreds of years for the boundary to move a significant distance (Figure 7.47).

The consequences of active saline water encroachment are considerably more severe, as the natural hydraulic gradient has been reversed and fresh water is actually moving away from the saltwater-freshwater boundary (Figure 7.48). This occurrence is mainly due to the concentrated withdrawals of groundwater creating a deep cone of depression. The boundary zone moves much more rapidly than it does during passive saline water encroachment. Furthermore, it will not stop until it has reached the low point of the hydraulic gradient: the center of pumping. It is this type of rapid encroachment that destroyed the aquifers beneath Brooklyn, New York, in the 1930s when the water table was lowered 10 to 15 meters (30 to 50 feet) below sea level.

Both types of saline water encroachment can occur in areas of inland connate saline waters as well as in coastal areas. In Baton Rouge, Louisiana, heavy industrial pumpage has lowered the potentiometric surface in the "600-foot sand" aquifer. This has caused connate saline groundwater in the aquifer to move northward toward the well field (61).

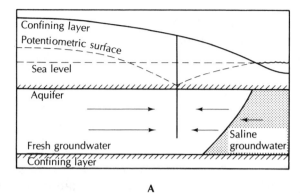

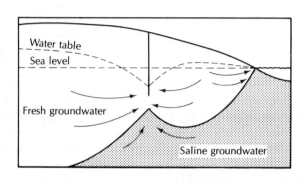

FIGURE 7.48.   A. Active saline water encroachment in a confined aquifer with the potentiometric surface below sea level; **B.** Active saline water encroachment in an unconfined aquifer with the water table drawn below sea level.

Saline water encroachment is also a serious problem in the Miami area of Florida. The unconfined Biscayne aquifer is the sole source of groundwater for this area, as the Floridan aquifer is salty there. The encroachment began with the 1907 installation of inland drainage canals to lower the water table. This diverted fresh water that had been flowing through the Biscayne aquifer to Biscayne Bay. Passive saline water encroachment resulted as the saltwater-freshwater boundary sought a new equilibrium position. Because of the high transmissivity of the aquifer, saline water occupied many areas of the Biscayne aquifer beneath Miami in only a few years (68). Sea water also moved up the canals toward the aquifer during high tides. Dams have now been built across the canals in order to keep salt water from traveling up them. Water impounded behind the dams is fresh, adding to the available recharge to the Biscayne aquifer. This fresh water had formerly been lost to the sea during flooding. New canals must be constructed with salinity-control dams.

In Savannah, Georgia, large quantities of groundwater are needed for both municipal and industrial use. When the first wells were drilled, the water was under artesian pressure. Today, a cone of depression some 45

243

meters (150 feet) below sea level exists. This reversed the original seaward hydraulic gradient, so that active saline water encroachment can occur. Saline water from the coastal area is now slowly moving toward the Savannah well-field area (69).

7.7 GROUNDWATER IN DESERT AREAS

Desert areas receive 25 centimeters (10 inches) of precipitation or less each year. Cold deserts, such as Antarctica, can have great accumulations of water. In warm deserts, however, the potential evapotranspiration may be many times the annual precipitation. Under such conditions, there is often virtually no local groundwater recharge. Yet, some warm deserts can have large volumes of fresh water stored beneath them.

Both bedrock and unconsolidated aquifers are known in desert areas. As the dry climate promotes mechanical weathering of rock, the uncon-solidated materials are often rather coarse and permeable. Alluvial fans and talus slopes at the base of mountains and escarpments can be productive. In South Yemen on the Arabian peninsula, groundwater is obtained from shallow wells in alluvial materials (70). Recharge occurs during flooding that accom-panies the infrequent precipitation.

Vast bedrock aquifers are known to occur in the sedimentary basins of Egypt, Jordan, and Saudi Arabia (71). The Sahara Desert is underlain by the Nubian aquifer, a sandstone up to 900 meters (3000 feet) thick. Several younger aquifers overlie it. The groundwater in the Nubian Sandstone is con-fined, and there are large initital heads. High-capacity wells now tap this aqui-fer. Several major aquifer systems, primarily sandstone, also underlie the Ara-bian peninsula. Carbonate aquifers may not have high-solution permeability due to the lack of actively circulating groundwater. Those that are permeable are a result of primary permeability or solution permeability developed during moister periods of the past.

Recharge to aquifers can occur through runoff from adjacent mountains which receive relatively more water. However, under much of North Africa and Arabia, the groundwater is old, exceeding 35,000 radiocarbon years B.P. (72). It has been shown by model studies that the water in these aquifers was probably recharged during wet climatic periods during the Pleistocene. Under such conditions, groundwater is mined from the aquifers, as there is no modern recharge.

Arid zones may have significant water quality problems. The slow circulation of groundwater results in mineralization. In addition, evapora-tion of groundwater from discharge areas results in salt deposition, with con-sequent high salinity in the soil and shallow groundwater.

GROUNDWATER REGIONS OF THE UNITED STATES 7.8

One useful generalization in the study of hydrogeology is the concept of groundwater regions. These are geographical areas of similar occurrence of groundwater. If an area is subdivided into several smaller regions, useful comparisons can be made between areas of well-known hydrogeology and similar areas that are not as well understood.

In delineating groundwater regions, the problem arises of how detailed to make them. Meinzer (73) divided the United States into twenty-one groundwater provinces that paralleled the physiographic provinces. Thomas (74) later consolidated and rearranged these into ten larger groundwater regions, some of which were renamed by McGuinness (62). Our description of these ten groundwater regions will follow the McGuinness system. A far more detailed description can be found in McGuinness's or Thomas's work. The groundwater regions are outlined in Figure 7.49.

None of the aforementioned authors accounted for the Hawaiian Islands or Alaska in designating the groundwater regions. The Hawaiian Islands could constitute Region 11, Volcanic Islands. A description of this region has previously been given in this chapter as a case study (see pages 230–36). Alaska is more difficult, with its great variety of bedrock types and physiographic provinces. It was glaciated and is affected by permafrost. Groundwater supplies are obtained primarily from unfrozen glacial or alluvial deposits. No doubt other groundwater sources exist in Alaska, but they have not yet been exploited to an appreciable degree.

1. WESTERN MOUNTAIN REGION

The mountains of the western United States are the headwater areas of many rivers, including the Columbia, Snake, San Joaquin, Sacramento, Missouri, Platte, Colorado, Arkansas, and Rio Grande. The mountains tend to be well watered, receiving higher levels of precipitation than the surrounding lowlands. The rocks in the western mountains are predominantly crystalline, although there are some sedimentary ranges. Intermontaine valleys may contain aquifers of alluvial deposits or glacial outwash. Most consolidated rock aquifers are of local rather than regional extent. Mountain ranges included in this region are the northern Coast Ranges, the Sierra Nevada and Cascade Range, parts of the physiographic Basin and Range Province, and the Rocky Mountains.

2. ALLUVIAL BASINS

Lying between the ranges of the Western Mountains are tectonic troughs which have been partially filled with erosional products from the adjacent uplands

245

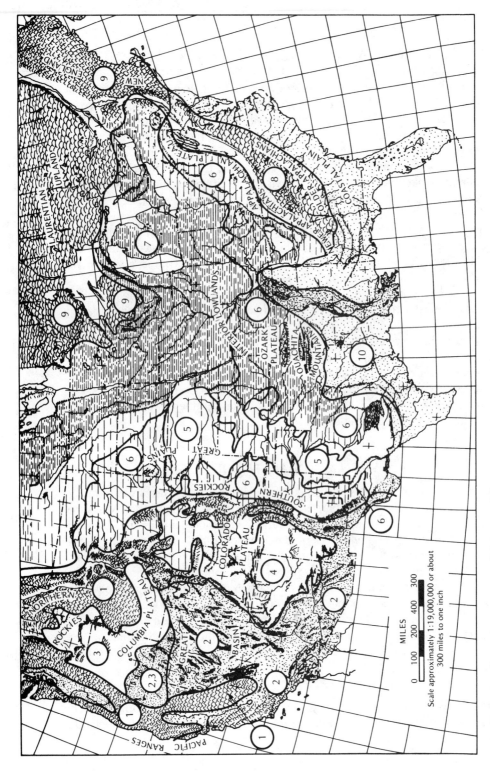

FIGURE 7.49. Groundwater areas of the United States.

and mountains. The sediment filling the valleys has been sorted into various size fractions by running water. Sand and gravel layers are found interspersed between silt and clay deposits. The valley-fill material is porous; hence, there are many permeable zones—especially, close to the mountains where coarser debris was deposited in coalescing alluvial fans.

Runoff coming from the mountains can soak into the coarse deposits; however, rainfall over the basins, themselves, may be rather low. The deep sedimentary basins hold vast volumes of water in storage. Many of the aquifers can yield large amounts of water to individual wells. Demands for water are high, and the amount of groundwater recharge is relatively low. As a result, the water in storage is being withdrawn and groundwater levels are falling in many of the basins.

3.   COLUMBIA LAVA PLATEAU

The Columbia Lava Plateau is formed of intrusive volcanic rock with interbedded alluvium and lake sediments. The plateau is somewhat arid, but some of the higher areas and surrounding mountains are more humid.

The lava flows are nearly horizontal, with some layers and interflow surfaces highly permeable. Shrinkage cracks create vertical permeability. The amount of water in storage is high; it is replenished by local precipitation, runoff from nearby mountains, and excess irrigation water applied to the surface.

The plateau is deeply dissected by the Columbia and Snake rivers and their tributaries. There are many large springs flowing from the permeable basalt layers into incised rivers; e.g., the Thousand Springs area of the Snake River. Groundwater is obtained from wells in the basalt, many of them 300 meters (1000 feet) deep or more. Glacial outwash gravel deposits along many of the rivers also are productive aquifers.

4.   COLORADO PLATEAU AND WYOMING BASIN

The Colorado Plateau and Wyoming Basin are areas of extensive sedimentary rocks. The strata are flat-lying for the most part, but some areas have been tilted, folded, or faulted. There are a number of plateaus standing at different levels which are deeply dissected by the Colorado River system or the Missouri River system. The sediments are chiefly shales and low-permeability sandstones.

The rate of groundwater recharge is very low, as the warm and arid climate causes most of the precipitation to be quickly evaporated. Some of the higher plateaus, such as the southern edge of the Colorado, do receive somewhat more precipitation. The most productive consolidated rock aquifer of the region, the Coconino Sandstone (Permian), outcrops here. Other aquifers include the Dakota Sandstone (Cretaceous), the Entrada and Junction Creek sandstones (Jurassic), and the Wingate Sandstone (Triassic). Small amounts of

247

water can be obtained from sandstone and limestone aquifers, which can receive local recharge, as well as from some alluvial deposits, such as the Uinta Basin at the edge of the Uinta Range. In general, the Colorado Plateau-Wyoming Basin is a water-poor region.

5. HIGH PLAINS

The High Plains groundwater region corresponds with the High Plains part of the physiographic Great Plains Province. This area is underlain by sedimentary rocks that dip gently to the east and are upturned at their contact with the Rocky Mountains and other places at which dome mountains, such as the Black Hills, have pushed through them. The entire surface is mantled with an alluvial apron representing material eroded from the Rocky Mountains. This alluvial cover, known as the Ogallala Formation, slopes to the east, but has been dissected in some areas.

The area of Nebraska between the Platte and Niobrara rivers is mantled with deposits of windblown sand—the Sand Hills. Both the Sand Hills and the Ogallala Formation (Pliocene) are highly porous and permeable. Rivers, which have cut all the way into bedrock, isolate these unconsolidated aquifers from recharge by streams draining from the Rocky Mountains. Recharge is by means of local precipitation, and the Sand Hills receives enough to be relatively well supplied compared with the rest of the High Plains. The Ogallala Formation is the major aquifer through most of the High Plains. It had a large volume of water in storage, but with heavy groundwater development for irrigation, water levels are rapidly falling—particularly in Texas and New Mexico.

Consolidated rock aquifers in Nebraska include units of the Arikaree Group (Miocene), Chadron Formation (Oligocene) and Dakota Group (Cretaceous). The Dakota Group is overlain by up to 1000 meters (3000 feet) of impermeable Pierre Shale and is recharged in the Black Hills outcrop area. Water is also obtained from Quaternary-age river alluvium along modern rivers, which can furnish a local source for recharge.

6. UNGLACIATED CENTRAL REGION

This region encompasses a large area of the interior United States. The bedrock underlying the region consists of level or gently tilted and folded sedimentary rocks ranging in age from Paleozoic to middle Tertiary. The land surface encompasses plains and plateaus and is gently rolling to sharply dissected. Climate ranges from semiarid to humid, but most of the area is subhumid to humid.

Consolidated rock aquifers include the Edwards Limestone (Cretaceous) of southwestern and central Texas, from which a flowing well in San Antonio yielded 1050 liters per second (24 million gallons per day) when first drilled. Sands of the Trinity Group (early Cretaceous) and Woodbine Sand (late

Cretaceous) are productive aquifers in north-central Texas and southern Oklahoma. These units include primarily clay, sand, and slightly consolidated sandstone. The Dakota Sandstone (late Cretaceous), Minnelusa Formation (Pennsylvanian and Permian), and Pahasapa Limestone (early Mississippian) are bedrock aquifers in South Dakota. In North Dakota, the Dakota Sandstone and Lakota Formation (early Cretaceous) are important. In eastern Montana, the Hell Creek Formation (late Cretaceous), Fox Hills Sandstone (late Cretaceous), and Fort Union Formation (Paleocene) are widespread. Other aquifers in Montana that are more restricted are the Judith River Formation (late Cretaceous) and the Kootenai, Dakota, and Lakota formations (Cretaceous). Fresh water may be available at depth from the Madison Limestone (Mississippian). The Roswell Basin of New Mexico includes the San Andres Limestone (Permian), which is confined by the Artesia Group. The initially high artesian heads in the Roswell Basin have long since declined due to the many flowing wells, as has the head in the Dakota Sandstone elsewhere. Limestone aquifers also occur in Arkansas, Missouri, Kentucky, and Tennessee. These units tend to be cavernous and many yield large springs, but generally they are not very productive because of the sporadic occurrence of water and the low storage capacity. River alluvium is an important source of groundwater throughout the region.

7.   GLACIATED CENTRAL REGION

This region encompasses the remainder of the interior lowland. Whether the so-called driftless area of Wisconsin should be included in this region or in Region 6 is debatable. The bedrock aquifers of the area are contiguous with the surrounding glaciated area, and the land was probably ice-covered during the Pleistocene.

The principal feature of the Glaciated Central Region is the yield of groundwater from glacial drift. Outwash plains and buried bedrock valleys are the principal sources. Most of the drift is fine-grained, but there are numerous sand and gravel deposits. The region ranges from semiarid in North Dakota to humid in the Great Lakes states.

Major bedrock aquifers include limestones and sandstones of Paleozoic age. Some of these are widespread and include the following: Mt. Simon Sandstone, Galesville Sandstone, Ironton Sandstone, Franconia Sandstone, Jordan Sandstone, and Trempealeau Formation (all Cambrian); Prairie du Chien Formation and St. Peter Sandstone (both Ordovician). These are the bedrock aquifers of eastern Iowa, southern Minnesota, southern Wisconsin, and northern Illinois. The bedrock aquifers are collectively known locally as the deep aquifer, sandstone aquifer, or Cambrian-Ordovician aquifer. Elsewhere in this region the Paleozoic bedrock aquifers are not regional in extent, and saline water in them tends to be a problem. In areas underlain by Precambrian crystalline rocks, the yield to wells is generally small. Alluvium, especially glacial-fluvial deposits, is a major source of groundwater.

249

8. UNGLACIATED APPALACHIAN REGION

The Appalachian Mountains and Appalachian Plateau are a part of this region. Bedrock in the area includes flat-lying sedimentary rocks in the Appalachian Plateau, folded sedimentary units in the Valley and Ridge Province, and metamorphic and crystalline rocks of the Piedmont and Blue Ridge. Precipitation ranges from 76 centimeters (30 inches) up to 190 centimeters (75 inches).

Weathered zones in the metamorphic rocks of the Piedmont yield small to moderate amounts of water almost anywhere, with larger yield wells possible on fracture traces. More-productive wells are usually in valleys rather than on hilltops. Some sandstone units which lie in belts parallel to the structural trend are also moderately good aquifers. There is little groundwater in the Blue Ridge; for the most part, the bedrock is impermeable.

In the Valley and Ridge Province, the limestone units are the most productive aquifers. The yield of wells located at random is erratic, but wells located on fracture traces, especially at intersections, can be very prolific. In the Appalachian Plateau, the sedimentary strata are elevated and dissected, so that the groundwater is drained from the higher elevations. Small to moderate amounts of water are available from wells there.

9. GLACIATED APPALACHIAN REGION

This is an area similar to the southern Appalachians, only mantled with a generally thin cover of glacial drift. It is humid, with 76 centimeters (30 inches) to 127 centimeters (50 inches) of precipitation. In most areas, the glacial drift is thin, suitable only for domestic groundwater supplies, as is the weathered zone at the bedrock surface. Bedrock wells are somewhat more productive in the Triassic sandstones of the Connecticut River Valley and parts of central New York State underlain by Paleozoic limestones and sandstones. Outwash plains and glacio-fluvial deposits form productive aquifers in some areas.

10. ATLANTIC AND GULF COASTAL PLAIN

The Atlantic Coastal Plain starts at Cape Cod and includes Long Island and southern New Jersey; most of each of the Atlantic coastal states; all of Florida; most of Alabama and Mississippi; parts of Missouri, Arkansas, and Oklahoma; all of Louisiana; and southeastern Texas. At the interior edge, the deposits thin to a feather edge and thicken toward the coast. They consist of unconsolidated to consolidated continental and marine sediments. Almost all of the Coastal Plain has a humid climate, with ample water available to recharge the aquifers.

Cape Cod and Long Island are covered with glacial deposits, including outwash, which is a good aquifer. Pleistocene sediments are also the most productive aquifer in Delaware. Elsewhere, the coastal plain deposits contain many excellent aquifers, including sands, sandstones, dolomites, and limestones. Saline water is present at the coasts and at depth in many of the areas. Otherwise, water quality is generally good.

250

Major aquifers include the Magothy Formation and Lloyd Sand Member of the Raritan Formation (both Cretaceous) of Long Island and northern New Jersey; Cheswold and Fredonia aquifers (Miocene) in Delaware; principal artesian aquifer (Floridan aquifer) of southeastern South Carolina, Georgia, Florida, and southern Alabama; and Biscayne aquifer of Florida. A number of good aquifers exist in Texas and include the Wilcox Group and Carrizo Sand (Eocene); Catahoula Sandstone, Oakville Sandstone and sand units of Lagarto Clay (Miocene); and Goliad Sand, Willis Sand, and sand units of the Lissie Formation (Pliocene to Pleistocene). Other states have equally productive aquifers, either unnamed or of local extent.

The Atlantic and Gulf Coastal Plain is one of the most productive groundwater regions of the United States. However, some areas, such as most of Florida, Savannah, Georgia, and Long Island, have experienced saltwater intrusion problems due to overdevelopment. Likewise Houston, Texas, has undergone land subsidence due to heavy groundwater withdrawals.

References

1. DOONAN, C. J., G. E. HENDRICKSON, and J. R. BYERLAY. *Ground Water and Geology of Keweenaw Peninsula, Michigan.* Michigan Department of Natural Resources, Geological Survey Division, Water Investigation 10, 1970, 40 pp.

2. NORRIS, S. E. *Permeability of Glacial Till.* U.S. Geological Survey Professional Paper 450-E, 1963, pp. 150–51.

3. Warzyn Engineering, Inc., Madison, Wisconsin. Personal communication from D. R. Viste, 1979.

4. CARR, J. R. "Tacoma's Well Field Might Be World's Most Productive." *Johnson Drillers Journal* (September and October 1976):1–4.

5. WINTER, T. C. *Hydrogeology of Glacial Drift, Mesabi Iron Range, Northeastern Minnesota.* U.S. Geological Survey Water Supply Paper 2029-A, 1973, 23 pp.

6. NORRIS, S. E. and A. M. SPIEKER. *Ground-Water Resources of the Dayton Area, Ohio.* U.S. Geological Survey Water Supply Paper 1808, 1966, 167 pp.

7. PLUMMER, P. M. *An Investigation of the Effects of Waste Disposal Practices on Groundwater Quality of Montgomery County, Ohio.* Miami Conservancy District Report, 1973, 68 pp. and appendices.

8. EAKIN, T. E., D. PRICE, and J. R. HARRILL. *Summary Appraisals of the Nation's Ground Water Resources—Great Basin Region.* U.S. Geological Survey Professional Paper 813-G, 1976, 37 pp.

9. DUTCHER, L. C. and A. A. GARRETT. *Geological and Hydrologic Features of the San Bernardino Area, California.* U.S. Geological Survey Water Supply Paper 1419, 1963, 114 pp.

10. HARDT, W. F. and C. B. HUTCHINSON. "Model Aids Planners in Predicting Rising Ground-Water Levels in San Bernardino, California." *Ground Water,* 16 (1978):424–31.

11. YOUNG, H. *Digital Computer Model of the Sandstone Aquifer in Southeastern Wisconsin.* Southeastern Wisconsin Regional Planning Commission, Technical Report 16, 1976, 42 pp.

12. WALTON, W. C. *Groundwater Recharge and Runoff in Illinois.* Illinois State Water Survey Report of Investigation 48, 1965, 55 pp.

13. SCHICHT, R. J., J. R. ADAMS, and J. B. STALL. *Water Resources Availability, Quality and Cost in Northeastern Illinois.* Illinois State Water Survey Report of Investigation 83, 1976, 90 pp.

14. TWENTER, F. R. *Geology and Promising Areas for Ground-Water Development in the Hualapai Indian Reservation, Arizona.* U.S. Geological Survey Water Supply Paper 1576-A, 1962, 38 pp.

15. HUNTOON, P. W. "Cambrian Stratigraphic Nomenclature and Ground-Water Prospecting Failures in the Hualapai Plateau, Arizona." *Ground Water,* 15 (1977):426–33.

16. MORRIS, D. A. and A. I. JOHNSON. *Summary of Hydrologic and Physical Properties of Rock and Soil Materials as Analyzed by the Hydrologic Laboratory of the U.S. Geological Survey 1948–1960.* U.S. Geological Survey Water Supply Paper 1839-D, 1967.

17. WALTON, W. C. and S. CSALLANY. *Yields of Deep Sandstone Wells in Northern Illinois.* Illinois State Water Survey Report of Investigation 43, 1962, 47 pp.

18. EWERS, R. O. et al. "The Origin of Distributary and Tributary Flow within Karst Aquifers." Geological Society of America, *Abstracts with Programs,* 10, no. 7 (1978):398–99.

19. WHITE, W. B. "Conceptual models for Carbonate Aquifers." *Ground Water,* 7, no. 3 (1969):15–22.

20. SHUSTER, E. T. and W. B. WHITE. "Seasonal Fluctuations in the Chemistry of Limestone Springs: A Possible Means for Characterizing Carbonate Aquifers." *Journal of Hydrology,* 14 (1971):93–128.

21. BRUCKNER, R. W., J. W. HESS, and W. B. WHITE. "Role of Vertical Shafts in the Movement of Groundwater in Carbonate Aquifers." *Ground Water,* 10, no. 6 (1972):5–13.

22. SHERRILL, M. G. *Geology and Ground Water in Door County, Wisconsin, with Emphasis on Contamination Potential in the Silurian Dolomite.* U.S. Geological Survey Water Supply Paper 2047, 1978, 38 pp.

23. FOX, I. A. and K. R. RUSHTON. "Rapid Recharge in a Limestone Aquifer." *Ground Water,* 14, no. 1 (1976):21–27.

24. BOATWRIGHT, B. A. and D. W. AILMAN. "The Occurrence and Development of Guest Sink, Hernando County, Florida." *Ground Water,* 13, no. 4 (1975):372–75.

25. RUHE, R. V. "Summary of Geohydrologic Relationships in the Lost River Watershed, Indiana, Applied to Water Use and Environment." In *Hydrologic Problems in Karst Regions,* ed. R. R. Dilamarter and S. Csallany. Bowling Green, Ky: Western Kentucky University, 1977, pp. 64–78.

26. BROWN, R. F. *Hydrology of the Cavernous Limestones of the Mammoth Cave Area, Kentucky*. U.S. Geological Survey Water Supply Paper 1837, 1966, 64 pp.

27. WHITE, W. B. et al. "The Central Kentucky Karst." *Geographical Review*, 60 (1970):88–115.

28. SWENSON, F. A. "New Theory of Recharge to the Artesian Basin of the Dakotas." *Bulletin, Geological Society of America*, 79 (1968):163–82.

29. LE GRAND, H. E. and V. T. STRINGFIELD. "Water Levels in Carbonate Rock Terranes." *Ground Water*, 9, no. 3 (1971):4–10.

30. CUSHMAN, R. V. *Recent Developments in Hydrogeologic Investigation in the Karst Area of Central Kentucky*. International Association of Hydrogeologists Memoirs, Congress of Istanbul, 1967, pp. 236–48. Cited in Reference 29.

31. LATTMAN, L. H. and R. R. PARIZEK. "Relationship between Fracture Traces and the Occurrence of Groundwater in Carbonate Rocks." *Journal of Hydrology*, 2 (1964):73–91.

32. PARIZEK, R. R. "On the Nature and Significance of Fracture Traces and Lineaments in Carbonate and Other Terranes." In *Karst Hydrology and Water Resources*, ed. V. Yevjevich. Fort Collins: Colo.: Water Resources Publications, 1976, pp. 47–108.

33. SIDDIQUI, S. H. and R. R. PARIZEK. "Hydrogeologic Factors Influencing Well Yields in Folded and Faulted Carbonate Rocks in Central Pennsylvania." *Water Resources Research*, 7 (1971):1295–1312.

34. CSALLANY, S. and W. C. WALTON. *Yields of Shallow Dolomite Wells in Northern Illinois*. Illinois State Water Survey Report of Investigation 46, 1963, 43 pp.

35. LAMOUREAUX, P. E. and W. J. POWELL. *Stratigraphic and Structural Guides to the Development of Water Wells and Well Fields in Limestone Terrane*. International Association of Scientific Hydrology, Publ. 52, 1960, pp. 363–75.

36. MARCHER, M. V., R. H. BINGHAM, and R. E. LOUNSBURY. *Ground Water Geology of the Dickson, Lawrenceburg and Waverly Areas of the Western Highland Rim, Tennessee*. U.S. Geological Survey Professional Paper 1764, 1964, 50 pp.

37. ZURAWSKI, A. *Summary Appraisals of the Nation's Groundwater Resources—Tennessee Region*. U.S. Geological Survey Professional Paper 813-L, 1978, 35 pp.

38. STONE, R. and D. F. SNOEBERGER. "Cleat Orientation and Areal Hydraulic Anisotropy of a Wyoming Coal Aquifer." *Ground Water*, 15 (1977):434–38.

39. REHM, B. W., G. H. GROENEWOLD, and S. R. MORAN. "The Hydraulic Conductivity of Lignite and Associated Geological Materials and Strip Mine Spoils, Western North Dakota." *Geological Society of America, Abstracts with Programs*, 10, no. 7 (1978):477.

40. KEEFER, W. R. and R. F. HADLEY. *Land and Natural Resource Information and Some Potential Environmental Effects of Surface Mining of Coal in the Gillette Area, Wyoming*. U.S. Geological Survey Circular 743, 1976, 27 pp.

253

41. DAVIS, S. N. and L. J. TURK. "Optimum Depth of Wells in Crystalline Rocks." *Ground Water,* 2, no. 2 (1964):6–11.

42. JOHNS, R. H. "Sheet Structure in Granite, Its Origin and Use as a Measure of Glacial Erosion in New England." *Journal of Geology,* 51 (1945):71–98.

43. LACHENBRUCH, A. H. "Depth and Spacing of Tension Cracks." *Journal of Geophysical Research,* 66 (1961):4273–92.

44. STEWART, J. W. *Relation of Permeability and Jointing in Crystalline Metamorphic Rocks Near Jonesboro, Georgia.* U.S. Geological Survey Professional Paper 450-D, 1962, pp. 168–70.

45. LE GRAND, H. E. "Perspective on Problems in Hydrogeology." *Bulletin, Geological Society of America,* 73 (1962):1147–52.

46. LE GRAND, H. E. *Geology and Groundwater in the Statesville Area, North Carolina.* North Carolina Department of Conservation and Development, Division of Mineral Resources Bulletin 68, 1954, 68 pp.

47. PETERSON, F. L. "Water Development on Tropic Volcanic Islands—Type Example: Hawaii." *Ground Water,* 10, no. 5 (1972):18–23.

48. LUZIER, J. E. and J. A. SKRVIAN. *Digital Simulation and Projection of Water Level Declines in Basalt Aquifers of the Odessa-Lind Area, East Central Washington.* U.S. Geological Survey Water Supply Paper 2036, 1975, 48 pp.

49. NEWCOMB, R. C. *Quality of the Ground Water in Basalt of the Columbia River Group, Washington, Oregon and Idaho.* U.S. Geological Survey Water Supply Paper 1999-N, 1972, 71 pp.

50. BROWN, J. C. "Definition of the Basalt Groundwater Flow System, Horse Heaven Plateau, Washington." Geological Society of America, *Abstracts with Programs,* 11, no. 7 (1979):395.

51. COX, D. C. "Water development for Hawaiian Sugar Cane Irrigation." *Hawaiian Planters Record,* 54 (1954):175–97.

52. VISHER, F. N. and J. F. MINK. Ground Water Resources, Southern Oahu, Hawaii. U.S. Geological Survey Water Supply Paper 1778, 1964, 133 pp.

53. TAKASAKI, K. J. *Summary Appraisals of the Nation's Ground Water Resources – Hawaii Region.* U.S. Geological Survey Professional Paper 813-M, 1978, 29 pp.

54. DALE, R. H. *Land Use and Its Effect on the Basal Water Supply, Pearl Harbor Area, Oahu, Hawaii, 1931–1965.* U.S. Geological Survey Hydrologic Investigations Atlas HA-267, 1967.

55. CEDERSTROM, D. J., P. M. JOHNSON, and S. SUBITZKY. *Occurrence and Development of Ground Water in Permafrost Regions.* U.S. Geological Survey Circular 275, 1953, 30 pp.

56. BRANDON, L. V. "Evidences of Ground-Water Flow in Permafrost Regions." *Proceedings of the 1963 International Conference on Permafrost,* Lafayette, Indiana, 1965, pp. 176–77.

57. WILLIAMS, JOHN R. *Ground Water in the Permafrost Regions of Alaska.* U.S. Geological Survey Professional Paper 696, 1970, 83 pp.

58. TERZAGKI, K., "Permafrost." *Boston Society of Civil Engineers Journal,* 39 (1950):1–50.

59. LACHENBRUCH, A. H., M. C. BREWER, G. W. GREENE, and B. V. MARSHALL. "Temperatures in Permafrost." In *Temperature and Its Measurement and Control in Science and Industry,* vol. 3, American Institute of Physics. New York: Reinhold Publishing Corporation, 1962, pp. 791–803.

60. CEDERSTROM, D. J. *Ground-Water Resources of the Fairbanks Area, Alaska.* U.S. Geological Survey Water Supply Paper 1590, 1963, 84 pp.

61. KAZMANN, R. G. *The Present and Future Ground Water Supply of the Baton Rouge Area.* Louisiana Water Resources Research Institute Bulletin 5, 1970, 44 pp. and appendices.

62. MC GUINNESS, C. L. *The Role of Groundwater in the National Water Situation.* U.S. Geological Survey Water Supply Paper 1800, 1963, 1121 pp.

63. ROLLO, J. R. *Ground Water in Louisiana,* Louisiana Geological Survey and Louisiana Department of Public Works Water Resources Bulletin 1, 1960, 84 pp.

64. KOHOUT, F. A. et al. "Fresh Groundwater Stored in Aquifers under the Continental Shelf: Implications from a Deep Test Well, Nantucket Island, Massachusetts. *Water Resources Bulletin,* 13, (1977):373–86.

65. COLLINS, M. A. "Discussion." *Water Resources Bulletin,* 14 (1978):484–85.

66. PLUHOWSKI, E. J. and I. H. KANTROWITZ. *Hydrology of the Babylon-Islip Area, Suffolk County, Long Island, New York.* U.S. Geological Survey Water Supply Paper 1768, 1964, 119 pp.

67. FETTER, C. W., JR. "Water Resources Management in Coastal Plain Aquifers." *Proceedings of the International Water Resources Association, First World Congress on Water Resources,* 1973, pp. 322–31.

68. PARKER, G. G., R. H. BROWN, D. G. BOGART, and S. K. LOVE. "Salt Water Encroachment." In *Water Resources of Southeastern Florida.* U.S. Geological Survey Water Supply Paper 1255, 1955, pp. 571–711.

69. COUNTS, H. B. and E. DONSKY. *Salt Water Encroachment, Geology and Ground Water Resources of the Savannah Area, Georgia and South Carolina.* U.S. Geological Survey Water Supply Paper 1611, 1963, 100 pp.

70. CEDERSTROM, D. J. "Ground Water in the Aden Sector of Southern Arabia." *Ground Water,* 9, no. 2 (1971):29–34.

71. HARSHBARGER, J. W. "Ground-Water Development in Desert Areas," *Ground Water,* 6, no. 5 (1968):2–4.

72. LLOYD, J. W. and M. H. FARAG. "Fossil Groundwater-Gradients in Arid Regional Sedimentary Basins." *Ground Water,* 16 (1978):388–93.

73. MEINZER, O. E. *The Occurrence of Groundwater in the United States, with a Discussion of Principles.* U.S. Geological Survey Water Supply Paper 489, 1923, 321 pp.

74. THOMAS, H. E. "Ground Water Regions of the United States—Their Storage Facilities." U.S. 83rd Congress, House Interior and Insular Affairs Committee, *The Physical and Economic Foundation of Natural Resources,* vol. 3, 1952, 78 pp.

Groundwater Flow to Wells

chapter

8.1 INTRODUCTION

Wells are one of the most important aspects of applied hydrogeology. Water wells are used for the extraction of groundwater to fill domestic, municipal, industrial, and irrigation needs. Wells have also been used to control saltwater intrusion, remove contaminated water from an aquifer, lower the water table for construction projects, relieve pressures under dams, and drain farmland.

Wells also function to inject fluids into the ground. On Long Island, New York, all groundwater pumped for cooling purposes must be returned to the same aquifer by an injection well. Still another function of wells is to dispose of wastewater into isolated aquifers. Finally, as a means of groundwater management, wells are sometimes used to artificially recharge aquifers at rates greater than natural recharge.

The same theoretical considerations apply to both wells that extract water and those that inject water. The only difference is that, for injection wells, the rate of pumping has a negative value.

8.2 UNSTEADY RADIAL FLOW

For purposes of analysis, the aquifers that we will consider will be assumed to be isotropic and homogeneous. It will be shown that solutions can be found for cases in which the value of the horizontal conductivity is different from that of the vertical conductivity. However, we will assume that the aquifer has **radial symmetry**; i.e., the value of the horizontal conductivity does not depend on the direction of flow in the aquifer.

Flow toward a well has been termed **radial flow**. It moves as if along the spokes of a wagon wheel toward the hub. We can deal with radial flow using a coordinate system called **polar coordinates**. The position of a point in a plane is specified according to its distance and direction from a fixed point or pole. The distance is measured directly from the pole to the point in the plane. The direction is determined by the angle between the line from the pole to the point and a fixed reference line—the **polar axis**. The polar axis is usually drawn horizontally and to the right of the pole, and the angle is measured by a counterclockwise rotation from the polar axis (Figure 8.1). Only the value of the angle (θ) and the radial distance (r) need be specified. If the aquifer is isotropic in a horizontal plane, then flow will have radial symmetry. Under these conditions, the equation for confined flow becomes

$$\frac{\delta^2 h}{\delta r^2} + \frac{1}{r}\frac{\delta h}{\delta r} = \frac{S}{T}\frac{\delta h}{\delta t} \tag{8-1}$$

where

h is the hydraulic head

S is the storativity

T is the transmissivity

t is the time

r is the distance from the well

If there is leakage through a confining layer, or recharge to the aquifer, the flow equation is then expressed as

$$\frac{\delta^2 h}{\delta r^2} + \frac{1}{r}\frac{\delta h}{\delta r} + \frac{e}{T} = \frac{S}{T}\frac{\delta h}{\delta t} \tag{8-2}$$

where e is the rate of vertical leakage.

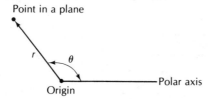
Point in a plane
r
θ
Origin
Polar axis

FIGURE 8.1.   The use of polar coordinates to describe the position of a point in a plane. It lies a distance (r) from the origin, and the angle between the polar axis and a line connecting the point and the origin is θ.

Solution of the flow equation for a variety of boundary values for radial flow to wells has yielded a number of extremely useful equations. The mathematics behind the basic solutions involve some rather esoteric mathematical functions. These include Laplace transforms, finite Fourier transforms, Bessel functions, and error functions.* The solutions can be used to determine the drawdown of the potentiometric surface or water table near a pumping well, if the formation characteristics are known. Conversely, if the formation characteristics of an aquifer are unknown, an aquifer test (pumping test) can be

---

*For a review of these functions as they apply to well hydraulics, see M. S. Hantush, "Hydraulics of Wells," in *Advances in Hydroscience*, vol. 1, ed. V. T. Chow (New York, Academic Press, 1964).

made. The well is pumped at a known rate, and the response of the potentiometric surface is measured. The equations can then be applied to determine the unknown aquifer characteristics.

8.3 WELL HYDRAULICS IN A COMPLETELY CONFINED AREALLY EXTENSIVE AQUIFER

When a well is pumped in a completely confined aquifer, the water is obtained from the elastic or specific storage of the aquifer. You will recall that the elastic storage is water that is released from storage by the expansion of the water as pressure in the aquifer is reduced and as the pore space is reduced as the aquifer compacts. The product of the specific storage, S_s, and the aquifer thickness is an aquifer parameter called **storativity**. For a confined aquifer, it is generally small (0.001 or less), and pumpage affects a relatively large area of the aquifer. Further, if there is no recharge, the area of drawdown of the potentiometric surface will expand indefinitely as the pumpage continues.

We will assume the following conditions in applying the solution: flow is in the range of Darcy's law; water is discharged instantaneously from storage; and the aquifer is homogeneous, isotropic, has a constant thickness, a negligible slope, and is of infinite extent. We will also assume that the pumping well and any observation wells fully penetrate the aquifer (i.e., water can enter at any level from the top to the bottom) and that the well diameter is infinitesimal (Figure 8.2).

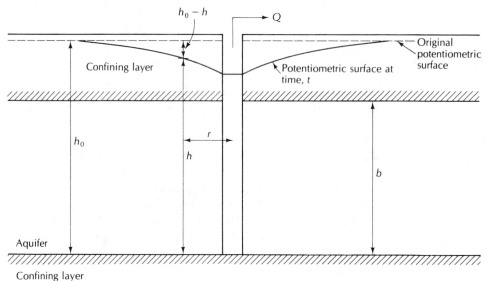

FIGURE 8.2. Fully penetrating well pumping from a confined aquifer.

8.3.1   THEIS METHOD

The flow equation for confined aquifers was first solved by C. V. Theis based on the analogy between flow of water in an aquifer and flow of heat in a thermal conductor (2). The original level of the potentiometric surface is h_0. The solution that Theis developed is

$$h_0 - h = \frac{Q}{4\pi T}\left[-0.5772 - \ln u + u - \frac{u^2}{2\cdot 2!} + \frac{u^3}{3\cdot 3!} - \frac{u^4}{4\cdot 4!} + \cdots\right] \tag{8-3}$$

where the argument u is given as

$$u = \frac{r^2 S}{4Tt} \tag{8-4}$$

and

Q is the constant pumping rate

h is the hydraulic head at time t since pumping began

h_0 is the hydraulic head before the start of pumping

r is the radial distance from the pumping well to the observation well

T is the aquifer transmissivity

S is the aquifer storativity

The infinite-series term in Equation (8-3) has been called the **well function** and is generally designated as $W(u)$. For a table of the value of $W(u)$ for various values of u, see Appendix 1 (3).

EXAMPLE PROBLEM

A well is located in an aquifer with a conductivity of 15 meters per day and a storativity of 0.005. The aquifer is 20 meters thick and is pumped at a rate of 2725 cubic meters per day. What is the drawdown at a distance of 7 meters from the well after one day of pumping?

$$T = Kb = 15 \text{ m/day} \times 20 \text{ m} = 300 \text{ m}^2/\text{day}$$

$$u = \frac{r^2 S}{4Tt} = \frac{(7 \text{ m})^2 \times 0.005}{4 \times 300 \text{ m}^2/\text{day} \times 1 \text{ day}} = 0.0002$$

261

From the table of $W(u)$ and u, if $u = 2 \times 10^{-4}$, $W(u) = 7.94$:

$$h_0 - h = \frac{Q}{4\pi T}W(u) = \frac{2725 \text{ m}^3/\text{day} \times 7.94}{4 \times \pi \times 300 \text{ m}^2/\text{day}} = 5.73 \text{ m}$$

The drawdown is 5.73 meters after one day.

Equations (8-3) and (8-4) are often called the **nonequilibrium** or **Theis equations**. A graph on full-logarithmic paper of $W(u)$ versus $1/u$, as shown in Figure 8.3, has the shape of the cone of depression near the pumping well. This graph is called the **reverse type curve**. This leads to a method for the solution of the nonequilibrium equation when drawdown and time are known but transmissivity and storativity are unknown. The procedure for collecting time-drawdown data involves making a pumping test of a well. (Pumping-test procedures are described in Section 8.10.) A pumping test results in a series of drawdown values at different times after the start of pumping for one or more nonpumping observation wells and the extraction well.

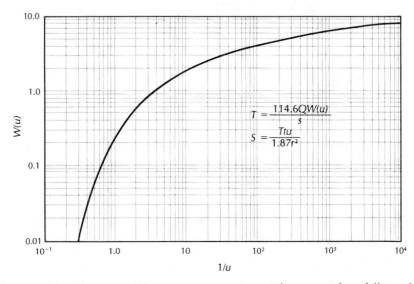

FIGURE 8.3. The nonequilibrium reverse type curve (Theis curve) for a fully confined aquifer.

The Theis graphical method of solution starts with the construction of the reverse type curve of $W(u)$ plotted against $1/u$ on logarithmic paper (Figure 8.3). Field data for drawdown $(h_0 - h)$ at all of the wells are plotted as a function of t/r^2 on logarithmic paper of the same scale as the type curve (Figure 8.4). By plotting t/r^2, data from observation wells located at differing distances from the pumping well can be used. If there is only one observation well, it is sufficient to plot $(h_0 - h)$ as a function of t. The graph paper with the data curve

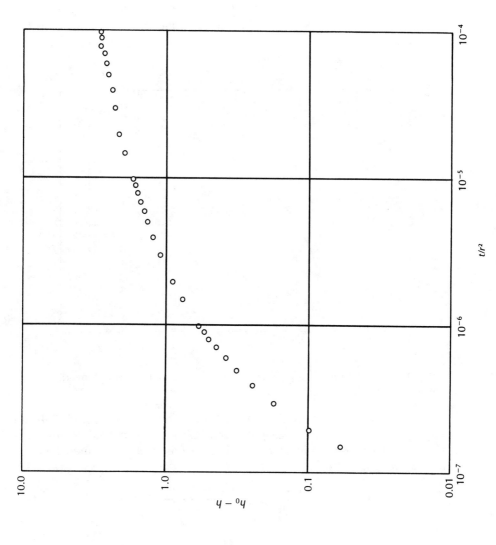

FIGURE 8.4. Field-data plot of drawdown ($h_0 - h$) as a function of t/r^2 for the Theis graphical method of analysis of pumping-test data.

is laid on a light table over a type curve drawn to the same scale. The position of the field-data sheet is adjusted, keeping the axes of both sheets parallel, until the data points fall on the type curve (Figure 8.5). Any arbitrary point (not necessarily on the type curve) is selected as a match point. A pin may be pushed through both sheets to mark the match point. From the match point, values of drawdown $(h_0 - h)$, t/r^2, $W(u)$, and $1/u$ are obtained. A very convenient match point is the intersection of the $W(u) = 1$ and $1/u = 1$ lines on the type curve. This simplifies later arithmetic computations. The value of u is the reciprocal of $1/u$.

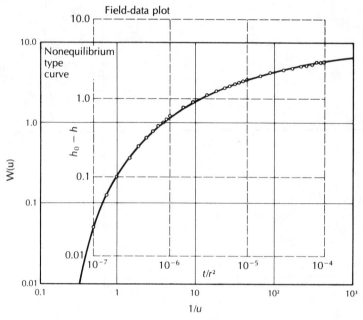

FIGURE 8.5. Field-data plot of drawdown as a function of t/r^2 overlain on a nonequilibrium type curve.

If consistent units are used, then Equations (8-3) and (8-4) are used directly. In common usage in the United States, the nonequilibrium equations are often presented in a form that uses inconsistent units (American practical hydrologic units):

$$h_0 - h = \frac{114.6QW(u)}{T} \tag{8-5}$$

$$S = \frac{Ttu}{2693r^2} \tag{8-6}$$

where

$h_0 - h$ is the drawdown (feet)

Q is the well discharge (gallons per minute)

T is the transmissivity (gallons per day per foot)

r is the distance to the observation well (feet)

S is the storativity (dimensionless)

t is the time since pumping began (minutes)

Using the data for $W(u)$ from the match point and the constant for pumping rate, Q, the value of transmissivity is computed using Equation (8-5).

After T has been determined, it is used along with the match-point values for $1/u$ and t/r^2 to compute the storativity from Equation (8-6).

EXAMPLE PROBLEM

A well in a confined aquifer was pumped at a rate of 220 gallons per minute for about 8 hours. The aquifer was 18 feet thick. Time-drawdown data for an observation well 824 feet away are given in Table 8.1. Find T, K, and S.

TABLE 8.1

Time After Pumping Started (min)	t/r^2	Drawdown (ft)
3	4.46×10^{-6}	0.3
5	7.46×10^{-6}	0.7
8	1.18×10^{-5}	1.3
12	1.77×10^{-5}	2.1
20	2.95×10^{-5}	3.2
24	3.53×10^{-5}	3.6
30	4.42×10^{-5}	4.1
38	5.57×10^{-5}	4.7
47	6.94×10^{-5}	5.1
50	7.41×10^{-5}	5.3
60	8.85×10^{-5}	5.7
70	1.03×10^{-4}	6.1
80	1.18×10^{-4}	6.3
90	1.33×10^{-4}	6.7
100	1.47×10^{-4}	7.0
130	1.92×10^{-4}	7.5
160	2.36×10^{-4}	8.3
200	2.95×10^{-4}	8.5
260	3.83×10^{-4}	9.2
320	4.72×10^{-4}	9.7
380	5.62×10^{-4}	10.2
500	7.35×10^{-4}	10.9

In the Theis solution, the data are plotted on logarithmic paper. When overlain on a type curve, the following match point is obtained:

$$W(u) = 1$$
$$1/u = 1$$
$$h_0 - h = 2.4$$
$$t/r^2 = 6.06 \times 10^{-6}$$

Transmissivity:

$$T = \frac{114.6QW(u)}{h_0 - h} = \frac{114.6 \times 220 \times 1.0}{2.4}$$
$$= 10,500 \text{ gpd/ft}$$

Hydraulic Conductivity:

$$K = T/b$$
$$= 10,500/18 = 580 \text{ gpd/ft}^2$$

Storativity:

$$S = \frac{Tu}{2693} \times t/r^2$$
$$= \frac{10,500 \times 1}{2693} \times 6.06 \times 10^{-6}$$
$$= 0.00002$$

8.3.2 JACOB STRAIGHT-LINE METHOD

C. E. Jacob (4) observed that after the pumping well had been running for some time, higher values of the infinite series became very small, and the nonequilibrium formula could be closely approximated by

$$h_0 - h = (2.3Q/4\pi T)\log_{10}(2.25Tt/Sr^2) \tag{8-7}$$

Equation (8-7) is obtained by truncating Equation (8-3) after the first two terms of the infinite series. If $t > 5r^2S/T$, a plot of drawdown on the arithmetic scale as a function of time (or t/r^2 for multiple observation wells) on the semilogarithmic log scale will be a straight line (Figure 8.6). Examination of the limiting condition, $t > 5r^2S/T$, shows that this can be met by a combination of large values of t and small values of r. In a pumping test with multiple observation wells, the closer wells will meet these conditions before the more distant ones.

In the **Jacob straight-line method**, a straight line is drawn through the field-data points and extended backward to the zero-drawdown axis. It

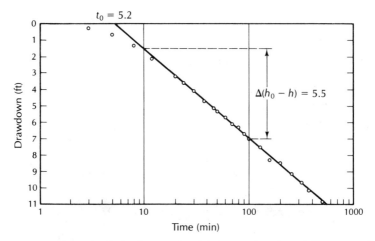

FIGURE 8.6. Jacob method of solution of pumping-test data for a fully confined aquifer. Drawdown is plotted as a function of time on semilogarithmic paper.

should intercept this axis at some positive value of time (or t/r^2 if there are multiple observation wells). This value is designated t_0 (or $(t/r^2)_0$). The value of the drawdown per log cycle $\Delta(h_0 - h)$ is obtained from the slope of the graph. With consistent units for the parameters, the values of transmissivity and storativity may be found from the equations

$$T = \frac{2.3Q}{4\pi\Delta(h_0 - h)} \tag{8-8}$$

and

$$S = \frac{2.25Tt_0}{r^2} \quad \text{or} \quad 2.25T(t/r^2)_0 \tag{8-9}$$

In American practical units, these equations are expressed as

$$T = \frac{264Q}{\Delta(h_0 - h)} \tag{8-10}$$

$$S = \frac{Tt_0}{4790r^2} \tag{8-11}$$

where

Q is the constant rate of pumpage (gallons per minute)

$\Delta(h_0 - h)$ is the drawdown per log cycle of time (feet)

t is the time since pumping began (minutes)

T is the transmissivity (gallons per day per foot)

r is the distance to the pumped well (feet)

267

EXAMPLE PROBLEM

Evaluate the pumping-test data of Table 8.1 by the Jacob method.

The field data are plotted on semilogarithmic paper (Figure 8.6). A straight line is fit to the later time data and extended back to the zero-drawdown axis. The value of t_0 is 5.2 minutes, while the drawdown per log cycle is 5.5 feet.

$$T = \frac{264Q}{\Delta(h_0 - h)} = \frac{264 \times 220}{5.5} = 10,560 \text{ gal/day/ft}$$

$$S = \frac{Tt_0}{4790r^2} = \frac{10,560 \times 5.2}{4790 \times 824^2} = 0.00002$$

As a check, the value of $5r^2S/T$ is 0.005 days, or 7.9 minutes. Inspection of Figure 8.6 will show that after about 8 minutes the drawdown data fall on a straight line.

In comparing the Jacob solution with the Theis solution in the preceding problems, we see that the resulting answers are almost the same. As these are graphical methods of solution, there will often be a slight variation in the answers, depending upon the accuracy of the graph construction and subjective judgments in matching field data to type curves.

An aquifer test may be made even if there are no observation wells. In this case, drawdown must be measured in the pumping well. There are energy losses as the water rushes into the well, so that the head in the aquifer is higher than the water level in the pumping well. For this reason, aquifer storativity cannot be determined. However, a plot of drawdown versus time for the pumping well can be used to determine aquifer transmissivity. Either the Theis or Jacob method can be used. It is important that the well be pumped at a constant rate, as any slight fluctuations will immediately affect the water level in the well. Likewise, drawdown data for the start of pumping is affected by the volume of water stored in the well casing. At the start of pumping, the water comes from the well casing rather than from the aquifer, especially when the well diameter is large and/or the pumping rate is small. The measured drawdown data should be adjusted to compensate for this factor (6).

8.3.3 DISTANCE-DRAWDOWN METHODS

If simultaneous observations are made of the drawdown in three or more observation wells, a modification of the Jacob straight-line method may be used. If drawdown is measured at the same time in several wells, it is found to vary with the distance from the pumping well in accordance with the Theis equation. Drawdown is plotted on the arithmetic scale as a function of the distance from the pumping well on the logarithmic scale. The drawdown in the wells closest to the pumping well should fall on a straight line.

 A line is drawn through the data points for the closest wells. It is extended until it intercepts the zero-drawdown line. This is the distance at which the well is not affecting the water level, and is designated r_0. The drawdown per log cycle is $\Delta(h_0 - h)$, as before. With consistent units for the parameters, the values of T and S may be found from

$$T = \frac{2.3Q}{2\pi\Delta(h_0 - h)} \qquad \text{(8-12)}$$

$$S = \frac{2.25Tt}{r_0^2} \qquad \text{(8-13)}$$

In American practical hydrologic units,

$$T = \frac{528Q}{\Delta(h_0 - h)} \qquad \text{(8-14)}$$

$$S = \frac{Tt}{4790r_0^2} \qquad \text{(8-15)}$$

where

 T is the transmissivity (gallons per day per foot)

 Q is the constant rate of pumpage (gallons per minute)

 $\Delta(h_0 - h)$ is the drawdown per log cycle of distance (feet)

 t is the time since pumping began (minutes)

 r_0 is the intercept of the straight line with the zero-drawdown axis (feet)

EXAMPLE PROBLEM

A well pumping at 400 gallons per minute has observation wells located 10, 40, 150, 300, and 400 feet away. After 200 minutes of pumping, the following drawdowns were observed:

Distance (ft)	Drawdown (ft)
10	15.1
40	9.4
150	4.4
300	1.7
400	0.2

The data are plotted in Figure 8.7. The drawdown per log cycle is 8.8 feet and r_0 is 460 feet. Find the values of T and S.

$$T = \frac{528Q}{\Delta(h_0 - h)}$$

$$= \frac{528 \times 400}{8.8}$$

$$= 24{,}000 \text{ gal/day/ft}$$

$$S = \frac{Tt}{4790r_0^2}$$

$$= \frac{24{,}000 \times 200}{4790 \times 460^2}$$

$$= 0.0047$$

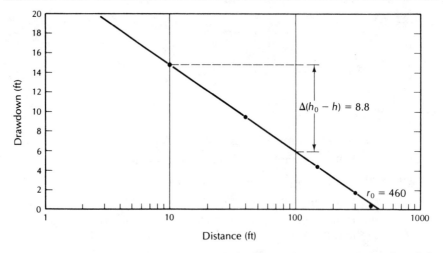

FIGURE 8.7. Variation of the Jacob method of solution of pumping-test data for a fully confined aquifer. Drawdown is plotted as a function of distance to observation well on semilogarithmic paper.

8.3.4 SLUG TESTS

In some field investigations, the practicing hydrogeologist may be working with low-permeability materials. This is especially likely in site studies for areas of potential waste storage or disposal. The soil materials in these areas may have a conductivity too small to conduct a pumping test. An alternative method of testing involves either injecting into or withdrawing a slug of water of known volume from the well. The rate at which the water level rises or falls is controlled by the formation characteristics. This is known as a **slug test** (7, 8, 9).

A well is drilled and cased above the confined aquifer. A well screen or open hole penetrates the aquifer (Figure 8.8). The well casing has a radius, r_c, and the well screen has a radius, r_s. Immediately after injection or withdrawal, the water level in the well has an elevation, H_0, above or below the initial head. As the water level rises or falls, the difference, H, in water-level elevation between that at time t and that at the original head is measured.

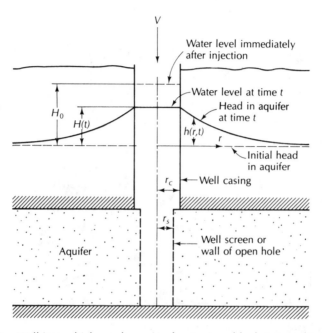

FIGURE 8.8. Well into which a volume, V, of water is suddenly injected for a slug test of a confined aquifer. SOURCE: H. H. Cooper, Jr., J. D. Bredehoeft, and I. S. Papadopulos, *Water Resources Research*, 3 (1967):263–69.

A plot of the ratio of the measured head to the head after injection (H/H_0) is made as a function of time. The ratio H/H_0 is on the arithmetic scale, and time is on the logarithmic scale of semilogarithmic paper:

$$H/H_0 = F(\eta, \mu) \tag{8-16}$$

where

$$\eta = Tt/r_c^2 \tag{8-17}$$

and

$$\mu = r_s^2 S/r_c^2 \tag{8-18}$$

$F(\eta, \mu)$ is a function, the values of which are given in tabulated form in Appendix 2 (9). The tabulated values are plotted as a series of type curves in Figure 8.9. The type curves should be on semilogarithmic paper with the same scale as field data.

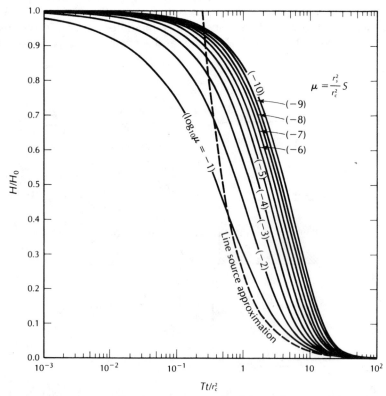

FIGURE 8.9. Type curves for slug test in a well of finite diameter. SOURCE: S. S. Papadopulos, J. D. Bredehoeft, and H. H. Cooper, Jr., *Water Resources Research*, 9 (1973):1087–89.

The field-data curve is placed over the type curves with the arithmetic axis coincident. That is, the value of $H/H_0 = 1$ for the field data lies on the horizontal axis of 1.0 on the type curve. The data are matched to the type curve (μ), which has the same curvature. The vertical time-axis, t_1, which overlays the vertical axis for $Tt/r_c^2 = 1.0$ is selected. The transmissivity is found from

$$T = \frac{1.0 r_c^2}{t_1} \qquad \textbf{(8-19)}$$

The value of storativity can be found from the value of the μ-curve for the field data. Since $\mu = (r_s^2/r_c^2)S$,

$$S = (r_c^2\mu)/r_s^2 \qquad \textbf{(8-20)}$$

For small values of μ, however, the curves are often very similar; therefore, in matching the field data, the question of which μ-value to use is often encountered. The use of this method to estimate storativity should be approached with caution. Likewise, the value of T that is determined is representative only of the formation in the immediate vicinity of the test hole.

Several other types of slug tests can be run. One useful procedure can be applied to both confined and unconfined aquifers and used for either partially penetrating or fully penetrating wells (10). This procedure is based on a bail-down-type test, in which the well has a volume of water suddenly removed by bailing. The rate of rise of water in the well is then observed. Auger holes can also be used for slug tests (11). These are shallow, unlined, cylindrical holes, augered by hand or machine for the purpose of determining transmissivity.

EXAMPLE PROBLEM

A casing with a radius of 7.6 centimeters is installed through a confining layer. A screen with a radius of 5.1 centimeters is installed in a formation with a thickness of 5 meters. A slug of water is injected, raising the water level 0.42 meter. The decline of the head is given in Table 8.2. Find the values of T, K, and S.

TABLE 8.2

Time (sec)	H (m)	H/H_0
2	0.37	0.89
5	0.34	0.81
10	0.27	0.65
21	0.18	0.43
46	0.09	0.21
70	0.05	0.11
100	0.02	0.06

A plot of H/H_0 as a function of t is made on semilogarithmic paper. It is overlain on the type curve (Figure 8.10). At the axis for $Tt/r_c^2 = 1.0$, the value t_1 is 13 seconds.

$$T = \frac{1.0 r_c^2}{t_1} = \frac{1.0 \times 7.6 \text{ cm}^2}{13 \text{ sec}} = 4.44 \text{ cm}^2/\text{sec}$$

$$K = T/b$$

$$= 4.44 \text{ cm}^2/\text{sec}/500 \text{ cm}$$

$$= 8.9 \times 10^{-3} \text{ cm/sec}$$

The μ-curve is 10^{-3}. With $r_c = 7.6$ centimeters and $r_s = 5.1$ centimeters, we find

$$S = (\mu r_c^2)/r_s^2$$

$$= (10^{-3} \times 7.6^2)/(5.1)^2$$

$$= 2 \times 10^{-3}$$

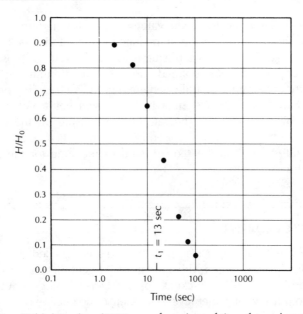

FIGURE 8.10. Field-data plot of H/H_0 as a function of time for a slug-test analysis.

8.4 FLOW IN A SEMICONFINED AQUIFER

Most confined aquifers are not toally isolated from sources of vertical recharge. Semipervious layers, either above or below the aquifer, can leak water into the aquifer if the direction of the hydraulic gradient is favorable (Figure 8.11).

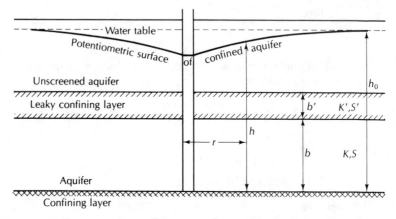

FIGURE 8.11. Fully penetrating well in an aquifer overlain by a semipermeable confining layer.

The flow equation for a confined aquifer with recharge is

$$\frac{\delta^2 h}{\delta r^2} + \frac{1}{r}\frac{\delta h}{\delta r} - \left(\frac{hK'^2}{T^2 b'^2}\right) = \frac{S}{T}\frac{\delta h}{\delta t} \tag{8-21}$$

where

    K' is the vertical hydraulic conductivity of the leaky layer

    b' is the thickness of the leaky layer

    h is the head

    r is the radial distance from the pumping well

    t is the time

    S is the storativity

    T is the transmissivity

8.4.1   NO STORAGE IN THE LEAKY CONFINING LAYER

The solution to this equation assumes conditions similar to the Theis solution, with the following additions: leakage through the confining bed is vertical and proportional to the drawdown; the head in the deposits supplying leakage is constant; and storage in the confining bed is negligible.

The solution (12) to Equation (8-21) is given by

$$h_0 - h = \frac{2.3Q}{4\pi T} W(u, r/B) \tag{8-22}$$

where

$$u = \frac{r^2 S}{4Tt} \tag{8-23}$$

and

$$B = (Tb'/K')^{1/2} \tag{8-24}$$

Equation (8-22) is known as the **Hantush-Jacob formula** for leaky aquifers. The factor B is termed the **leakage factor** (13). The equation is valid for all values of r_s (radius of well screen), providing that

$$t > (30r_s^2 S/T)[1 - (10r_s/b)^2] \quad \text{and} \quad r_s/B < 0.1$$

The rate of flow from storage in the main aquifer, q_s, is found from the formula

$$q_s = Q \exp(-Tt/SB^2) \tag{8-25}$$

The rate of induced leakage, q_L, may be found from

$$q_L = Q - q_s \tag{8-26}$$

Values of $W(u,r/B)$ are given in Appendix 3.

8.4.2 STORAGE IN THE LEAKY CONFINING LAYER

If there is significant storage in the leaky confining layer, then part of the flow during the initial time period will come from storage in the confining layer (2, 14, 15). It is necessary to use several equations to completely describe the drawdown, depending upon the length of time since pumping began.

The assumptions will be the same as before, except that here the confining layer has finite storativity (S') and there is no drawdown in the unpumped aquifer. The latter assumption may lead to errors for large values of time (15); hence, we will examine only the case in which the time is short (less than $(S'b'/10K')$. This is the most useful solution, as hydraulic parameters for both the aquifer and the confining layer can be obtained.

The drawdown equation is

$$h_0 - h = \frac{Q}{4\pi T} H(u,\beta) \tag{8-27}$$

where $H(u,\beta)$ is a function with values tabulated in Appendix 4 and

$$\beta = \frac{r}{4B}(S'/S)^{1/2} \tag{8-28}$$

$$B = [T/(K'/b')]^{1/2} \tag{8-29}$$

where

K' is the vertical conductivity of the confining layer

b' is the thickness of the confining layer

S' is the storativity of the confining layer

and

$$u = \frac{r^2 S}{4Tt}$$

The rate of flow from storage in the main aquifer, q_s, is given by

$$q_s = Q \exp(\eta t)\text{erfc}(\sqrt{\eta t}) \tag{8-30}$$

where

$$\eta = (K'/b')(S'/S^2) \tag{8-31}$$

and erfc is the complimentary error function, providing that $t < S'b'/10K'$.

The error function $\text{erf}(x)$ is a function with tabulated values. The complimentary error function $\text{erfc}(x)$ is defined as $1 - \text{erf}(x)$. A table of the error function is given in Table 8.3.

TABLE 8.3. Values of the error function, erf(x)

X	erf(x)
0.00	0.000
0.20	0.223
0.40	0.428
0.60	0.604
0.80	0.742
0.90	0.797
1.00	0.843
1.10	0.880
1.20	0.910
1.30	0.934
1.40	0.952
1.50	0.966
1.60	0.976
1.70	0.984
1.80	0.989
1.90	0.993
2.00	0.995
∞	1.000

A plot of drawdown as a function of time on semilogarithmic paper for several types of confined aquifers is given in Figure 8.12. Other than at very early times, the nonleaky aquifer will decline on a straight line. As there is no recharge, drawdown per log cycle of time will remain constant.

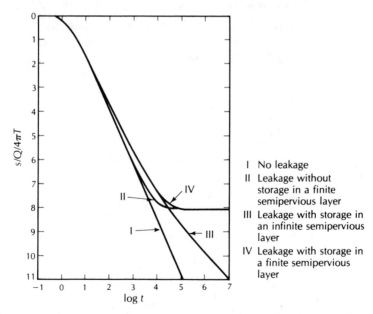

I   No leakage
II Leakage without storage in a finite semipervious layer
III Leakage with storage in an infinite semipervious layer
IV Leakage with storage in a finite semipervious layer

FIGURE 8.12. Plots of log of dimensionless drawdown as a function of time for an aquifer with various types of overlying confining layer. SOURCE: M. S. Hantush, *Journal of Geophysical Research*, 65 (1960):3713–25.

In a leaky aquifer, the drawdown curve will initially follow the nonleaky curve. However, after a finite time interval, the lowered hydraulic head in the aquifer will induce leakage from the confining layer. As part of the well discharge is now coming through the impervious layer, the rate of decline of head will decrease. If there is storage in the confining layer, the rate of drawdown will be lower than if there were no storage. Eventually, the drawdown cone will be large enough so that the pumped water will be coming entirely from leakage through the confining layer, and there will be no further drawdown with time. As water is no longer coming from storage in the leaky layer, the two curves coincide.

EXAMPLE PROBLEM

An aquifer 10 meters thick is penetrated by a well. It is overlain by a semipervious layer 1 meter thick with a K' of 10^{-5} centimeter per second. There is no storage in the leaky confining layer. The aquifer has a K of 10^{-2} centimeter per second and an S of 0.0005. If a well pumps at 500 cubic meters per day, compute values of drawdown at 1, 5, 10, 50, 100, 500, and 1000 meters in order to describe the cone of depression after 1 day.

$$K = 10^{-2} \text{ cm/sec} \times 60 \text{ sec/min} \times 1440 \text{ min/day}$$
$$\times 10^{-2} \text{ m/cm}$$
$$= 8.64 \text{ m/day}$$
$$K' = 10^{-5} \text{ cm/sec} \times 60 \text{ sec/min} \times 1440 \text{ min/day}$$
$$\times 10^{-2} \text{ m/cm}$$
$$= 8.64 \times 10^{-3} \text{ m/day}$$
$$b' = 1 \text{ m}$$
$$b = 10 \text{ m}$$
$$T = Kb = 86.4 \text{ m}^2/\text{day}$$
$$B = (Tb'/K')^{1/2}$$
$$= (86.4 \text{ m}^2/\text{day} \times 1 \text{ m}/8.64 \times 10^{-3} \text{ m/day})^{1/2}$$
$$= (10^4)^{1/2} = 100$$
$$u = \frac{r^2 s}{4Tt} = \frac{r^2 \times 0.0005}{4 \times 86.4 \times 1} = 1.44 \times 10^{-6} r^2$$
$$r/B = r/10 = 10^{-2} r$$
$$h_0 - h = \frac{2.6Q}{4\pi T} W(u, r/B) = 1.06 \, W(u, r/B)$$

As $u = 1.44 \times 10^{-6} r^2$, we can find the value of u for each r-value. Then, from the values of u and r/B, a value of $W(u, r/B)$ can be found from Appendix 3.

r	u	r/B	$W(u,r/B)$
1 m	1.44×10^{-6}	0.01	9.44
5 m	$3.6 \ \times 10^{-5}$	0.05	6.23
10 m	1.44×10^{-4}	0.1	4.83
50 m	$3.6 \ \times 10^{-3}$	0.5	1.85
100 m	1.44×10^{-2}	1	0.824
500 m	$3.6 \ \times 10^{-1}$	5	0.007
1000 m	1.44	10	0.0001

From the computed values of $W(u,r/B)$ at each observation point, the drawdown can be computed from $h_0 - h = 1.06 \, W(u,r/B)$:

r	$h_0 - h$
1 m	10.00 m
5 m	6.60 m
10 m	5.12 m
50 m	1.96 m
100 m	0.873 m
500 m	0.007 m
1000 m	0.0001 m

8.4.3   PUMPING TESTS FOR A LEAKY ARTESIAN AQUIFER WITH NO STORAGE IN THE CONFINING LAYER

Pumping tests may be used to determine the formation constants for both an aquifer and an overlying or underlying semipervious layer of finite thickness. Leaky artesian aquifers are also known as **semiconfined aquifers**.

W. C. Walton (16, 17) devised a graphical method based on type curves for $W(u,r/B)$. The type curves are plots on logarithmic paper of $W(u,r/B)$ as a function of $1/u$ for various values of $1/u$ and r/B (Figure 8.13). The type curve for $r/B = 0$ is identical to the Theis nonequilibrium reverse type curve.

Field data are plotted as drawdown versus time, or t/r^2 if there are several observation wells. The data curve is placed over the type curve with the axes parallel. The data curve should match one of the type curves for r/B; it may have to be matched to an imaginary line interpolated between two r/B-lines.

The early drawdown data will tend to fall on the nonequilibrium portion of the type curve. As leakage starts to contribute to flow from the well, the drawdown will follow an r/B-type curve.

A match point is picked on the graph. The coordinates on both the data plot and the type curve yield the values of $W(u,r/B)$, $1/u$, t, and $h_0 - h$. In addition, the type curve matched yields a value for r/B. These values are substituted into the Hantush-Jacob equations to find formation constants for both the aquifer and the semipervious layers.

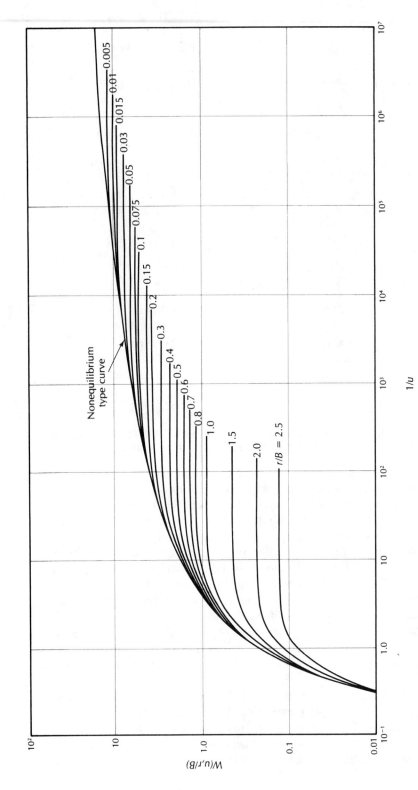

FIGURE 8.13. Type curves of leaky artesian aquifer in which no water is released from storage in the confining layer. SOURCE: W. C. Walton, Illinois State Water Survey Bulletin 49, 1962.

In American practical hydrologic units, the Hantush-Jacob formulas are

$$h_0 - h = \frac{114.6Q}{T} W(u, r/B) \qquad \textbf{(8-32)}$$

$$S = \frac{Ttu}{2693r^2} \qquad \textbf{(8-33)}$$

$$r/B = r/(Tb'/K')^{1/2} \qquad \textbf{(8-34)}$$

$$K' = [Tb'(r/B)^2]/r^2 \qquad \textbf{(8-35)}$$

where

Q is the pumping rate (gallons per minute)

T is the transmissivity of the main aquifer (gallons per day per foot)

t is the time since pumping began (minutes)

S is the storativity of the main aquifer

r is the distance from the pumping well to the observation well (feet)

K' is the vertical conductivity of the semipervious layer (gallons per square foot)

b' is the thickness of the semipervious layer (feet)

and

$$B = (Tb'/K')^{1/2}$$

EXAMPLE PROBLEM

Time drawdown data (17) for a well confined by a stratum of silty fine sand 14 feet thick are given in Table 8.4.

TABLE 8.4

Time (min)	Drawdown (ft)
5	0.76
28	3.30
41	3.59
60	4.08
75	4.39
244	5.47
493	5.96
669	6.11
958	6.27
1129	6.40
1185	6.42

281

The well was pumped at 25 gallons per minute. Using the Hantush-Jacob method, find the values of T, S, and K.

A plot of drawdown versus time must be made. This is shown in Figure 8.14. The match-point values are

$$W(u,r/B) = 1.0$$
$$1/u = 10, \quad u = 0.1$$
$$h_0 - h = 1.9 \text{ ft}$$
$$t = 33 \text{ min}$$

Substituting the match-point values in Equations (8-32), (8-33), (8-34), and (8-35),

$$T = \frac{114.6Q}{h_0 - h}W(u,r/B)$$

$$= \frac{114.6 \times 25 \times 1}{1.9}$$

$$= 1510 \text{ gal/day/ft}$$

$$S = \frac{Ttu}{2693r^2}$$

$$= \frac{1510 \times 0.1 \times 33}{2693 \times 96^2}$$

$$= 0.0002$$

$$K' = [Tb'(r/B)^2]/r^2$$

$$= \frac{1510 \times 14 \times 0.22^2}{96^2}$$

$$= 0.11 \text{ gal/day/ft}^2$$

M. S. Hantush (2) developed an alternative method that does not require the plotting and use of type curves. As can be seen from Figure 8.15, a plot of time versus drawdown on semilogarithmic paper has the shape of an elongated reverse S. At some point, the curve has an inflection. In the **Hantush inflection-point method**, the solution is based on finding this inflection point. For a pumping test with one observation well, the following procedure is used:

1. Plot the drawdown on the arithmetic scale as a function of time since pumping began on the logarithmic scale. If the test has reached equilibrium, then determine the maximum drawdown, $(h_0 - h)_{max}$. If drawdown is still occurring, extrapolate the curve to find $(h_0 - h)_{max}$.

2. The drawdown at the inflection point $(h_0 - h)_i$ is defined as being equal to one-half the maximum drawdown.

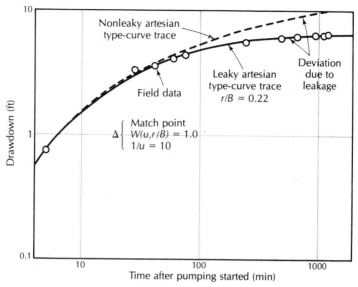

FIGURE 8.14.   Field-data plot of drawdown as a function of time for a leaky confined aquifer. SOURCE: W. C. Walton, Illinois State Water Survey Bulletin 49, 1962.

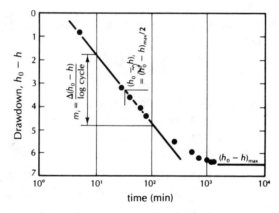

FIGURE 8.15.   Plot of drawdown in a confined aquifer as a function of time on semilogarithmic paper for use in the Hantush inflection-point method of analysis.

3. From the graph, determine the time, t_i, when $(h_0 - h)_i$ occurs; also graphically find the slope of the drawdown curve at the inflection point (m_i). This is generally equal to the slope of the straight portion of the drawdown curve. The slope is expressed as drawdown per log-cycle.

The following relations hold true for the inflection point:

$$u_i = r^2 S/4t_i T = r/2B \qquad\qquad\qquad (8\text{-}36)$$

$$m_i = (2.3Q/4\pi T)\exp(-r/B) \qquad\qquad (8\text{-}37)$$

$$(h_0 - h)_i = 0.5(h_0 - h)_{max} = \frac{Q}{4\pi T} K_0(r/B) \qquad (8\text{-}38)$$

$$B = [T/(K'/b')]^{1/2} \qquad\qquad\qquad\qquad (8\text{-}29)$$

283

$$f(r/B) = 2.3(h_0 - h)_i/m_i = \exp(r/B)K_0(r/B) \tag{8-39}$$

where K_0 is a function with values tabulated in Appendix 5 as $K_0(x)$ and $\exp(x)K_0(x)$.

From the drawdown and slope at the inflection point, the value of $f(r/B)$ may be found:

$$f(r/B) = 2.3(h_0 - h)_i/m_i \tag{8-40}$$

Knowing the value of $f(r/B)$, the function tables may be used to find the value of r/B, since $f(x) = \exp(x)K_0x$. Since r is known, the value of B may be easily found.

The transmissivity may be found from the relation

$$T = \frac{QK_0(r/B)}{2\pi(h_0 - h)_{max}} \tag{8-41}$$

where K_0 is a Bessel function, known as a zero-order modified Bessel function of the second kind (see Appendix 5).

The value of the storativity is found from

$$S = 4t_iT/2rB \tag{8-42}$$

The conductivity of the semipervious layer may be determined if its thickness, b', is known:

$$K' = \frac{Tb'}{B^2} \tag{8-43}$$

A computer program for the Hantush inflection-point method has been developed for use on a minicomputer (18). Field data are fed into the computer and the program computes T, S, and K'/b'.

8.4.4 PUMPING TEST FOR A LEAKY ARTESIAN AQUIFER WITH STORAGE IN THE CONFINING LAYER

A type-curve method can also be used for the leaky confined aquifer with storage in the confining layer. A set of type curves on logarithmic paper are prepared from the tabulated values of $H(u,\beta)$ (Figure 8.16). On logarithmic paper of similar scale, drawdown is plotted against time. The field-data sheet is overlain on the type curves with the $H(u,\beta)$-axis parallel to the drawdown axis, and it is matched to a β-curve. A match point is selected and values of $H(u,\beta)$, $1/u$, drawdown, and time are obtained. The value of the curve is also noted.

From the match-point value of $H(u,\beta)$ and $h_0 - h$, the value of T is found from Equation (8-27):

$$T = \frac{Q}{4\pi(h_0 - h)}H(u,\beta) \tag{8-44}$$

and S may be found from Equation (8-23) from the match-point values of t, $1/u$, the measured value of r, and the computed value of T:

$$S = \frac{4Tut}{r^2} \tag{8-45}$$

The value of β can be used to compute the product $K'S'$ from Equation (8-28):

$$\beta^2 = \frac{r^2 S'}{16B^2 S} \tag{8-46}$$

$$B^2 = T/(K'/b') \tag{8-47}$$

Combining Equations (8-46) and (8-47),

$$K'S' = \frac{16\beta^2 Tb'S}{r^2}$$

If one of the values, either K' or S', is known, the value of the other can be found.

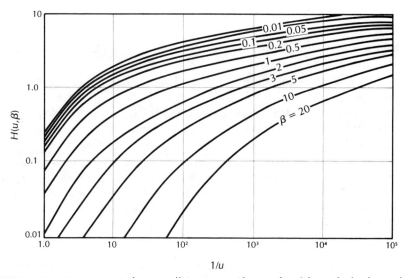

FIGURE 8.16. Type curves for a well in an aquifer confined by a leaky layer that releases water from storage. SOURCES: Data from M. S. Hantush, *Journal of Geophysical Research*, 65 (1960):3713-25; type curves from W. C. Walton, *Groundwater Resource Evaluation* (New York: McGraw-Hill Book Company, 1970).

EFFECT OF PARTIAL PENETRATION OF WELLS

In many cases, the open hole or well screen of a pumping well does not extend from the top to the bottom of an aquifer. In all of the cases considered thus far, we assumed the wells to penetrate the entire thickness of the aquifer. This caused flow in the aquifer to be essentially horizontal. However, the flow toward a partially penetrating well will be three-dimensional due to vertical

285

flow components (Figure 8.17). In addition, if the aquifer is anisotropic, the value of the vertical conductivity, K_v, and the horizontal conductivity, K_h, are important. This will affect both the amount of water pumped from the well and the potential field caused by drawdown.

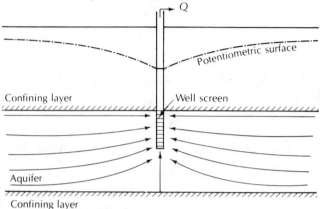

FIGURE 8.17. Flow lines toward a partially penetrating well in a confined aquifer.

If an observation well completely penetrates an aquifer, or if a partially penetrating well is located more than $1.5b\sqrt{K_h/K_v}$ distance units from the pumped well, the effect of a partially penetrating pumping well is negligible (1). The drawdown is described by either the Hantush-Jacob formula for leaky aquifers or the Theis formula for nonleaky aquifers.

If, however, the pumping well is partially penetrating, and the observation wells are also partially penetrating and located closer to the pumping well than $1.5b\sqrt{K_h/K_v}$ distance units, then the drawdown formula is different and quite complex. Hantush (1) makes the following observations about the drawdown in observation wells in such cases:

1. If two observation wells equidistant from the pumping well are screened in different parts of the aquifer, the time-drawdown curves may be different.

2. Depending upon the length and relative position of observation-well screens, it is possible for a more distant well to have a greater drawdown than a closer well.

3. The effects of partial penetration produce a time-drawdown curve similar in shape to one produced when there is a downward leakage from storage through a thick, semipervious layer.

4. Partial-penetration effects may produce a time-drawdown curve that resembles the effect of a recharge boundary, a fully penetrating well in either a sloping water-table aquifer or an aquifer of nonuniform thickness.

The preceding observations suggest that the hydrogeologist must always use extreme care in collecting and interpreting pumping-test data. Test drilling should be used to delineate the aquifer system before the pumping test. If a pumping test is to be made on a partially penetrating well, fully penetrating observation wells or wells located at a distance of more than $1.5b\sqrt{K_h/K_v}$ would be desirable. The analysis of partial-penetration effects is beyond the scope of this book. The hydrogeologist who must deal with this problem can use the **Hantush partial-penetration method** as described in Reference 19. Using partially penetrating wells, it is also possible to run pumping tests that determine the value of both horizontal and vertical conductivity of the confining layer and the storativity of both the aquifer and the confining layer (20–23).

The pumping-test solutions given in the preceding sections are all based on some simplifying assumptions that may not be valid in certain circumstances. For leaky artesian aquifer flow, the assumption that the storage in the aquifer is negligible may be erroneous (14, 15). The effect of storage in the confining layer occurs early in the pumping test, as leakage water is being furnished from storage in the confining layer. With increasing time, more of the leakage is being contributed by the aquifer above the leaky confining layer, and the amount of water from storage diminishes. If analysis of pumping-test data for a leaky artesian aquifer does not follow an r/B-curve for early drawdown data, significant flow from storage should be suspected and an alternative analysis made (14, 15, 23, 24, 25).

The Theis solution may overestimate the value of the storativity of shallow elastic aquifers. If the ratio of the average depth of a confined aquifer to its thickness is less than 0.5, the overestimation can be as great as 40 percent due to three-dimensional consolidation of the aquifer (15). If the average depth-to-thickness ratio is 1.0 or more, then the Theis solution is valid even for deforming aquifers.

WATER-TABLE AQUIFER

A well pumping from a water-table aquifer extracts water by two mechanisms. As with confined aquifers, the decline in pressure in the aquifer yields water due to the elastic storage of the aquifer storativity (S_s). The declining water table also yields water as it drains under gravity from the sediments. This is termed specific yield (S_y). The flow equation has been solved for radial flow in compressible unconfined aquifers under a number of different conditions using a variety of mathematical gambits (23, 26–37). The many and various equations in these solutions can lead to confusion; however, a qualitative description of the response of a water-table well to pumping may be helpful.

There are three distinct phases of time-drawdown relations in water-table wells. We will examine the response of any typical annular region of the aquifer located a constant distance from the pumping well. Sometime after pumping has begun, the pressure in the annular region will drop. As the pressure first drops, the aquifer responds by contributing a small volume of water due to the expansion of water and compression of the aquifer. During this time, the aquifer behaves as an artesian aquifer, and the time-drawdown data follow the Theis nonequilibrium curve for S equal to the elastic storativity of the aquifer. Flow is horizontal during this period, as the water is being derived from the entire aquifer thickness.

Following this initial phase, the water table begins to decline. Water is now being derived primarily from the gravity drainage of the aquifer, and there are both horizontal and vertical flow components. The drawdown-time relationship is a function of the ratio of horizontal-to-vertical conductivities of the aquifer, the distance to the pumping well, and the thickness of the aquifer.

As time progresses, the rate of drawdown decreases and the contribution of the particular annular region to the overall well discharge diminishes. Flow is again essentially horizontal, and the time-discharge data again follow a Theis type curve. The Theis curve now corresponds to one with a storativity equal to the specific yield of the aquifer. The importance of the vertical flow component as it affects the average drawdown is directly related to the magnitude of the ratio of the specific yield to the elastic storage coefficient (S_y/S_s). As the value of S_s approaches zero, the time duration of the first stage of drawdown also approaches zero. As S_y approaches zero, the length of the first stage increases, so that if $S_y = 0$, the aquifer behaves as an artesian aquifer of storativity S_s (37).

A number of type-curve solutions have been developed (39). The one that we will consider is based on a fully penetrating production well (35). The drawdown response is measured in observation wells that also fully penetrate the aquifer, and the well is pumped at a constant rate.

The flow equation for unconfined aquifers is given by (38):

$$T = \frac{Q}{4\pi h_0 - h} W(u_A, u_B, \Gamma) \tag{8-48}$$

where $W(u_A, u_B, \Gamma)$ is the well function for the water-table aquifer.

$$u_A = \frac{r^2 S}{4Tt} \quad \text{(for early drawdown data)} \tag{8-49}$$

$$u_B = \frac{r^2 S_y}{4Tt} \quad \text{(for later drawdown data)} \tag{8-50}$$

$$\Gamma = \frac{r^2 K_v}{b^2 K_h} \tag{8-51}$$

$h_0 - h$ is the drawdown
Q is the pumping rate
T is the transmissivity

r is the radial distance from the pumping well

S is the storativity

S_y is the specific yield

t is the time

K_h is the horizontal hydraulic conductivity

K_v is the vertical hydraulic conductivity

b is the initial saturated thickness of the aquifer

Two sets of type curves are used. Type-A curves are good for early drawdown data, when instantaneous release of water from storage is occurring. As time elapses, the effects of gravity drainage and vertical flow cause deviations from the nonequilibrium type curve, which is accounted for in the family of Type-A curves. The Type-B curves are used for late drawdown data, when effects of gravity drainage are becoming smaller. The Type-B curves end on a Theis curve. Figure 8.18 shows the two sets of type curves for fully penetrating wells. Values of $W(u_A, \Gamma)$ and $W(u_B, \Gamma)$ are found in Appendix 6.

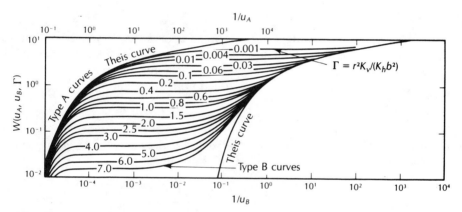

FIGURE 8.18. Type curves for drawdown data from fully penetrating wells in an unconfined aquifer. SOURCE: S. P. Neuman, *Water Resources Research,* 11 (1975):329–42.

The type curves are then used to evaluate the field data for time and drawdown, which are plotted on logarithmic paper of the same scale as the type curve. The following procedure can be used (37):

1. Superpose the latest time-drawdown data on the Type-B curves. The axes of the graph papers should be parallel and the data matched to the curve with the best fit. At any match point, the values of $W(u_B, \Gamma)$, u_B, t, and $h_0 - h$ are determined. The value of Γ comes from the type curve. The value of T is found using these values and Equation (8-48). The specific yield is found from Equation (8-50).

289

2. The early drawdown data are then superposed on the Type-A curve for the Γ-value of the previously matched Type-B curve. A new set of match points are determined. The value of T calculated from Equation (8-48) should be approximately equal to that computed from the Type-B curve. Equation (8-49) can be used to compute the storativity.

3. The value of the horizontal hydraulic conductivity can be determined from

$$K_h = T/m \qquad \text{(8-52)}$$

4. The value of the vertical hydraulic conductivity can also be computed using the Γ-value of the matched type curve. Rearrangement of Equation (8-51), yields the following formula:

$$K_v = \frac{\Gamma b^2 K_h}{r^2} \qquad \text{(8-53)}$$

The preceding analysis is based upon a very low value of drawdown compared with the saturated thickness of the aquifer. If the drawdown is substantial, some authorities (36) suggest that the drawdown data be corrected. The corrected drawdown $(h_0 - h)'$ is found from the relation (12)

$$(h_0 - h)' = (h_0 - h) - [(h_0 - h)^2/2h_0] \qquad \text{(8-54)}$$

This is normally only necessary for the later time-drawdown data; if $h_0 - h$ is small compared with h_0, correction will not be needed.

EXAMPLE PROBLEM

A well pumping at 1000 gallons per minute fully penetrates an unconfined aquifer with an initial saturated thickness of 100 feet. Time-drawdown data for a well located 200 feet away are plotted on log paper (Figure 8.19). Find T, S_y, S, K_h, and K_v.

The later time-drawdown data fit best on the $\Gamma = 0.1$ type curve. With the axes of the two sheets of graph paper parallel, the selected match point has values of $W(u_B, \Gamma) = 0.1$, $1/u_B = 10$, $h_0 - h = 0.043$ feet, and $t = 128$ minutes. The value of u_y is 0.1, the pumping rate is equal to 1.9×10^5 cubic feet per day, and $t = 0.089$ days. From Equation (8-48),

$$T = \frac{Q}{4\pi(h_0 - h)} W(u_B, \Gamma)$$

$$T = \frac{1.9 \times 10^5 \text{ ft}^3/\text{day} \times 0.1}{4\pi \times 0.043 \text{ ft}}$$

$$= 3.56 \times 10^4 \text{ ft}^2/\text{day}$$

The value of the specific yield can be determined from Equation (8-50):

$$S_y = \frac{4Ttu_B}{r^2}$$

$$= \frac{4 \times 3.56 \times 10^4 \text{ ft}^2/\text{day} \times 0.1 \times 0.089 \text{ days}}{(200 \text{ ft})^2}$$

$$= 0.32$$

The early time-drawdown data are now matched to the Type-A curve for $\Gamma = 0.1$. The selected match point is $W(u_A,\Gamma) = 0.1$, $1/u_A = 1.0$, $h_0 - h = 0.041$ foot and $t = 0.9$ minute. The value of u_A is 1.0 and $t = 6.25 \times 10^{-4}$ day. From Equation (8-48),

$$T = \frac{1.9 \times 10^5 \text{ ft}^3/\text{day} \times 0.1}{4\pi \times 0.041 \text{ ft}}$$

$$= 3.73 \times 10^4 \text{ ft}^2/\text{day}$$

The storativity value is found from Equation (8-49):

$$S = \frac{4Ttu_A}{r^2}$$

$$= \frac{4 \times 3.73 \times 10^4 \text{ ft}^2/\text{day} \times 0.1 \times 6.25 \times 10^{-4} \text{ days}}{(200 \text{ ft})^2}$$

$$= 0.0023$$

The value of the horizontal hydraulic conductivity can be found from Equation (8-52). The average of T is 3.65×10^4 square feet per day:

$$K_h = T/b$$

$$= (3.65 \times 10^4 \text{ ft}^2/\text{day})/100 \text{ ft}$$

$$= 365 \text{ ft/day}$$

and the value of K_v is determined using Equation (8-53):

$$K_v = \frac{\Gamma b^2 K_h}{r^2}$$

$$= \frac{0.1 \times (100 \text{ ft})^2 \times 365 \text{ ft/day}}{(200 \text{ ft})^2}$$

$$= 9 \text{ ft/day}$$

291

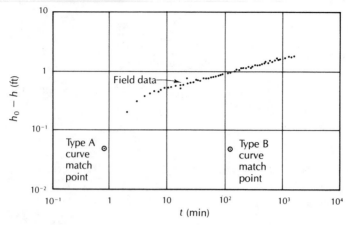

FIGURE 8.19. Field data for example problem of analysis of aquifer test for an unconfined aquifer.

 STEADY-STATE RADIAL FLOW

Prior to the development of the Theis nonequilibrium formula, people doing well analysis assumed a condition of steady flow. That is, the drawdown rate in the pumped well was assumed to be so low as to be essentially zero. This assumption is valid for both nonleaky artesian and water-table aquifers if a long period of time has elapsed since the start of pumping (41). For nonleaky artesian aquifers, the appropriate equation is

$$T = \frac{Q}{2\pi(h_2 - h_1)} \ln(r_2/r_1) \qquad \textbf{(8-55)}$$

where

Q is the pumping rate

h_1 is the head at distance r_1 from the pumping well

h_2 is the head at distance r_2 from the pumping well

For an unconfined aquifer,

$$K = \frac{Q}{\pi(h_2^2 - h_1^2)} \ln(r_2/r_1) \qquad \textbf{(8-56)}$$

To find values of T or K from steady-state equations, there must be at least two observation wells at different distances from the pumping well. The well must be pumped long enough for the drawdown to approach a steady-state condition.

The usefulness of steady-state analysis is limited, as values of storativity or specific yield are not obtained. However, transmissivity or hydraulic conductivity values obtained by the above equations are likely to be more accurate than those obtained from a transient analysis.

There may be circumstances in which an exising well which pumps constantly and cannot be shut down must be analyzed. If the well has reached a steady-state condition and discharges at a constant rate, and there are at least two observation wells, the aquifer transmissivity can be found from either Equation (8-55) or (8-56), depending on whether it is a water-table or a confined aquifer.

INTERSECTING PUMPING CONES AND WELL INTERFERENCE

The cases we have considered thus far have involved only one well pumping from an aquifer. However, there are often several wells tapping the same aquifer with intersecting pumping cones. At any given point in a confined aquifer, the total drawdown is the sum of the individual drawdowns for each well. Because the Laplace equation is linear, the superposition of drawdown effects is found by simple addition. In Figure 8.20, the well interference for a multiple-aquifer well field is presented graphically. Linear superposition is valid only for confined aquifers, in which the value of the transmissivity does not change with drawdown. In water-table aquifers, if the drawdown is significant in relation to the total saturated thickness, the use of linear superposition will result in a predicted composite drawdown that is less than the actual composite drawdown. As a decrease in saturated thickness reduces the transmissivity, the multiple-well system will result in a composite hydraulic gradient greater than that of an equivalent confined system in order to compensate for a reduced value of aquifer transmissivity.

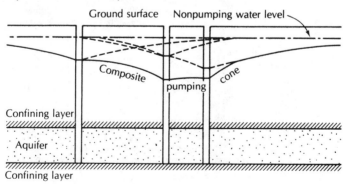

FIGURE 8.20. Composite pumping cone for three wells tapping the same aquifer. Each well is pumping at a different rate; thus, the pumping level of each is different.

In designing well-field layouts, it is necessary to take into account well interference. The level of the water in the well during pumping determines the length of pipe necessary to carry water to the surface. The characteristics of the well pump and the horsepower requirements of the motor

293

also depend upon the depth to the pumping level. If wells are spaced too closely together, the amount of well interference could be excessive. Aligning wells parallel to a line source of recharge, such as a river, would result in less well interference than a perpendicular configuration.

8.9 EFFECT OF HYDROGEOLOGIC BOUNDARIES

If a well is not located in an aquifer of infinite areal extent, as is the case with all real wells in real aquifers, the drawdown cone will extend until either the well is supplied by vertical recharge or a hydrogeologic boundary is reached. A hydrogeologic boundary could be the edge of the aquifer, a region of recharge to a fully confined artesian aquifer, or a source of recharge, such as a stream or lake.

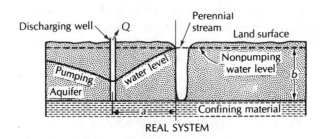

REAL SYSTEM

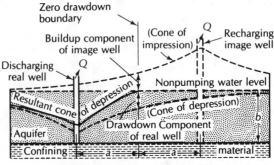

NOTE: Aquifer thickness *b* should be very large compared to resultant drawdown near real well

HYDRAULIC COUNTERPART OF REAL SYSTEM

FIGURE 8.21A. Idealized cross section of a well in an aquifer bounded on one side by a stream. SOURCE: J. G. Ferris et al., U.S. Geological Survey Water Supply Paper 1536-E, 1962.

Boundaries are considered to be either recharge or barrier boundaries. A **recharge boundary** is a region in which the aquifer is replenished. A **barrier boundary** is an edge of the aquifer, where it terminates, either by thinning or abutting a low-permeability formation, or has been eroded away. Parts A and B of Figure 8.21 show the effect of boundaries on the cross section of the potentiometric surface of a pumped aquifer. Boundaries have the most dramatic impact on the drawdown of a pumped well for the aquifer with no source of vertical recharge. As the well withdraws water only from storage in the aquifer, drawdown proceeds as a function of the logarithm of time.

Figure 8.22 shows a theoretical straight-line plot of drawdown as a function of time on semilogarithmic paper. The effect of a recharge boundary is to retard the rate of drawdown. Drawdown can become zero if the well comes to be supplied entirely with recharged water. The effect of a barrier to flow in some region of the aquifer is to accelerate the drawdown rate. The water level declines faster than the theoretical straight line.

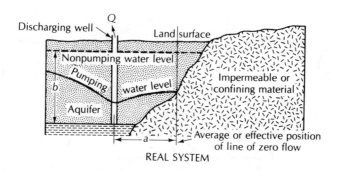

REAL SYSTEM

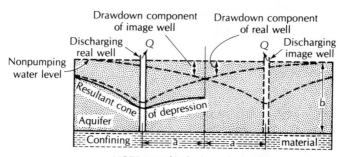

NOTE: Aquifer thickness b should be very large compared to resultant drawdown near real well

HYDRAULIC COUNTERPART OF REAL SYSTEM

FIGURE 8.21B.   Idealized cross section of a well in an aquifer bounded on one side by an impermeable boundary. SOURCE: J. G. Ferris et al., U.S. Geological Survey Water Supply Paper 1536-E, 1962.

295

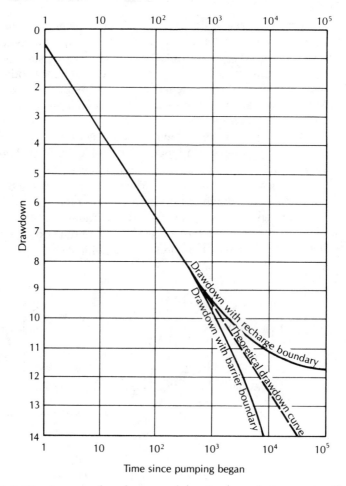

FIGURE 8.22. Impact of recharge and barrier boundaries on semilogarithmic drawdown-time curves.

8.10 PUMPING-TEST DESIGN

Adequate design and execution of a pumping test involves considerable planning and attention to detail. An understanding of fundamental well hydraulics is necessary, not only for the interpretation of data, but also for the experimental design by which valid and usable data are obtained. The purpose of the pumping test must be established first. Determining the yield of a new well involves simply pumping the well. This type of test, as it is generally conducted, yields only the scantiest information about the aquifer, itself. With careful planning, the pumping-well test can yield data to compute the aquifer transmissivity. It can also indicate the general type of aquifer.

The amount of information gained from a pumping test expands greatly if one or more observation wells are involved, in addition to the pumping well. Both transmissivity and storativity of the aquifer can be determined, as can the vertical hydraulic conductivity of any overlying semipervious layers. More eloquent tests can be used to determine the value of the vertical anisotropy of the formation. Radial anisotropy and recharge or barrier boundaries can also be detected.

8.10.1 SINGLE-WELL PUMPING TESTS

The basics of a single-well pumping test are also applicable to pumping tests involving multiple wells. The first step is to determine the location of the well to be drilled. This is best done on the basis of detailed exploration using geological, geophysical, and perhaps aerial photo techniques. However, the location of the well is often dictated by economic or engineering factors. If economic or engineering factors predominate, the hydrogeologist should determine if there is a reasonable chance of success based on the known hydrogeology of the site.

A test well may be bored as the first step, or the production well may be drilled immediately. The geologist should make a log of the geologic formations encountered. The water level in the drilled hole should be recorded as a function of the depth of the hole; however, this might not be possible if certain drilling techniques such as rotary and reverse rotary are used. Based on the test hole and selected borehole geophysical studies, the hydrogeologist can determine the depth and thickness of potential aquifer zones. An aquifer is selected, and a test or permanent well is installed. If at all feasible, the well should be open throughout the entire thickness of the aquifer. The physical dimensions of the well should be recorded, along with the depth, thickness, and type of aquifer. A description of the aquifer material should be included. An inventory of nearby wells should be made, and it should be determined whether any other wells will be running when an aquifer test is planned. Intersecting cones of depression during an aquifer test should be avoided.

A pump is installed in the completed well. The pump, engine, wiring, piping, and assorted equipment must be reliable. If a pumping test is terminated due to mechanical failure prior to the planned time, the data may not be sufficient. The discharge pipe from the pump must be equipped with a valve to control the volume of flow and with a means of measuring the flow.

Small pumping rates may be measured by means of a water meter in the line or by filling a container of known volume in a measured time. These methods are generally useful for flows of 6 liters per second (100 gallons per minute) or less. For larger flows, a common method is to use a circular orifice weir on the discharge pipe (42). Generally, the well-drilling contractor furnishes the equipment and makes the discharge measurements. The hydrogeologist should always check the apparatus and measurements.

There must be a means of conveying the pumped water away from the test site. This is especially true for shallow, unconfined aquifers, where

297

the water could recharge the aquifer and render any pumping-test results useless. If a pumping test runs for several days, the quantity of pumped water that must be conveyed from the site can be considerable.

A means to measure the water level in the well before as well as during pumping must be available. A steel tape, air line, or electrical tape can serve this function (42).

The pumping tests we have studied all have been based on a constant discharge rate. In reality, the water level in the well falls with pumping, the pumping lift increases, and the discharge of the pump tends to decline. To avoid this, the valve on the discharge pipe should be partially closed to restrict the initial discharge. During the course of the test, the valve can be opened as necessary to keep the pumping rate constant. There should be no more than a 10 percent variation in rate during the test, with a smaller pumping-rate variation if possible. After construction, a new well is usually developed by on-and-off pumping, which causes the water to surge back and forth through the well screen or open hole. This increases the yield of the well by washing out fine particles and mud used for drilling. The well should be fully developed before pumping tests are made.

It is necessary to select a measuring point on the well to serve as a fixed reference for water-level measurements. The measuring point should be marked, and its elevation measured and recorded. Prior to the pumping test, the water level should be measured. Other production wells that may be nearby should either be shut down for the duration of the test or pumped at a constant rate. It is difficult to correct pumping-test data for the effect of wells starting and stopping. For the most accurate results, the nonpumping water elevation should be measured several times before pumping begins. Groundwater levels may have a long-term trend of rising or falling. They may also be affected by tides or changes in the barometric pressure. If the static level is found to fluctuate, then detailed pretest measurements must be made for at least twice the expected length of the pumping test. If a long-term linear trend is observed, the drawdown observed during the pumping test must be corrected. The corrected drawdown is the difference between the measured depth to water and the projected static level based on the long-term trend. Tidal fluctuations and fluctuations due to barometric pressure require measurements of the water level in the well and either the tide or air pressure prior to pumping, in order to establish a relationship between them. Measurements of tides or air pressure must then be made during the pumping test to find a corrected static level at the time of each drawdown measurement.

Prior to the test, the discharge of the well should be measured and a planned pumping rate selected. The valve should be preset for this rate. When the pump is started for the test, the valve is adjusted to yield the desired pumping rate. Periodic measurements of discharge and corrections using the valve should be made about every half hour. The time and measurement of pumping rate should be recorded in the field notes.

Water-level readings in the well commence after one minute of pumping. The usual procedure is to record the depth to water below the

298

measuring point and the time of measurement. Computation of drawdown is made later. On the order of ten readings per log cycle of time are made. The first reading could be at one minute, then at one and one-half minutes, and then at every minute up to ten. Between ten and twenty minutes, readings are taken every two minutes; then every five minutes between twenty minutes and one hour; and every ten minutes between one and two hours. After two hours have elapsed, the recommended ten readings per log cycle of time are made. The most work occurs during the first minutes of the pumping test.

There must be some advance planning for the length of time the test is to proceed. For an artesian or leaky artesian aquifer, the test may last twenty-four hours or less. This is often sufficient to delineate values for the formation constants and to determine whether there are any recharge or barrier boundaries. Water-table wells must be pumped for a sufficient duration to preclude any significant effects of vertical flow near the well. The time period is a function of the distance from the pumping well, the conductivity of the formation, and the degree of anisotropy. The time is naturally greater the further the distance from the pumping well, the lower the horizontal conductivity, and the greater the anisotropy. Pumping tests of unconfined aquifers normally run for several days to several weeks.

Periodically during the pumping test, a sample of the water is collected for chemical analysis. A series of samples will reveal any trend in chemical or bacteriological quality with continued pumping. It is possible to predict final well quality fairly accurately, even on the basis of chemical analysis of water from temporary test wells (43).

Following the collection of time and drawdown data in the pumped well, an analysis is made using the appropriate type curves. It should be remembered that with only one well, the value of T, but not of S, can be determined. In another computation usually made at the end of the pumping test, the well yield is divided by the maximum drawdown to obtain the specific capacity, which is widely used as an index of the capacity of the well.

8.10.2 PUMPING TESTS WITH OBSERVATION WELLS

If it is feasible, drawdown should be measured in one or more nonpumping wells situated close to the pumping well. These observation wells are often constructed especially for the pumping tests. However, domestic-supply wells, abandoned wells, other wells in a well field, or other wells under construction are sometimes used. The use of one or more observation wells can enable the hydrogeologist to compute the storativity or specific yield of the aquifer. Under some circumstances, the anisotropy of the pumping well and the leakage factor for leaky confined aquifers may be determined. In most cases, the hydrogeologist will be able to employ only one observation well—especially if the aquifer is deep and the area is undeveloped with no existing wells available.

The selection of the location of the single observation well is critical. It should be located at a radial distance such that the time-drawdown data collected during the planned pumping period will fall on a type curve of unique curvature. If the curvature is too flat, selection of the proper type curve is difficult. After a test boring is made, the hydrogeologist should know the type of aquifer system to be confronted. The formation characteristics are then estimated, and using the correct formula with the planned pumping rate, time-drawdown curves for several hypothetical observation wells at different distances are plotted. On the basis of these curves, the location of the single observation well is selected. As a general rule, it will be closer for a water-table well than for a confined well.

If there are two observation wells, the second should be in a radial line with the first, but at ten times the distance. If is there are more than two wells, they should form two or more radial lines from the pumping well. This will indicate any radial anisotropy in the aquifer. A map should be carefully made in the field showing the relative locations of the pumping well and all observation wells. The distance from the pumping well to each observation well should be measured with a steel tape.

Ideally, observation wells should fully penetrate the aquifer, so that they measure the average head in the formation at that location. Plastic pipe through which holes are drilled at the depth of the aquifer serves as a suitable observation well. The annular space between the drilled hole and the plastic pipe can be backfilled with coarse sand. If the well goes through a confining layer, the hole must be backfilled with clay and tamped to prevent leakage around the pipe.

If the observation well has a short well screen, then the screen should be placed such that the head it measures is representative of the average head in the formation at that location. For a confined aquifer, it should be at the mid-depth of the stratum. For a water-table aquifer, it should be one-third of the distance from the static water table to the bottom of the aquifer.

The observation well should have a rapid response to changes in the water level in the aquifer. One way to test response time is to pour water into the observation well. The induced head should drain away in a fairly short time, usually in a few hours or less in most aquifers. If the observation well does not show a good response, an effort should be made to unclog it. Pouring water in the well may clear it. If the water level is within a few meters of the surface, a plunger on a stick can be used to surge the well to clear it.

Prior to the pumping test, a measuring point should be chosen for each observation well. Usually this is the top of the casing. The elevation of each measuring point should be determined. The depth to the static water level should also be measured prior to pumping. This will be useful in mapping the potentiometric surface.

Depth-to-water measurements are made in the observation wells on the same schedule as the pumping-well measurements. For the first minutes, this probably will necessitate one observer for each well. After twenty minutes, when the readings go to a frequency of every ten minutes, fewer people will be needed. After two and one-half hours, readings are made only every hundred minutes, so that one observer usually can do everything. Also, at the start of the test, one or two additional people should be on hand to measure pump discharge and adjust the flow.

After the end of the scheduled pumping test, recovery measurements can be made in the wells. The water levels will recover at the same rate they fill. In some cases, the drawdown data are affected by uncontrolled variations in the pumping rate. This does not affect the recovery rate. The flow rate for recovery data is equal to the mean discharge for the entire pumping period. In using recovery data, the difference between the water level at the end of pumping and that after a given time since pumping stopped is plotted as a function of the time since pumping stopped. The standard methods of well-test analysis are used. Recovery measurements are a standard part of the aquifer test. In many cases, the recovery data prove to be more useful than the drawdown data; for example, if there was a short period when the well shut down during the test, or if the rate of pumping was extremely variable during the drawdown phase of the test.

References

1. HANTUSH, M. S. "Hydraulics of Wells." In *Advances in Hydroscience*, vol. 1, ed. V. T. Chow. New York: Academic Press, 1964, pp. 281–432.

2. THEIS, C. V. "The Lowering of the Piezometric Surface and the Rate and Discharge of a Well Using Ground-Water Storage." *Transactions, American Geophysical Union*, 16 (1935):519–24.

3. WENZEL, L. K. *Methods for Determining Permeability of Water-Bearing Materials with Special Reference to Discharging Well Methods*. U.S. Geological Survey Water Supply Paper 887, 1942, p. 192.

4. COOPER, H. H., JR. and C. E. JACOB. "A Generalized Graphical Method for Evaluating Formation Constants and Summarizing Well-Field History." *Transactions, American Geophysical Union*, 27 (1946):526–34.

5. JACOB, C. E. "Flow of Ground-Water." In *Engineering Hydraulics*, ed. H. Rouse. New York: John Wiley & Sons, 1950, pp. 321–86.

6. SCHAFER, D. C. "Casing Storage Can Affect Pumping Test Data." *Johnson Drillers Journal* (January-February 1978):1–5, 10–11.

7. FERRIS, J. G. et al. *Theory of Aquifer Tests*. U.S. Geological Survey Water Supply Paper 1536-E, 1962.

8. COOPER, H. H., JR., J. D. BREDEHOEFT, and I. S. PAPADOPULOS. "Response of a Finite Diameter Well to an Instantaneous Charge of Water." *Water Resources Research*, 3 (1967):263–69.

9. PAPADOPULOS, I. S., J. D. BREDEHOEFT, and H. H. COOPER, JR. "On the Analysis of 'Slug Test' Data." *Water Resources Research*, 9 (1973):1087–89.

10. BOUWER, H. and R. C. RICE. "A Slug Test for Determining Hydraulic Conductivity of Unconfined Aquifers with Completely or Partially Penetrating Wells." *Water Resources Research,* 12 (1976):423–28.

11. BOAST, C. W. and D. KIRKHAM. "Auger Hole Seepage Theory." *Soil Science Society of America, Proceedings,* 35 (1971):365–73.

12. HANTUSH, M. S. "Analysis of Data from Pumping Tests in Leaky Aquifers." *Transactions, American Geophysical Union,* 37 (1956):702–14.

13. HANTUSH, M. S. and C. E. JACOB. "Plane Potential Flow of Ground-Water with Linear Leakage." *Transactions, American Geophysical Union,* 35 (1954):917–36.

14. HANTUSH, M. S. "Modification of the Theory of Leaky Aquifers." *Journal of Geophysical Research,* 65 (1960):3713–25.

15. NEUMAN, S. P. and P. A. WITHERSPOON. "Applicability of Current Theories of Flow in Leaky Aquifers." *Water Resources Research,* 5 (1969):817–29.

16. WALTON, W. C. *Selected Analytical Methods for Well and Aquifer Evaluation.* Illinois State Water Survey Bulletin 49, 1962, p. 81.

17. WALTON, W. C. *Leaky Artesian Aquifer Conditions in Illinois.* Illinois State Water Survey Report of Investigation 39, 1960.

18. HOLZSCHUH, J. C., III. "A Simple Computer Program for the Determination of Aquifer Characteristics from Pump Test Data." *Ground Water,* 14 (1976):283–85.

19. HANTUSH, M. S. "Aquifer Test on Partially Penetrating Wells." *Proceedings of the American Society of Civil Engineers,* 87 (1961):171–95.

20. HANTUSH, M. S. "Analysis of Data from Pumping Tests in Anisotropic Aquifers." *Journal of Geophysical Research,* 71 (1960):421–26.

21. HANTUSH, M. S. "Wells in Homogeneous Anisotropic Aquifers." *Water Resources Research,* 2 (1966):273–79.

22. WEEKS, E. P. "Determining the Ratio of Horizontal to Vertical Permeability by Aquifer Test Analysis." *Water Resources Research,* 5 (1969):196–214.

23. BOULTON, N. S. and T. D. STRELTSOVA. "New Equations for Determining the Formation Constants of an Aquifer from Pumping Test Data." *Water Resources Research,* 11 (1975):148–53.

24. STRELTSOVA, T. D. "Analysis of Aquifer-Aquitard Flow." *Water Resources Research,* 12 (1976):415–22.

25. GAMBOLATI, G. "Deviations from the Theis Solution in Aquifers Undergoing Three-Dimensional Consolidation." *Water Resources Research,* 13 (1977):62–68.

26. BOULTON, N. S. "The Drawdown of the Water Table Under Non-Steady Conditions Near a Pumped Well in an Unconfined Formation." *Proceedings of the Institute of Civil Engineers* (London), 3, no. 3 (1954):564–79.

27. BOULTON, N. S. *Unsteady Radial Flow to a Pumped Well Allowing for Delayed Yield from Storage.* International Association of Scientific Hydrology Publication 37, 1955, pp. 472–77.

28. BOULTON, N. S. "Analysis of Data from Non-Equilibrium Pumping Tests Allowing for Delayed Yield from Storage." *Proceedings of the Institute of Civil Engineers* (London), 26 (1963):269–82.

29. BOULTON, N. S. "The Influence of the Delayed Drainage on Data from Pumping Tests in Unconfined Aquifers." *Journal of Hydrology,* 19 (1973):157–69.

30. BOULTON, N. S. and J. M. A. PONTIN. "An Extended Theory of Delayed Yield from Storage Applied to Pumping Tests in Unconfined Anisotropic Aquifers." *Journal of Hydrology,* 14 (1971):53–65.

31. STRELTSOVA, T. D. "Unsteady Radial Flow in an Unconfined Aquifer." *Water Resources Research,* 8 (1972)1059–66.

32. STRELTSOVA, T. D. "Unsteady Radial Flow in an Unconfined Aquifer." *Water Resources Research.* 9 (1973):236–42.

33. DAGAN, G. "A Method of Determining the Permeability and Effective Porosity of Unconfined Anisotropic Aquifers." *Water Resources Research,* 3 (1967):1059–71.

34. NEUMAN, S. P. "Theory of Flow in Unconfined Aquifers Considering Delayed Response of the Water Table." *Water Resources Research,* 8 (1972):1031–45.

35. NEUMAN, S. P. "Analysis of Pumping Test Data from Anisotropic Unconfined Aquifers Considering Delayed Gravity Response." *Water Resources Research,* 11 (1975):329–42.

36. NEUMAN, S. P. "Effect of Partial Penetration on Flow in Unconfined Aquifers Considering Delayed Gravity Response," *Water Resources Research,* 10 (1974):303–12.

37. GAMBOLATI, G. "Transient Free Surface Flow to a Well: An Analysis of Theoretical Solutions." *Water Resources Research,* 12 (1976):27–39.

38. PRICKETT, T. A. "Type Curve Solution to Aquifer Tests under Water Table Conditions." *Ground Water,* 3, no. 3 (1965):5–14.

39. WALTON, W. C. "Comprehensive Analyses of Water-Table Aquifer Test Data." *Ground Water,* 16 (1978):311–17.

40. STRELTSOVA, T. D. "Comments on 'Analysis of Pumping Test Data from Anisotropic Unconfined Aquifers Considering Delayed Gravity Response,' by Shlomo P. Neuman." *Water Resources Research,* 12 (1976):113–14.

41. THIEM, G. *Hydrologische Methoden.* Leipzig: Gebhardt, 1906, p. 56.

42. Johnson Division, Universal Oil Products Company. *Ground Water and Wells.* St. Paul: Minn., 1966, p. 440.

43. FETTER, C. W., JR. "Use of Test Wells as Water-Quality Predictors." *Journal American Water Works Association,* 67 (1975):516–18.

Water Chemistry

chapter

INTRODUCTION

For most of its uses, the chemical properties of water are as important as the physical properties and available quantity. In the next chapter, we will consider water quality, which involves the type and amount of substances dissolved in the water. In this chapter, our focus will be the chemical reactions that occur between water and the solids and gases it contacts.

Natural waters are never pure; they always contain at least small amounts of dissolved gases and solids. The composition of the aqueous solution is a function of a multiplicity of factors; for example, the initial composition of the water, the partial pressure of the gas phase, the type of mineral matter the water contacts, and the pH and oxidation potential of the solution. Water containing a biotic assemblage has an even more complex chemistry due to the life processes of the biota.

The detailed study of water chemistry is far beyond the scope of this chapter. We will concentrate on the aspect of solubility of gases and liquids in dilute aqueous solutions. Further, we will assume that all reactions take place at a temperature of 25° C and a pressure of 1 atmosphere. Small deviations from this assumption (a few atmospheres pressure and ±10 to 15° C) will not lead to significant error (1). The systems considered will be presumed to be abiotic.

9.2 **UNITS OF MEASUREMENT**

Chemical analysis of an aqueous solution yields the amount of solute in a specified amount of solvent. Conventionally, this is reported as milligrams of solute (weight) per liter of solvent (volume), or mg/ℓ. Analyses may also be expressed as milliequivalents per liter. The **equivalent weight** of a substance is the concentration in parts per million multiplied by the valence change and then divided by the formula weight in grams.

In chemical thermodynamics, the concentrations should be expressed in terms of **molality**. A one-molal solution has one mole of solute in 1000 grams of water. One mole of a compound is its formula weight in grams.

EXAMPLE PROBLEM

Part A: What is the weight of NaCl in a 0.01-molal solution?

Atomic weight of sodium = 22.991 gm

Atomic weight of chlorine = 35.457 gm

One mole of NaCl = 58.448 gm

0.01 mole = 0.01×58.448 = 0.58448 gm

0.01-molal solution = 0.58448 gm NaCl in 1000 gm H_2O

Part B: What is the concentration of NaCl in a 0.01-molal solution at 25° C?

At 25° C, the density of water is 0.99707 gram per milliliter. The volume of 1000 grams is 997.07 milliliters. The concentration is 0.58448 gram per 0.99707 liter, or 586.1 milligrams per liter.

For dilute solutions, it is not necessary to make density corrections; the following conversion factors may be used (2):

$$\text{Molality} = \frac{\text{milligrams per liter} \times 10^{-3}}{\text{formula weight in grams}} \qquad \text{(9-1)}$$

$$\text{Molality} = \frac{\text{milliequivalents per liter} \times 10^{-3}}{\text{valence of ion}} \qquad \text{(9-2)}$$

TYPES OF CHEMICAL REACTIONS IN WATER

Chemical reactions in an aqueous solution are either **reversible** or **irreversible**. Those that are reversible can reach equilibrium with their hydrochemical environment and are amenable to study by kinetic and thermodynamic methods.

The simplest aqueous reaction is the dissociation of an inorganic salt. If the salt is present in excess, it will tend to form a saturated solution:

$$NaCl \rightleftharpoons Na^+ + Cl^- \qquad \text{(9-3)}$$

Natural systems always tend toward equilibrium; thus, if the solution is under-saturated, more salt will dissolve. If it is supersaturated, salt will crystallize, although for kinetic reasons the solution may remain supersaturated. Notice that the water molecule does not actively participate in this reaction.

Water molecules can actively bond with either a gas or solid in a reversible reaction:

$$CaCO_3 + H_2O \rightleftharpoons Ca^{++} + HCO_3^- + OH^- \tag{9-4}$$

$$CO_2 + H_2O \rightleftharpoons HCO_3^- + H^+ \tag{9-5}$$

In this type of reaction, the water molecule breaks into H^+ and OH^- radicals when combining with the species in solution.

Reversible reactions may also involve the transfer of electrons from one ion to another. When this happens, the species undergo a valence change:

$$4Fe^{++} + 3O_2 + 8e^- \rightleftharpoons 2Fe_2O_3 \tag{9-6}$$

In this example, the ferrous iron is oxidized to ferric iron by the transfer of an electron from the iron to the oxygen.

A gas or solid may also dissolve in an aqueous solution without dissociation:

$$O_{2\ (gas)} \rightleftharpoons O_{2\ (aqueous)} \tag{9-7}$$

IDEAL GAS LAW

The **ideal gas** is described in terms of the pressure, P, volume, V, and tempera-ture, T, of n moles of gas by the relation

$$PV = nRT \tag{9-8}$$

where R is the thermodynamic gas constant. The value of R is 1.987 calories per degree Kelvin per mole at 25° C and 1 atmosphere pressure. Under temper-ature and pressure conditions expected in hydrogeological studies, gases will tend to behave as ideal gases, thus Equation (9-8) may be used.

The thermodynamic activity, a, of an ideal gas is defined as unity when the gas is in its standard state (1 atmosphere P and specified T). For pressure other than 1 atmosphere, the activity may be found from the equation

$$a = \frac{P_m^2}{P_c} \tag{9-9}$$

where P_m is the measured pressure of the gas and P_c is the pressure calculated from the ideal gas law ($P_c = nRT/V$). High pressures are sometimes encountered in deep groundwater flow systems, so that this correction occasionally might be necessary.

If a reversible reaction can go in either of two directions, which way will it go? The answer to this basic question is found in the **law of mass action**, which suggests that the reaction will strive to reach equilibrium. In an aqueous mixture, both reactions are occurring simultaneously:

$$A + B \rightarrow C + D \tag{9-10a}$$

and

$$C + D \rightarrow A + B \tag{9-10b}$$

At chemical equilibrium, the two rates are equal; thus, if the mixture is not at chemical equilibrium, it will proceed in the direction which produces equilibrium. A chemical reaction may be expressed as

$$cC + dD \rightleftharpoons xX + yY \tag{9-11}$$

where capital letters represent chemical constituents and lowercase letters represent coefficients. The **equilibrium concentration** of each chemical formula is $[X]$, and the **equilibrium constant**, K, for the given reaction is

$$K = \frac{[X]^x [Y]^y}{[C]^c [D]^d} \tag{9-12}$$

where $[X]$ represents the molal concentration of the X-ion. An equilibrium constant is valid only for a specific chemical reaction. It is either experimentally determined or calculated from thermodynamic properties. In equilibrium studies, the value of the concentration of a pure liquid or solid is defined as 1.

If AgCl is dissolved in water, it will eventually saturate the water and no more will dissolve. The reaction is

$$AgCl \rightleftharpoons Ag^+ + Cl^- \tag{9-13}$$

The equilibrium reaction is given by

$$K_{sp} = \frac{[Ag^+][Cl^-]}{[AgCl]} \tag{9-14}$$

The equilibrium constant for a slightly soluble salt is termed the **solubility product**, K_{sp}. The experimentally determined value of K_{sp} for the reaction is $10^{-9.8}$. Since $[AgCl]$ is defined as 1,

$$K_{sp} = [Ag^+][Cl^-] = 10^{-9.8}$$

309

EXAMPLE PROBLEM

What is the solubility of Ag^+ at equilbrium?

The two ions have equal solubility:

$$[Ag^+] = [Cl^-] = \text{Solubility}$$

The product of $[Ag^+][Cl^-]$ is the square of the solubility of either ion; thus, the solubility of either ion is the square root of the equilibrium constant:

$$\text{Solubility} = \sqrt{K_{sp}} = \sqrt{10^{-9.8}} = 1.26 \times 10^{-5} \text{ mol}$$

The solubility of Ag+ is 1.26×10^{-5} mole per liter.

The situation is more complex if it involves a salt, such as $PbCl_2$. The reaction is

$$PbCl_2 \rightleftharpoons Pb^{++} + 2Cl^-$$

and the solubility product is given by

$$K_{sp} = \frac{[Pb^{++}][Cl^-]^2}{[PbCl_2]} \tag{9-15}$$

One mole of $PbCl_2$ yields one mole of Pb^{++} and two moles of Cl^-. To solve the equation for the solubility, X, of $PbCl_2$, use the expression

$$K_{sp} = [X][2X]^2 \tag{9-16}$$

The value of K_{sp} is $10^{-4.8}$ and, X, the solubility of $PbCl_2$, is found from

$$K = 4X^3$$
$$X = \sqrt[3]{K/4} = \sqrt[3]{0.25 \times 10^{-4.8}}$$
$$= 0.0158 \text{ mol}$$

EXAMPLE PROBLEM

One thousand grams of water will dissolve 1.0×10^{-4} mole of $PbSO_4$. Calculate K_{sp}.

$$[Pb^{++}] = [SO_4^=] = 1 \times 10^{-4} \text{ mol}$$
$$K_{sp} = [Pb^{++}][SO_4^=] = 10^{-8}$$

COMMON ION EFFECT 9.6

If the solvent contains another source for an ion also present in a salt, the **common ion effect** will reduce the solubility of the salt. This applies to any salt in equilibrium with its saturated solution. If we dissolve AgCl in two solutions, one pure water and one containing 0.1 mole of NaCl, less of the AgCl will dissolve in the solution of NaCl. The solubility of NaCl is many orders of magnitude greater than that of the AgCl, and is not affected. The total amount of the common ion in solution controls the amount of the less soluble salt that can dissolve. For example, consider the solution of AgCl in the 0.1-molal solution of NaCl. There will be X moles of AgCl and 0.1 mole of Cl^- from the NaCl. Thus, there are X moles of Ag^+ and X + 0.1 moles of Cl^-:

$$K_{sp} = [Ag^+][Cl^-] = [X][X + 0.1] = 10^{-9.8}$$

and

$$[0.1X] + [X^2] = 10^{-9.8}$$

Since [X] is small, $[X^2]$ is very small and can be ignored; hence,

$$[X] = 10^{-8.8} = 1.58 \times 10^{-9}$$

The solubility of AgCl in pure water is 1.26×10^{-5} mole, while in a 0.1-molal solution of NaCl, it is only 1.58×10^{-9} mole. In general, ground and surface waters contain ions from many sources, so that the common ion effect must be considered.

CHEMICAL ACTIVITIES 9.7

In very dilute aqueous solutions, the molal concentrations can be used to determine equilibrium and solubility. For the general case, chemical activities must be computed from the concentration before the law of mass action can be applied. This is due to the fact that electrostatic forces cause the behavior of the solutes to be nonideal.

The **chemical activity** of an ion is equal to the molal concentration times a factor known as the **activity coefficient**:

$$\alpha = \gamma m \qquad\qquad \text{(9-17)}$$

where

α is the chemical activity
m is the molal concentration
γ is the activity coefficient

In order to compute the activity coefficient of an individual ion, the **ionic strength** of the solution must first be determined. For a mixture of electrolytes in solution, the ionic strength is given by

$$I = \frac{1}{2} \sum m_i z_i^2 \qquad (9\text{-}18)$$

where

I is the ionic strength

m_i is the molality of ith ion

z_i is the charge of ith ion

The ionic strength of 0.2-molal solution of $CaCl_2$ is

$$I = \tfrac{1}{2}(m_{Ca^{++}} \times 2^2 + m_{Cl^-} \times 1^2)$$
$$= \tfrac{1}{2}(0.2 \times 2^2 + 0.4 \times 1^2) = 0.6$$

EXAMPLE PROBLEM

Compute the ionic strength of groundwater from a Cambrian-age sandstone in Neenah, Wisconsin.

Chemical Analysis (mg/ℓ) (major ions only)

Ca^{++}	Mg^{++}	HCO_3^-	$SO_4^=$
234	39	290	498

The concentrations must be converted to molality by Equation (9-1):

Chemical Analysis (molalities)

Ca^{++}	Mg^{++}	HCO_3^-	$SO_4^=$
0.00584	0.0016	0.00475	0.00518

The ionic strength is then computed using Equation (9-18):

$$I = \tfrac{1}{2}(0.00584 \times 2^2 + 0.0016 \times 2^2 + 0.00475 \times 1^2 + 0.00518 \times 2^2)$$
$$= 0.0276$$

Once the ionic strength of a solution of electrolytes is known, the activity coefficient of the individual ion can be determined from the **Debye-Hückel equation**:

$$-\log \gamma_i = \frac{Az_i^2 \sqrt{I}}{1 + a_i B \sqrt{I}} \tag{9-19}$$

where

γ_i is the activity coefficient of ionic species i

z_i is the charge of ionic species i

I is the ionic strength of the solution

A is a constant equal to 0.5085 at 25° C

B is a constant equal to 0.3281 at 25° C

a_i is the effective diameter of the ion (Table 9.1)

The Debye-Hückel equation is valid for solutions with an ionic strength of 0.1 or less (approximately 5000 milligrams per liter).

TABLE 9.1.   Values of the parameter a_i in the Debye-Hückel equation

a_i	Ion
11	Th^{+4}, Sn^{+4}
9	Al^{+3}, Fe^{+3}, Cr^{+3}, H^+
8	Mg^{+2}, Be^{+2}
6	Ca^{+2}, Cu^{+2}, Zn^{+2}, Sn^{+2}, Mn^{+2}, Fe^{+2}, Ni^{+2}, Co^{+2}, Li^+
5	$Fe(CN)_6^{-4}$, Sr^{+2}, Ba^{+2}, Cd^{+2}, Hg^{+2}, S^{-2}, Pb^{+2}, CO_3^{-2}, SO_3^{-2}, MoO_4^{-2}
4	PO_4^{-3}, $Fe(CN)_6^{-3}$, Hg_2^{+2}, SO_4^{-2}, SeO_4^{-2}, CrO_4^{-3}, HPO_4^{-2}, Na^+, HCO_3^-, $H_2PO_4^-$
3	OH^-, F^-, CNS^-, CNO^-, HS^-, ClO_4^-, K^+, Cl^-, Br^-, I^-, CN^-, NO_2^-, NO_3^-, Rb^+, Cs^+, NH_4^+, Ag^+

SOURCE: J. Kielland, "Individual Activity Coefficients of Ions in Aqueous Solutions," *American Chemical Society Journal*, 59 (1937):1676–78.

Whereas Equation (9-12) is valid only for very dilute solutions where $\gamma \approx 1$, the law of mass action, expressed in terms of activity coefficients, is valid for solutions with any ionic strength:

$$K = \frac{(\alpha_X)^x (\alpha_Y)^y}{(\alpha_C)^c (\alpha_D)^d} \tag{9-20}$$

where $cC + dD \rightleftharpoons xX + yY$ and α_X is the activity of the X-ion.

313

EXAMPLE PROBLEM

Determine γ_i and α for Ca^{++} in a solution where the molal concentration of Ca^{++} is 0.00584 and $I = 0.0276$ at 25° C. The value of a_i for Ca^{++} is 6.

The activity coefficient can be determined using Equation (9-19):

$$-\log \gamma_i = \frac{Az_i^2 \sqrt{I}}{1 + a_i B \sqrt{I}}$$

$$\log \gamma_i = -\frac{0.5085(2)^2\sqrt{0.0276}}{1 + (6)(0.3281)\sqrt{0.0276}}$$

$$= -\frac{(0.5085)(4)(0.166)}{1 + (6)(0.3281)(0.166)}$$

$$= -0.254$$

$$\gamma_i = 0.557$$

The activity of calcium is then found from Equation (9-17):

$$\alpha = \gamma m$$

$$= (0.557)(0.00584)$$

$$= 0.00325$$

The **ion activity product** (K_{iap}), which is the product of the measured activities, can be calculated for any aqueous solution in order to test for saturation. The value of K_{iap} for a mineral equilibrium reaction in a natural water may be compared with the value of K_{sp}, the solubility product of the mineral. If the value of K_{iap} is equal to or greater than K_{sp}, the natural water is saturated or supersaturated with respect to the mineral. If K_{iap} is less than K_{sp}, the solution is undersaturated with respect to the mineral, and the mineral may be actively dissolving. For the case where the mineral C is being dissolved according to the reaction $cC \rightleftharpoons xX + yY$, K_{iap} is given by

$$K_{iap} = (\alpha_X)^x(\alpha_Y)^y \qquad (9\text{-}21)$$

 IONIZATION CONSTANT OF WATER AND WEAK ACIDS

Water undergoes a dissociation into two ionic species:

$$H_2O \rightleftharpoons H^+ + OH^-$$

In reality, a hydrogen ion (H^+) cannot exist; it must be in the form H_3O^+, the hydronium ion, formed by the interaction of water with the hydrogen ion. For convenience, however, we will represent it as H^+. The equilibrium constant for water is

$$K_{eq} = \frac{\alpha_{H^+}\alpha_{OH^-}}{\alpha_{H_2O}} \tag{9-22}$$

For water that is neutral, there are exactly the same number of H^+ and OH^- radicals, 10^{-7}. The negative logarithm of the number of H^+ ions in an aqueous solution is called the pH of the solution. For all aqueous solutions, either acidic or basic, the product $\alpha_{H^+}\alpha_{OH^-}$ is always 10^{-14} (at about $25°$ C). Since a neutral solution has equal amounts of H^+ and OH^- radicals, the pH is 7. If there are more H^+ ions than OH^- ions, the solution is acidic and the pH is less than 7.0. Basic solutions have more OH^- than H^+ ions and a pH between 7 and 14.

EXAMPLE PROBLEM

What is the $[H^+]$ and $[OH^-]$ of an aqueous solution of pH 3.2?

Since pH is the negative logarithm of $[H^+]$, the value of $[H^+]$ is $10^{-3.2}$. Since the product $[H^+][OH^-] = 10^{-14}$,

$$[OH^-] = 10^{-14}/[H^+] = 10^{-14}/10^{-3.2} = 10^{-10.8}$$

It is apparent that by measuring the pH of an aqueous solution, we can obtain the numerical value of both $[H^+]$ and $[OH^-]$. If the solution is nonideal, the pH meter measures the activity of H^+, since $\alpha_{H^+}\alpha_{OH^-} = 10^{-14}$.

An acid is a substance that can add H^+ (more properly, H_3O^+) ions to aqueous solutions. Strong acids will completely disassociate in water to release H^+ ions. Since the $[H^+][OH^-]$ product is constant at a given temperature, the concentration of OH^- ions decreases. A 1-molal solution of HCl will have a pH of 0 and a $[H^+]$ of 1. A 0.01-molal solution will have a pH of 2, and a $[H^+]$ of 10^{-2}. On the other hand, a 0.01-molal solution of H_2CO_3 will have a higher pH, as it is a weak acid. In dilute aqueous solution, the H_2CO_3 is only slightly broken down into ions. Weak acids with more than one H^+ per molecule ionize in steps; for example,

$$H_2CO_3 \rightleftharpoons H^+ + HCO_3^- \tag{9-23a}$$
$$HCO_3^- \rightleftharpoons H^+ + CO_3^= \tag{9-23b}$$

The equilibrium constants at $25°$ C are

$$\frac{[H^+][HCO_3^-]}{[H_2CO_3]} = K_1 = 10^{-6.4} \text{ (First ionization constant)} \tag{9-24a}$$

and

$$\frac{[H^+][CO_3^=]}{[HCO_3^-]} = K_2 = 10^{-10.3} \text{ (Second ionization constant)} \tag{9-24b}$$

The value of K_{eq} for water varies significantly with temperature. Table 9.2 lists the equilibrium constants for the dissociation of water at temperatures between 0 and 60° C. At 0° C, a neutral solution has a pH of 7.5; at 60° C, neutrality occurs at pH 6.6.

TABLE 9.2. Equilibrium constants for dissociation of water

Temperature	K_{eq}	pH of a Neutral Solution
0	0.1139×10^{-14}	7.47
5	0.1846×10^{-14}	7.37
10	0.2920×10^{-14}	7.27
15	0.4505×10^{-14}	7.17
20	0.6809×10^{-14}	7.08
25	1.008×10^{-14}	7.00
30	1.469×10^{-14}	6.92
35	2.089×10^{-14}	6.84
40	2.919×10^{-14}	6.78
45	4.018×10^{-14}	6.70
50	5.474×10^{-14}	6.63
55	7.297×10^{-14}	6.57
60	9.614×10^{-14}	6.51

EXAMPLE PROBLEM

What is the pH of a 0.01-molal solution of H_2CO_3 at 25° C?

There are five ionic species present: H^+, OH^-, H_2CO_3, HCO_3^-, and $CO_3^=$. The total of the carbonate species, H_2CO_3, HCO_3^-, and $CO_3^=$, is 0.01 mole. The 0.01-molal solution is obtained by dissolving 0.01 mole of CO_2 in one liter of water. For most geologic applications, some assumptions can be made to simplify the problem. From the values of K_1 and K_2, we see that K_2 is 10^5 smaller than K_1, so almost all of the H^+ ions come from $H_2CO_3 \rightleftharpoons H^+ + HCO_3^-$. There will also be a very small value of $CO_3^=$, since K_2 is so small. Likewise, there will be relatively few OH^- ions since it is an acid solution.

From the dissociation reactions, we must balance the electrical charges:

$$[H^+] = [OH^-] + [HCO_3^-] + 2[CO_3^=]$$

Since OH^- and $CO_3^=$ are relatively small,

$$[H^+] \approx [HCO_3^-]$$

From the equilibrium equation,

$$\frac{[H^+][HCO_3^-]}{[H_2CO_3]} = K_1 = \frac{[H^+]^2}{[H_2CO_3]} = 10^{-6.4}$$

Since the solution has 0.01 mole CO_2, total,

$$[HCO_3^-] + [H_2CO_3] + [CO_3^=] = 0.01 \text{ mol/1000gm}$$

With a small value for $CO_3^=$, and HCO_3^- equal to H^+,

$$[H_2CO_3] + [H^+] \simeq 0.01$$

Since this is a weak acid, $[H^+]$ is very small compared with $[H_2CO_3]$, so that $[H_2CO_3] \simeq 0.01$. Putting these two results together,

$$\frac{[H^+]^2}{[H_2CO_3]} = 10^{-6.4} \quad \text{and} \quad [H_2CO_3] = 0.01$$

$$[H^+]^2 = 0.01 \times 10^{-6.4} = 0.01 \times 3.98 \times 10^{-7} = 3.98 \times 10^{-9}$$

$$[H^+] = 6.31 \times 10^{-5} = 10^{-4.2}$$

Thus, pH = 4.2. Concentrations of other ions would be

$$[HCO_3^-] = [H^+] = 10^{-4.2}$$

$$[OH^-] = \frac{10^{-14}}{[H^+]} = 10^{-9.8}$$

$$[CO_3^=] = \frac{10^{-10.3}[HCO_3^-]}{[H^+]} = 10^{-10.3}$$

These values are accurate to ± 1 percent (3).

While this type of problem can promote understanding of weak acids, in a real-world situaton there may be many other ionic species present, increasing the ionic strength. This would necessitate the use of chemical activities, and might also introduce the common ion effect.

CARBONATE EQUILIBRIUM 9.9

In hydrogeologic studies, the equilibrium of calcium carbonate in contact with natural water, either surface or groundwater, is one of the most important geochemical reactions. Neutral water exposed to CO_2 in the atmosphere will dissolve CO_2 equal to the partial pressure. The CO_2 will react with H_2O to form H_2CO_3, a weak acid, and the resulting solution will have a pH of about 5.7. Soil CO_2 from organic decomposition is another source of even more importance in groundwater studies. As calcite and dolomite are soluble in acid solution, even rainwater will dissolve carbonate rocks. Likewise, a change in pH can result in a precipitation of $CaCO_3$ from solution that was at equilibrium prior to the pH shift. The equilibrium constants for the various reactions in carbonate equilibrium are given in Table 9.3.

317

TABLE 9.3. Carbonate equilibrium constants at 25° C (4)

1.* $CaCO_3 \rightleftharpoons Ca^{++} + CO_3^=$

$$K_{CaCO_3} = \frac{\alpha_{Ca^{++}}\,\alpha_{CO_3^=}}{\alpha_{CaCO_3}} = 10^{-8.35}$$

2. $H_2CO_3 \rightleftharpoons H^+ + HCO_3^-$

$$K_{H_2CO_3} = \frac{\alpha_{H^+}\,\alpha_{HCO_3^-}}{\alpha_{H_2CO_3}} = 10^{-6.4}$$

3. $HCO_3^- \rightleftharpoons H^+ + CO_3^=$

$$K_{HCO_3^-} = \frac{\alpha_{H^+}\,\alpha_{CO_3^=}}{\alpha_{HCO_3^-}} = 10^{-10.3}$$

4.† $H_2O + CO_2 \rightleftharpoons H_2CO_3$

$$K_{CO_2} = \frac{\alpha_{H_2CO_3}}{P_{CO_2}} = 10^{-1.47}$$

*The solubility product for calcite is $10^{-8.35}$. For aragonite, it is $10^{-8.22}$.

†P_{CO_2} is the partial pressure of carbon dioxide, which for most hydrogeologic conditions is equal to the gas activity.

9.9.1 CARBONATE EQUILIBRIUM IN WATER WITH FIXED PARTIAL PRESSURE OF CO_2

Water in streams and lakes is in contact with the atmosphere, in which CO_2 is present. The gas is dissolved in the water, adding to the carbonate content of the water. The system is described by

$$H_2O + CO_2 \rightleftharpoons H_2CO_3 \tag{9-25a}$$
$$H_2CO_3 \rightleftharpoons H^+ + HCO_3^- \tag{9-25b}$$
$$HCO_3^- \rightleftharpoons H^+ + CO_3^= \tag{9-25c}$$
$$CaCO_3 \rightleftharpoons Ca^{++} + CO_3^= \tag{9-25d}$$
$$H_2O \rightleftharpoons H^+ + OH^- \tag{9-25e}$$

The system is electrically neutral, so that

$$2m_{Ca^{++}} + m_{H^+} = 2m_{CO_3^=} + m_{HCO_3^-} + m_{OH^-} \tag{9-25f}$$

The partial pressure of CO_2 in the atmosphere is $10^{-3.5}$. The value of the activity of H_2CO_3 can be computed from the activity product of CO_2:

$$K_{CO_2} = \frac{\alpha_{H_2CO_3}}{P_{CO_2}} = 10^{-1.5}$$

Therefore,

$$\alpha_{H_2CO_3} = P_{CO_2} \times 10^{-1.5} = 10^{-3.5} \times 10^{-1.5} = 10^{-5.0}$$

The activity values for the remaining ionic species can be determined in relationship to the α_{H^+}. This enables computation of the pH of a solution of calcite that is in equilibrium with atmospheric CO_2. The H_2CO_3 will dissociate into H^+ and HCO_3^-:

$$K_{H_2CO_3} = \frac{\alpha_{H^+}\alpha_{HCO_3^-}}{\alpha_{H_2CO_3}} = 10^{-6.4}$$

As we know, $\alpha_{H_2CO_3} = 10^{-5.0}$:

$$\alpha_{HCO_3^-} = (10^{-6.4} \times 10^{-5.0})/\alpha_{H^+} = 10^{-11.4}/\alpha_{H^+}$$

The HCO_3^- will further dissociate into H^+ and $CO_3^=$:

$$K_{HCO_3^-} = \frac{\alpha_{H^+}\alpha_{CO_3^=}}{\alpha_{HCO_3^-}} = 10^{-10.3}$$

This can be rearranged to give $\alpha_{CO_3^=}$:

$$\alpha_{CO_3^=} = \frac{(\alpha_{HCO_3^-})(10^{-10.3})}{\alpha_{H^+}}$$

$$= \frac{10^{-11.4}}{\alpha_{H^+}} \times \frac{10^{-10.3}}{\alpha_{H^+}} = \frac{10^{-21.7}}{(\alpha_{H^+})^2}$$

The equilibrium constant for calcite is $10^{-8.3}$. For a substance at saturation, $\alpha = 1$; therefore $\alpha_{CaCO_3} = 1$:

$$K_{CaCO_3} = \frac{\alpha_{Ca^{++}}\alpha_{CO_3^=}}{\alpha_{CaCO_3}} = 10^{-8.3}$$

The activity of Ca^{++} can be found as

$$\alpha_{Ca^{++}} = \frac{10^{-8.3}}{\alpha_{CO_3^=}} = \frac{10^{-8.3}}{10^{-21.7}/(\alpha_{H^+})^2} = 10^{13.4}(\alpha_{H^+})^2$$

By definition, $\alpha_{OH^-} = 10^{-14}/\alpha_{H^+}$.

For very dilute solutions $\gamma_i \simeq 1$; therefore $m_i = \alpha_i$, and the equation for electrical neutrality can be expressed as

$$2\alpha_{Ca^{++}} + \alpha_{H^+} = 2\alpha_{CO_3^=} + \alpha_{HCO_3^-} + \alpha_{OH^-}$$

Expressions for $\alpha_{Ca^{++}}$, $\alpha_{CO_3^=}$, $\alpha_{HCO_3^-}$, and α_{OH^-}, have been determined with respect to α_{H^+}. The equation for electrical neutrality can be determined to be

$$2[10^{13.4} \, (\alpha_{H^+})^2] + \alpha_{H^+} = 2[10^{-21.7}/(\alpha_{H^+})^2] + 10^{-11.4}/\alpha_{H^+} + 10^{-14}/\alpha_{H^+}$$

Solution of the above yields $\alpha_{H^+} = 10^{-8.4}$.

Thus, the pH of a solution open to atmospheric pH which is in equilibrium with calcite is 8.4. This is lower than the pH of a solution of calcite with no external source of CO_2, which is about 9.9. This suggests that field measurements of pH should always be made, especially if the water is from a source with no external CO_2. Exposure of such a sample to the atmosphere for more than a few minutes would result in a lowering of the pH. Groundwater samples should always be tested for pH in the field.

9.9.2 CARBONATE EQUILIBRIUM WITH EXTERNAL pH CONTROL

In most groundwater and many surface-water bodies, there are ionic species other than H_2CO_3 that influence or control the pH. The hydrogeologist often has a set of chemical analyses and a measured pH for the total solution. If pH, total calcium, total carbonate, and ionic strength are known, the ion activity product, K_{iap}, can be calculated and compared with the solubility product, K_{sp}, to determine whether or not the water is in equilibrium with calcite. If K_{iap}/K_{sp} is greater than 1, the solution is supersaturated; if less than 1, it is undersaturated; and if equal to 1, the solution is in equilibrium.

EXAMPLE PROBLEM

Determine whether the sample of groundwater represented by the following analysis is saturated with respect to calcite. The field pH is 7.15. The total dissolved solids (TDS) is 371 milligrams per liter.

	Ca^{++}	Mg^{++}	Na^+	K^+	HCO_3^-	$SO_4^=$	Cl^-	$NO_3^=$
Concentration (mg/ℓ)	82	9	25	7.6	252	17	40	38
Molality $\times 10^3$	2.046	0.16	1.087	0.194	4.13	0.177	1.128	0.613

1. Calculate the ionic strength.

$I = \frac{1}{2} (0.002046 \times 2^2 + 0.00016 \times 2^2 + 0.001087 + 0.000194$
$\quad + 0.00413 + 0.000177 \times 2^2 + 0.001127 + 0.000613 \times 2^2)$

$= 0.0193$

2. Compute γ_i for Ca^{++}, HCO_3^-, and $CO_3^=$. For water at 25° C, the Debye-Hückel equation is

$$\log \gamma_i = -\frac{0.5085z_i^2 \sqrt{I}}{1 + 0.3281a_i\sqrt{I}}$$

Values of a_i (from Table 9.1) are

$$Ca^{++} = 6 \qquad HCO_3^- = 4 \qquad CO_3^= = 5$$

$$\log \gamma_{Ca++} = -\frac{0.5085(2)^2\sqrt{0.0193}}{1 + 0.3281(6)\sqrt{0.0193}}$$

$$= -0.223$$

$$\gamma_{Ca++} = 0.598$$

$$\alpha_{Ca++} = m_{Ca++}\gamma_{Ca++} = 0.00246 \times 0.598 = 0.00147 = 10^{-2.83}$$

$$\log \gamma_{HCO_3^-} = -\frac{0.5085(1)\sqrt{0.0193}}{1 + 0.3281(4)\sqrt{0.0193}}$$

$$= -0.0597$$

$$\gamma_{HCO_3^-} = 0.871$$

$$\alpha_{HCO_3^-} = m_{HCO_3^-}\gamma_{HCO_3^-} = 0.00413 \times 0.871 = 0.003597 = 10^{-2.45}$$

$$\log \gamma_{CO_3^=} = -\frac{0.5085(2)^2\sqrt{0.0193}}{1 + 0.3281(5)\sqrt{0.0193}}$$

$$= -0.230$$

$$\gamma_{CO_3^=} = 0.589$$

The molality of $CO_3^=$ is below detectable limits, but activity can be computed since

$$\frac{\alpha_{H+}\,\alpha_{CO_3^=}}{\alpha_{HCO_3^-}} = 10^{-10.3}$$

From the pH, $\alpha_{H+} = 10^{-7.15}$

$$\alpha_{CO_3^=} = \frac{10^{-10.3} \times 10^{-2.45}}{10^{-7.15}} = 10^{-5.6}$$

3. The calculated ion activity product is

$$K_{iap} = \alpha_{Ca++}\alpha_{CO_3^=}$$
$$= 10^{-2.83} \times 10^{-5.6} = 10^{-8.43}$$

4. The value of K_{sp} for calcite is $10^{-8.35}$. By comparing the ratio K_{iap}/K_{sp}, we determine whether or not the solution is saturated:

$$K_{iap}/K_{sp} = 10^{-8.43}/10^{-8.35} = 10^{-0.08} = 0.83$$

The water is slightly undersaturated with respect to calcite.

9.10 FREE ENERGY

One of the functions in chemical thermodynamics is termed **free energy** (also known as **Gibbs free energy**). It is a measure of the driving energy of a reaction. At standard conditions, the standard Gibbs free energy of a reaction ΔG_r° is the difference between the sum of the free energy of the reactants and the sum of the free energy of the products:

$$\Delta G_r^\circ = \Sigma \Delta G_r^\circ \text{ products} - \Sigma \Delta G_r^\circ \text{ reactants} \tag{9-26}$$

It is related to the equilibrium constant by the formula

$$\Delta G_r^\circ = -RT \ln K_{sp} \tag{9-27}$$

This is a useful relationship, since the values of ΔG_r° have been measured for many reactions. The value of K_{sp} can be computed if ΔG_r° is known. At 1 atmosphere pressure, 25° C, and in base 10 logs,

$$\log K_{sp} = -\frac{\Delta G_r^\circ}{1.364} \tag{9-28}$$

9.11 OXIDATION POTENTIAL

For chemical reactions in which electrons are transfered from one ion to another (**oxidation-reduction reactions**), the oxidation potential of an aqueous solution is called the *Eh*. A transfer of electrons is an electrical current, therefore a redox equation has an electrical potential. At 25° C and 1 atmosphere pressure, the standard potential, $E°$ (in volts), has been measured for many reactions. The sign of the potential is positive if the reaction is oxidizing and negative if it is reducing. The absolute value of $E°$ is a measure of the oxidizing or reducing tendency.

The oxidation potential of a reaction is given by the **Nernst equation:**

$$Eh = E° + \frac{RT}{nF} \ln K_{sp} \tag{9-29}$$

where

R is the universal gas constant

T is the temperature (degrees Kelvin)

F is the Faraday constant

n is the number of electrons

Oxidation potential is measured with a specific ion electrode meter. A positive value indicates that the solution is oxidizing; a negative value indicates that it is chemically reducing.

If the pH and Eh of an aqueous solution are known, the stability of minerals in contact with the water may be determined. This stability relationship is best represented on an Eh-pH diagram. Water, itself, is only stable in a certain part of the Eh-pH field. Figure 9.1 shows the framework of aqueous Eh-pH fields. Water in nature at near-surface environments is usually between pH 4 and pH 9, although values that are more acid or more basic can occur.

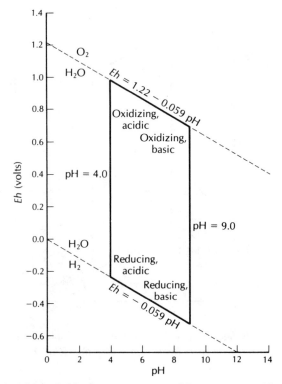

FIGURE 9.1.   The Eh-pH field where water is a stable component. The usual limits of Eh and pH for near-surface environments are also indicated by solid lines. SOURCE: K. Krauskopf, *Introduction to Geochemistry* (New York: McGraw-Hill Book Company, 1967). Used with permission.

The Eh-pH diagram can be used to show the fields of stability for both solid and dissolved ionic species. It has been used very effectively for iron. The Eh-pH diagram depends upon the concentrations of all ionic species present. For the simple ions and hydroxides of iron, the fields depend upon the molality of the iron in solution. Figure 9.2 shows the stability-field diagram for a 10^{-7}-molal solution of iron. The iron may be either in the Fe^{+++} or Fe^{++} valence state, depending upon its position in the stability field. As the iron concentration increases, the line separating the ferrous and ferric state shifts to the left. This is demonstrated by a dashed line in Figure 9.2, which represents a 1-molal iron concentration. Ferrous iron can exist as Fe^{++}, $Fe(OH)^+$, or $Fe(OH)_2$, depending upon the Eh and pH of the solution, while ferric iron can

323

be in the forms Fe^{+++}, $Fe(OH)_2^+$, and $Fe(OH)_3$. The pH at which these species change is also a function of the total amount of iron present. Procedures are available to compute an *Eh*-pH field for iron of any molality (5).

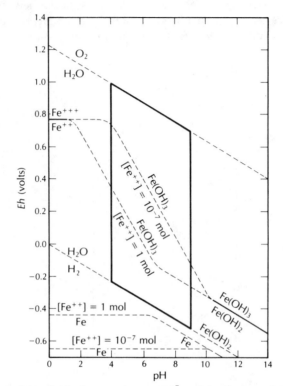

FIGURE 9.2. Stability-field diagram for a 10^{-7}-molal solution of iron. SOURCE: K. Krauskopf, *Introduction to Geochemistry* (New York: McGraw-Hill Book Company, 1967). Used with permission.

Of practical concern is the great difficulty in measuring the *Eh* of groundwater under field conditions. Even for spring water, the measured *Eh* has been shown to be too great for the amount of ferrous iron in the sample (5). With very careful work, oxygen can be excluded from the sampling procedure and field *Eh* measured (6).

High *Eh* is generally the direct result of dissolved oxygen in the water. For deep groundwater systems, the *Eh* is usually sufficiently low that, for a pH of less than about 8, iron is present as the soluble Fe^{++} ion. Near a recharge zone, the groundwater may have sufficient dissolved oxygen to elevate the *Eh*. As the water travels through the aquifer, the oxygen is chemically reduced by contact with reducing species, and the *Eh* is lowered. The oxygen can react with the small amount of ferrous iron to form ferric hydroxide, $Fe(OH)_3$. Interestingly, the ferric hydroxide thus formed may be colloidal, and can move through the aquifer with the groundwater. In the *Eh*-pH range where Fe^{++} exists, large amounts of dissolved iron can be present.

The total amount of dissolved iron that may be present in an aqueous solution is a function of the Eh and pH, as is related in Figure 9.3. For example, with an Eh of 0.20 volt and a pH of 6, the activity of dissolved iron would be 14 parts per million. The activity is, of course, less than the molal concentration. At an ionic strength of 0.01, the activity coefficient of Fe^{++} is 0.68, and the molal concentration of the given solution as computed by Equation (9-17) is 20.6.

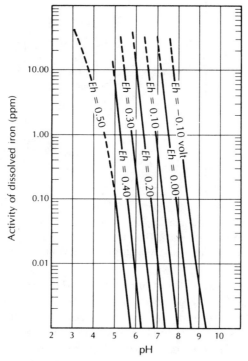

FIGURE 9.3.   Relation of total activity of iron in water to Eh and pH. SOURCE: J. D. Hem and W. H. Cropper, U.S. Geological Survey Water Supply Paper 1459-A, 1959.

Natural waters contain many ionic species. Again, using iron as an example, an Eh-pH diagram can be used to show the stable iron minerals in a mixed aqueous solution with iron, sulfur, and carbonate present. The given activities are: iron, 10^{-4} mole, sulfur, 10^{-6} mole, and carbonate, 1 mole (Figure 9.4). Stable species include Fe^{+++}, Fe_2O_3, Fe_3O_4, Fe^{++}, $FeCO_3$, and FeS_2. In reducing environments, pyrite, siderite, or magnetite may be the stable mineral, while under oxidizing conditions, hemitite is stable. For a thorough discussion of Eh-pH, diagram references 3, 4, 5, 7, and 8 are suggested.

Because of the difficulty of measuring *in situ* Eh in groundwater, there is not a great deal of information on the Eh-pH range of natural groundwaters. Some data do suggest that a range of Eh from -0.2 to $+0.7$ volt can occur (6, 8). In one study, the measured Eh ranged from -0.04 to $+0.7$ volt in

325

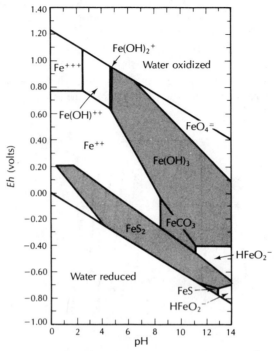

FIGURE 9.4. Stable iron minerals in an aqueous solution of 10^{-4}-molal iron, 10^{-6}-molal sulfur and 1-molal carbonate. The pressure is 1 atmosphere and the temperature 25° C. SOURCE: J. D. Hem, U.S. Geological Survey Water Supply Paper 1473, 1970.

groundwater found in a single county in Maryland (6). The groundwater was in a coastal plain aquifer with a regional flow pattern toward the sea. The highest oxidation potentials were in shallow groundwater of recharge areas. *Eh* was found to decrease with increasing length of flow from the aquifer recharge area. As might be expected, an inverse relationship was found between the oxidation potential and the amount of iron in solution. In the same study, field pH was found to range from 3.20 to 7.79, although, in general, higher and lower values are possible. For example, water draining from mineral deposits or mines can have a pH as low as 2 (8).

Surface waters are usually oxidizing, although low *Eh* can occur in the anerobic depths of some lakes. The pH of surface waters typically is in the range of 4 to 10 (8).

9.12 SURFACE PHENOMENA

The ionic species present in aqueous solution can react with the surfaces of solids of the rock and soil particles. The surface area of a soil or sediment increases with decreasing grain size, so that the clays are the most reactive.

There are two separate processes: **ion exchange**, in which an ion in the mineral lattice is replaced by one of the ions in the aqueous solution; and **adsorption**, in which the solid surface attracts and retains a layer of ions from the solution. These processes may be important in determining the chemical composition of natural groundwaters. In addition, potential contaminants from such sources as septic tanks, sanitary landfills, and wastewater may be attenuated, at least in part, by passage through soil (9).

9.12.1 ADSORPTION

The surfaces of solids, especially clays, have an electrical charge due to either isomorphous replacement, broken bonds, or lattice defects (10). The electrical charge is imbalanced, and may be satisfied by absorbing a charged ion. The absorption may be relatively weak, essentially a physical process caused by van der Waals forces. It may be stronger if chemical bonding occurs between the surface and the ion. Clays tend to be strong adsorbers, since they have both a high surface area per unit volume and significant electrical charges at the surface.

Most clay minerals have an excess of imbalanced negative charges in the crystal lattice. Adsorptive processes in soils thus favor the adsorption of cations. Divalent cations are usually more strongly adsorbed than monovalent ions. Some positively charged sites exist, but they are not as abundant as negative sites. In addition, some common negatively charged ions, such as HCO_3^-, $SO_4^=$, and NO_3^-, are too large to be effectively adsorbed. Chloride ions are larger than the common cations (1).

The adsorptive capacity of specific soils or sediments is usually determined experimentally. It is a function of minerology, particle size, ambient temperature, soil moisture, tension, pH, Eh, and activity of the ion (11–15). Equal weights of air-dried soil are shaken in solutions of varying activities of a particular ion. The amount of the ion absorbed will be proportional to the activity. Analysis of the aqueous solution in equilibrium with the soil yields an equilibrium concentration, C. Knowing the initial concentration, the weight of ion removed, X, per unit weight of soil, m, can be determined.

The **Langmuir adsorption isotherm** (16) is given by

$$C/X/m = \frac{1}{\beta_1 \beta_2} + \frac{C}{\beta_2} \tag{9-30}$$

where

$\quad$ C $\quad$ is the equilibrium concentration of the ion in contact with the soil

$\quad$ X/m $\quad$ is the amount of the ion absorbed per unit weight of soil

$\quad$ β_1 $\quad$ is an adsorption constant related to the binding energy

$\quad$ β_2 $\quad$ is the adsorption maximum for the soil

327

A plot of $C/X/m$ as a function of C is made on rectilinear scales. The data points will fall on a straight line. Some experiments yield two straight-line segments—one at lower concentrations of ion and one at higher concentrations with a lower slope (17). In some cases, the soil may have adsorbed some of the ion under natural conditions prior to the laboratory test. If this is the case, a correction must be made (18).

The maximum ion adsorption, β_2, is the reciprocal of the slope of the straight line. The binding energy constant, β_1, is the slope of the line divided by the intercept. The experimental procedure usually tests for only a single ion at a time. Natural waters are more complex, and field reactions may differ from those determined by laboratory study. The Langmuir adsorption isotherm can be used for both anions and cations.

EXAMPLE PROBLEM

A calcareous glacial outwash sediment was present at the proposed site for an artificial recharge basin for wastewater (9). The phosphorus-removal capacity of the soil was determined by laboratory adsorption studies. Equal weights of soil were shaken in various concentrations of disodium phosphate. The soil already had 0.016 milligram of phosphorus per gram sorbed prior to the test, and this was added to the value of X/m sorbed during the test to determine total X/m in equilibrium with the solution. The equilibrium concentration of the solute is C.

A plot of $C/X/m$ as a function of C is given in Figure 9.5. There are two straight-line segments, indicating that one type of adsorption is taking place at low activities of phosphorus and another type of adsorption, with a higher bonding energy, is occurring at greater concentrations. The slope of the straight line at lower concentrations is $\dfrac{20 \text{ mg/}\ell\text{/mg/gm}}{1 \text{ mg/}\ell}$, so the reciprocal is 0.05 milligram of phosphorus per gram of soil, which is the adsorption maximum (β_2). For higher concentrations, the adsorption maximum is 0.16 milligram of phosphorus per gram of soil.

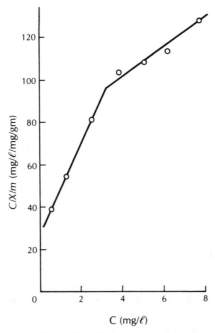

FIGURE 9.5. Adsorption isotherm for phosphorus sorbed on calcareous glacial outwash. SOURCE: C. W. Fetter, Jr., *Ground Water*, 15 (1977):365–71.

9.12.2 ION EXCHANGE

Under certain conditions, the ions attracted to a solid surface may be exchanged for other ions in aqueous solution. This process is known as ion exchange. Both cation exchange and anion exchange can occur, but in some natural soils cation exchange is the dominant process. The presence of exchange sites is a function of the same general conditions affecting adsorption sites. The ion-exchange process can be conceptualized as the preferential adsorption of selective ions with concommitant loss of other ions. Ion-exchange sites are found primarily on clays and soil organic materials (19), although all soils and sediments have some ion-exchange capacity.

The ion-exchange reactions of different soils must be studied individually in the laboratory. Results are reported in terms of milliequivalents per 100 grams of soil. In one study of exchange capacities for stream sediments (20), the following results were reported:

TABLE 9.4. Ion-exchange values for stream sediments (20)

Size Fraction	Ion Exchange (meq/100 grams soil)
4 microns	14–65
4–61 microns	4–30
61–1000 microns	0.3–13

A general ordering of cation exchangeability for common ions in groundwater is

$$Na^+ > K^+ > Mg^{++} > Ca^{++}$$

The divalent ions are more strongly bonded and tend to replace monovalent ions. However, it is a reversible reaction and, at high activities, the monovalent ions can replace divalent ions. This is the concept behind the home water softener. The divalent Ca^{++} and Mg^{++} ions replace the monovalent Na^+ ions on the exchange media. The exchange medium is regenerated when a brine solution with very high Na^+ activity is forced through the softener. The Na^+ replaces the Ca^{++} and Mg^{++} at the exchange sites. Ion-exchange capacities of organic colloids and clays can also remove heavy metal cations, and thus provide some protection to groundwater supplies (21); but enough cases of groundwater pollution from heavy metals have been documented to demonstrate that such protection is limited.

One particularly well-studied ion-exchange reaction is the replacement of calcium in the soil with sodium. If water used for irrigation is high in sodium and low in calcium, the cation-exchange complex may become saturated with sodium. This can destroy the soil structure due to dispersion of the clay particles. A simple method of evaluating the danger of high-sodium water is the **sodium-adsorption ratio, SAR** (23).

$$SAR = \frac{(Na^+)}{\left[\dfrac{(Ca^{++}) + (Mg^{++})}{2}\right]^{0.5}} \tag{9-31}$$

A low SAR (2 to 10) indicates little danger from sodium; medium hazards are between 7 and 18, high hazards between 11 and 26, and very high hazards above that. The lower the ionic strength of the solution, the greater the sodium hazard for a given SAR. Anions present in the water can affect calcium replacement (24, 25).

If the ion-exchange process is controlled by a reversible equilibrium process, the following equation applies:

$$b[\bar{A}] + a[B] \rightleftharpoons a[\bar{B}] + b[A] \tag{9-32}$$

where A and B are chemically exchanging species, A with a valence of a and B with a valence of b.

330

[X] is the concentration of the solute in terms of mass per unit volume of liquid.

[$\bar{X}$] is the amount of the solute adsorbed by ion exchange on a unit mass of sediment or soil.

When the exchanged ions are in equilibrium, the concentration of products and reactants at equilibrium is described by the ion-exchange selectivity coefficient, K_s:

$$K_s = \frac{[\bar{B}]^a [A]^b}{[\bar{A}]^b [B]^a}$$
(9-33)

The **cation-exchange capacity** (CEC) is defined as $[\bar{A}] + [\bar{B}]$ and the total solute concentration, C_0, is equal to $[A] + [B]$. When the concentration of one of the exchanging ions is very low, the adsorbed phase of the other (dominant) ion is approximately equal to CEC, and the total solute concentration is almost entirely that of the dominant ion. Equation (9-33) can be rewritten under these conditions, if B is the major species, as

$$K_s = \frac{[\bar{B}]^a C_0^b}{CEC^b [B]^a}$$
(9-34)

The **ion-exchange distribution coefficient**, K_d, is the ratio of the adsorbed species concentration to the concentration of the solute:

$$K_d = \frac{[\bar{B}]}{[B]}$$
(9-35)

COLLECTION OF WATER SAMPLES

Collecting water samples for chemical analysis is sometimes done in a rather cavalier manner. A bottle is grabbed from the shelf, rinsed under the tap, and the water sample taken at the edge of the stream, usually at a place the collector can reach and still maintain dry feet. Such a sample may well be worthless for purposes of obtaining meaningful data. A water-chemistry sampling program must be carefully planned.

Ideally, containers for sample collection should be new. Plastic bottles are generally preferable to glass ones. The bottles should be labeled with a waterproof marker. If they have been previously used, they should first be washed with 10 percent HCl, rinsed with tap water, and then rinsed again with distilled water. A detergent or similar product should never be used, as such products can contaminate the container. In the field, the container should be rinsed twice in the water to be collected. Any distilled water left from a prior rinsing would dilute the natural water being sampled. Bottles used for bacteriological or viral testing should be sterilized and kept capped.

Measurements of pH, *Eh*, temperature, and electrical conductivity are best made in the field. The first three parameters are of less (if of any) value if measured in the laboratory. Under many pH-*Eh* conditions, it is also necessary to measure such constituents as bicarbonate or soluble iron immediately. Exposure to atmospheric oxygen and carbon dioxide can affect the water chemistry enough to initiate changes in the equilibrium of these dissolved species. Acidity and alkalinity should also be determined in the field.

Samples for certain types of analysis are treated upon collection to stabilize them (26). Field filtration through 0.45-micrometer membrane filter is recommended for samples to be analyzed for boron, chloride, fluoride, hardness, lithium, nitrate and nitrite nitrogen, dissolved phosphorus, potassium, selenium, silica, sodium, dissolved solids, and sulfate. An aliquot of the field sample which has been both microfiltered and acidified to a pH under 3.0 is used for analysis of aluminum, arsenic, barium, cadmium, calcium, chromium, cobalt, copper, total iron, lead, lithium, magnesium, manganese, molybdenum, nickel, potassium, silver, sodium, strontium, vanadium, and zinc. An unfiltered aliquot, but one that is well mixed, is used for analysis of ammonia nitrogen, organic nitrogen, chemical oxygen demand, biochemical oxygen demand, cyanide, total phosphorus, suspended solids, volatile solids, and turbidity. An unfiltered sample, but one in which the sediments have been allowed to settle, is used for analyzing acidity, alkalinity, dissolved oxygen, pH, *Eh*, and color.

It is good laboratory practice to analyze samples as soon as possible after collection. Some species, such as dissolved phosphorus, nitrite nitrogen, ammonia nitrogen, and organic nitrogen require analysis within twenty-four hours of collection. Furthermore, samples should be kept at 0.2° C as soon as collected to minimize biological activity. This also applies for bacteriological and viral samples and for those used in biochemical oxygen-demand testing.

9.14 DESIGNING WATER-SAMPLING PROGRAMS

9.14.1 GROUNDWATER

In a groundwater-sampling program, as in any experimental program, the intended purpose will dictate the program's design. If there is a cogent purpose, such as a health threat, all wells in an area might be tested. Such might be the case if a chemical pollutant were accidentally spilled.

In many instances, only a representative sample of the wells and springs in an area will be sampled. The nature of the geological formations tapped by the wells almost always must be known. Some wells might be eliminated from a sampling program for lack of such information. Likewise, if a well draws water from more than one aquifer, it is often not usable. Since most groundwater-sampling programs involve a study of the geographical distribution of ions in a single aquifer, only wells drawing water from that aquifer alone are tested. A good geographical distribution of wells is necessary. In studies of chemical geohydrology, both wells located in the recharge areas and those located in the discharge areas are critical.

If several aquifers in an area are to be studied, then ideal experimental conditions demand a series of wells screened in different aquifers. If a test-well drilling program is included in the study, temporary test wells can be used to obtain water samples from different depths. A test hole is drilled and the drill column and bit are pulled from the hole at progressively greater depths. A temporary well casing and screen are lowered to the bottom of the hole and an inflatable packer (seal) or clay plug is used to seal off the upper part of the drill hole. The temporary well is pumped, usually until the water is clear. Replicate samples of water from the aquifer are collected, whereupon the packer and temporary screen and casing are removed and the well deepened into the next aquifer. It has been shown that water samples from such temporary test wells are very similar to those from high-capacity municipal wells later constructed in the same aquifer (27).

The field-sampling procedure should be planned to yield water that accurately reflects the field conditions. For example, wells designed for collection of shallow groundwater should be made of plastic or fiberglass casing. Water collected from wells with metal casing should be pumped for a long enough time prior to sample collection so that the water standing in the casing is removed. This will lessen the chances of the water sample being affected by contact with the metal of the well casing.

Groundwater quality is relatively constant with time because of the low rate of groundwater flow. In steady-state groundwater domains, in an absence of aquifer contamination, the water chemistry remains constant at a point. Water chemistry can vary with time under nonequilibrium groundwater conditions if the flow field is altered. Reversals of the flow direction near a well field could cause abrupt changes in the water chemistry. Such abrupt changes could also be caused by groundwater contamination. For example, when a railway tank car derailed and spilled 34,000 liters of phenol in Walworth County, Wisconsin, nearby wells tested one week later were found to yield water contaminated by phenol (28).

For deep aquifers, one-time sampling of water should be sufficient for a study. This is not true of shallow aquifers and the recharge areas of deep aquifers. It has been shown that the water chemistry of shallow water, in both recharge and discharge areas, can undergo seasonal variations (29). For shallow groundwater, seasonal sampling, including both dry and wet periods, is necessary.

333

9.14.2 SURFACE WATER

The chemical composition of surface water varies with time to a much greater extent than does that of groundwater. Water in streams comes from overland runoff and baseflow in varying proportions. As the chemistry of each of these sources may be vastly different, the composite stream quality can vary simply by a change in proportion of each. The chemistry of the overland-runoff component will also vary with meteorological and seasonal conditions of the drainage basin. Land-use patterns of the drainage basin have a direct impact on water chemistry. Biotic factors of surface water can create diurnal variations of water chemistry. Some important parameters can be mobile as either dissolved or suspended particles. Sediment moving in a stream may also carry with it sorbed ions of ecological importance. As the sampling of waters in lakes and reservoirs is a highly specialized field, our discussion will concentrate on stream sampling; however, many of the same principles apply to sampling the outlets of lakes and reservoirs.

If a sample from a particular reach of a river is being collected, it must not be assumed that the water chemistry will be the same at each part of the cross section. Some streams contain plumes of water, either from tributaries or from wastewater discharges. The plumes may be chemically, biologically, or thermally different from the rest of the stream. If only one sampling point is used, either in or out of a plume, results could be erroneous. When a sampling program is being established, it is necessary to take several samples across a stream, and at different depths, to determine whether the water is homogeneous. Stream reaches that are chemically homogeneous are more likely to occur in turbulent sections far downstream from major tributaries or wastewater sources. The fact that springs may create nonhomogeneous sections in streams should not be overlooked.

As water chemistry can vary diurnally, it is good practice to collect several samples over a period of twenty-four hours. This may be done by hand or by using an automatic sampler. Depending upon the purpose of the sampling program, the samples collected over the twenty-four hours may be either analyzed separately or combined for a composite sample. A biochemically oriented study would suggest the former; a geochemical study, the latter.

An annual sampling of stream water has little meaning. There can be such variation that a regular program of sampling over at least a one-year period is indicated. Samples should be collected under both low-flow and runoff conditions. The sampling program should be planned so that the frequency of sampling is proportional to the stream discharge. If samples are to be collected ten times over a one-year period, each sample should represent about 10 percent of the annual discharge of the stream. If stream-discharge measurements are made at the same time as samples are collected, the mass flux of each parameter may be determined. Otherwise, only the concentration is known.

For some pollution-monitoring studies, continuous sampling is required. This can be accomplished by a proportional sampler or by continuous-reading, analytical sensors.

334

The geographic distribution of sampling stations is a function of the project budget and purpose. To determine the water chemistry of a proposed water-supply intake, only one station may be necessary. If the geochemistry of a drainage basin is being studied, then the sampling stations should be distributed across the basin. For example, they could be placed on major tributaries and areas where bedrock lithologies or soil types change.

AGE DATING OF GROUNDWATER

Radiocarbon dating methods can be applied to obtain the age of groundwater. Carbon exists in several naturally occurring isotopes, ^{12}C, ^{13}C, and ^{14}C. Carbon 14 is formed in the atmosphere by the bombardment of ^{14}N by cosmic radiation (30). The ^{14}C forms CO_2, so that the atmospheric CO_2 has a constant radioactivity due to modern ^{14}C. If the CO_2 is incorporated into a form in which it is isolated from modern ^{14}C, age determinations can be made from the ^{14}C radioactivity as a percent of the original. The half-life of ^{14}C is 5570 years, so that if one-fourth of the original activity is present, two half-lives, or 11,140 years, have elapsed. When precipitation soaks into the ground, it is saturated with respect to CO_2, with a known ^{14}C activity. Once the water has entered the soil, additional carbon may come from soil CO_2 and the solution of carbonate minerals. The modern carbon is diluted by the inactive carbon from carbonate minerals. The raw dates obtained must be adjusted for this dilution.

If A is the measured ^{14}C radioactivity, and A_0 is the activity at the time the sample was isolated, then the following equation may be used:

$$A = QA_0 2^{-t/T} \qquad (9\text{-}36)$$

where

t   is the age

T   is the half-life of ^{14}C

Q   is an adjustment factor to account for dilution by **dead carbon*** (31).

The equation requires an estimation of initial value, A_0, and the adjustment factor, Q.

The value of A_0 will depend on the carbonate equilibria established under an open system in which the groundwater was exposed to an infinite reservoir of CO_2. This occurs in nature in the soil zone and in shallow groundwater. When the groundwater system becomes closed with respect to CO_2, then any added carbon would be only from carbonate rocks; i.e., dead carbon. The value of Q is generally in the range of 0.5 to 0.9. Carbon 14 dates of groundwater thus tend to be somewhat less than raw dates as a result of the

---

*Dead carbon is carbon from a source old enough for any ^{14}C to have decayed below measurable limits.

dilution by dead carbon from carbonate minerals. The determination of A_0 and Q is somewhat complex, and several different methods are available (31, 32).

Tritium is sometimes used to date young groundwater. It is a radio-isotope of hydrogen, ^3H, and has a half-life of 12.4 years. It occurs as a natural isotope in the atmosphere from cosmic radiation. The amount of atmospheric tritium was greatly increased by atmospheric testing of nuclear weapons, starting about 1954. If groundwater is free of tritium, it is at least 50 to 90 years old. Significant tritium activity indicates fairly young groundwater.

9.16 PRESENTATION OF RESULTS OF CHEMICAL ANALYSES

Tables of data are the most common form in which the results of an analysis of water chemistry are reported. The data can be expressed in milligrams per liter (mg/ℓ), milliequalents per liter (meq/ℓ), or millimoles per liter. For many purposes, the data may be also displayed in graphical form. There have been a great number of graphical forms proposed, including bar graphs, vectors, pie diagrams, and nomographs, to name but a few (1). One of the more widely used is the **trilinear diagram**. Inasmuch as this method also forms the basis for a common classification scheme for natural waters, it will be described in detail.

The major ionic species in most natural waters are Na^+, K^+, Ca^{++}, Mg^{++}, Cl^-, $CO_3^=$, HCO_3^-, and $SO_4^=$. A trilinear diagram can show the percentage composition of three ions. By grouping Ca^{++} and Mg^{++} together, the major cations can be displayed on one trilinear diagram. Likewise, if $CO_3^=$ and HCO_3^- are grouped, there are also three groups of the major anions. Figure 9.6

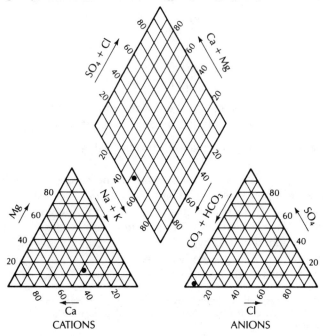

FIGURE 9.6. Trilinear diagram of the type used to display the results of water-chemistry studies.

shows the form of a trilinear diagram that is commonly used in water-chemistry studies (33). Analyses are plotted on the basis of the percent of each cation (or anion).

Each apex of a triangle represents a 100 percent concentration of one of the three constituents. If a sample has two constituent groups present, then the point representing the percentage of each would be plotted on the line between the apexes for those two groups. If all three constituent groups are present, the analyses would fall in the interior of the field. The diamond-shaped field between the two triangles is used to represent the composition of water with respect to both cations and anions. The cation point is projected parallel to the magnesium axis onto the center field, and the anion point is projected parallel to the sulfate axis. The intersection of the two points is then plotted.

EXAMPLE PROBLEM

Plot the results of the following analysis on a trilinear diagram:

	Ca^{++}	Mg^{++}	Na^+	K^+	HCO_3^-	$CO_3^=$	$SO_4^=$	Cl^-
mg/ℓ	23	4.7	35	4.7	171	0	1.0	9.5
meq/ℓ	1.15	0.39	1.52	0.12	2.80		0.02	0.27

The first step is to find the percent of each cation and anion group as a percentage of the total:

Cations	meq/ℓ	% of Total	Anions	meq/ℓ	% of Total
Ca^{++}	1.15	36	Cl^-	0.27	9
Mg^{++}	0.39	12	SO_4^-	0.02	1
$Na^+ + K^+$	1.64	52	$CO_3^= + HCO_3^-$	2.80	90
TOTAL	3.18		TOTAL	3.09	

NOTE: Due to analytical error and unreported minor constituents, the total equivalents of anions and cations do not exactly match. Theoretically, the total equivalent weight of the anions should be exactly that of the cations, as equivalent weights are based on the amount of the ion which would combine with 0_2.

Referring back to Figure 9.6, the points for both the cations and anions are plotted on the appropriate triangle diagrams. The positions of the points are projected parallel to the magnesium and sulfate axes, respectively, until they intersect in the center field.

As water flows through an aquifer it assumes a diagnostic chemical composition as a result of interaction with the lithologic framework. The term **hydrochemical facies** is used to describe the bodies of groundwater in an aquifer which differ in their chemical composition. The facies are a function of the lithology, solution kinetics, and flow patterns of the aquifer (34, 35). Hydrochemical facies can be classified on the basis of the dominant ions in the facies by means of the trilinear diagram (Figure 9.7).

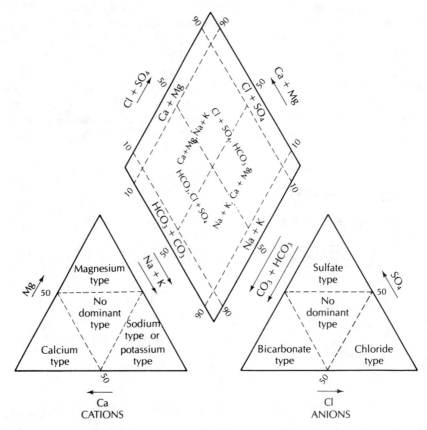

FIGURE 9.7. Hydrogeochemical classification system for natural waters using the trilinear diagram.

9.17   SOLUTE MOVEMENT IN GROUNDWATER

As water flows through the ground, solute is carried along with it at a rate equal to the actual groundwater flow velocity. This process is known as **advection**. In a porous medium the flow paths are tortuous, so that different flow paths have various lengths. Figure 9.8 illustrates a granular aquifer on a microscopic scale. Although the shortest distance from A to B is fixed, it is apparent that some flow paths are longer than others. Solute transported along a shorter path would arrive at the end point sooner than solute following a longer path. The result is **hydrodynamic dispersion**.

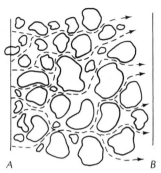

FIGURE 9.8. Flow paths in a porous medium which cause hydrodynamic dispersion.

We can establish a simple experiment to demonstrate hydrodynamic dispersion by means of a tube filled with sand. Distilled water is flowing through the tube at a steady rate. We then change the fluid to a 1 percent salt solution and begin to monitor the effluent water for chloride. The effluent has zero chloride initially, as the distilled water is still flushing from the tube. The solute does not first break through at the original concentration. Rather, a small amount is detected at first, and then this amount gradually increases until the original concentration is reached (Figure 9.9). The first salt to emerge followed the shortest flow paths. However, it was diluted by distilled water emerging at the same time from longer flow paths. This dilution continues in decreasing amounts until salt water in the longest flow paths reaches the end point, and all of the transported water is a 1 percent salt solution.

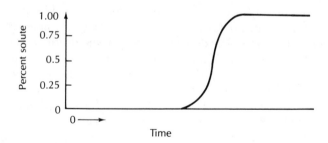

FIGURE 9.9. The concentration of an invading solute as a function of time as it is affected by hydrodynamic dispersion.

If the solute does not chemically react with other constituents in the groundwater, and is not adsorbed on the particle surfaces, it can be described by the one-dimensional hydrodynamic dispersion equation for homogeneous, isotropic aquifers (36, 37):

$$D_L \frac{\delta^2 C}{\delta x^2} - V_x \frac{\delta C}{\delta x} = \frac{\delta C}{\delta t} \tag{9-37}$$

339

where

D_L is the longitudinal dispersion coefficient

C is the solute concentration

V_x is the average groundwater velocity in the x-direction

t is the time since start of solute invasion

If the original solute concentration is C_0 at a point in the flow path L distance from the point of solute invasion, the concentration of solute C at time t is given by (38):

$$C = \frac{C_0}{2}\left[\mathrm{erfc}\left(\frac{L - V_x t}{2\sqrt{D_L t}}\right) + \exp\left(\frac{V_x}{D_L}\right)\mathrm{erfc}\left(\frac{L + V_x t}{2\sqrt{D_L t}}\right)\right] \qquad \textbf{(9-38)}$$

where erfc is the complimentary error function (see Table 8.3, page 277).

The value of D_L is determined experimentally for a given soil or aquifer. The travel-time concentration curve for an induced tracer is determined, and the value of D_L obtained by rearrangement of Equation (9-38). The dispersion coefficient is in units of square centimeters per second, with values in the general range of 0.0001 to 0.01 square centimeter per second. The value increases with increasing groundwater velocity. It also varies with the size and configuration of the pore materials. The determination of D_L must be made for a given aquifer over a range of velocities.

In natural flow situations, there can be dispersion both longitudinally and laterally. A solute being introduced into a groundwater flow field from a continuous point source results in a plume extending parallel to the direction of flow. The solute is less concentrated at the margins of the plume, but concentration increases toward the source. An example of this would be wastewater from a cesspool. Figure 9.10A shows a contaminant plume from a continuous point source. The diagram represents the distribution of solute at a given time. Since the value of the dispersion increases as groundwater velocity increases, the plume, or slug, will be larger with more rapid groundwater flow.

If there is a one-time point source, such as a contaminant spill, the solute will move by advection as a slug with the flow of groundwater. Due to hydrodynamic dispersion, the slug will expand in three dimensions to fill a larger volume, but with a lower solute concentration. Figure 9.10B illustrates the movement of a contaminant slug, showing the relative size initially (t_0), and then the growth of the slug at three separate times (t_1, t_2, and t_3).

The rate of advance of a contaminant front is retarded if there is a reaction between the solute and other groundwater constituents, or if adsorption of the solute occurs. The plume or slug of solute will spread more slowly, and the concentrations will be lower than those of an equivalent nonreactive solute. Hydrodynamic dispersion affects all solutes equally, while adsorption and reaction can affect various solutes at different rates. Thus, a contaminant with a number of solutes can have each moving at a different rate due to attenuation processes.

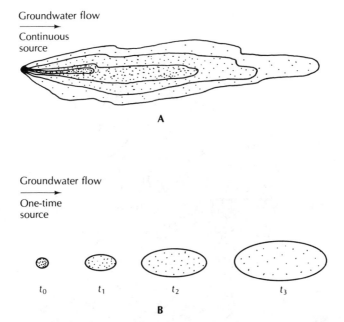

Groundwater flow

Continuous
source

A

Groundwater flow

One-time
source

t_0 t_1 t_2 t_3

B

FIGURE 9.10. **A.** The development of a contamination plume from a continuous point source; **B.** The travel of a contaminant slug from a one-time point source.

CASE STUDY: CHEMICAL GEOHYDROLOGY OF THE FLORIDAN AQUIFER SYSTEM

The regional geohydrology of the Floridan aquifer was discussed in Section 6.7. The chemical geohydrology of this aquifer system is also well known. We will consider the chemical changes that take place as the water flows from the central recharge area at Polk City to the south. Data from five wells form the basis for the hydrochemical cross sections (39, 40, 41). Figure 9.11 indicates the location of the wells on the potentiometric map of central Florida, with Polk City located at the southern edge of the recharge area. The five wells all tap the Floridan aquifer and lie approximately along the same flow path. Chemical analyses of water from the wells are given in Table 9.4. As water travels down the flow path, it increases in total dissolved solids, from 138 to 726 milligrams per liter. All ions except bicarbonate show a progressive increase along the flow path (Figure 9.12). Computation of ion-activity products from the analyses shows that both dolomite and calcite saturation increases along the flow path. For the most part, the K_{iap}/K_{sp} of the water is greater than 1, indicating supersaturation. Hydrochemical cross sections along the flow path are shown in Figure 9.12. A trilinear plot (Figure 9.13) of the well analyses indicates that the chemical composition of the water is

341

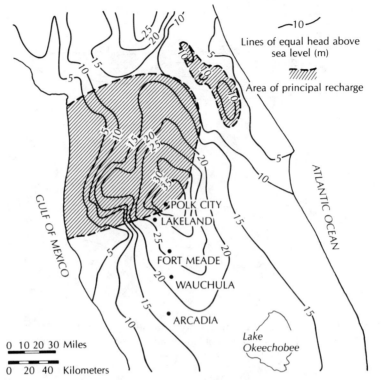

FIGURE 9.11. Potentiometric surface of the Floridan aquifer in central Florida. SOURCE: Adapted from L. N. Plummer, *Water Resources Research*, 13 (1977): 801–12.

TABLE 9.4. Chemical analysis of water from Floridan aquifer in central Florida

Well	Location	Temp. °C	Field pH	Milligrams per Liter								
				SiO₂	Ca	Mg	Na	K	HCO₃	SO₄	Cl	TDS
1	Polk City	23.8	8.0	12	34	5.6	3.2	0.5	124	2.4	4.5	138
2W	Lakeland	26.3	7.62	18	54	14	6.9	1.0	253	3.6	8.5	238
2S	Ft. Meade	26.6	7.75	16	58	17	6.1	0.7	163	71	9.0	272
3S	Wauchula	25.4	7.69	18	66	29	8.3	2.0	168	155	10	392
4S	Arcadia	26.3	7.44	31	106	60	21	3.7	206	344	28	726

SOURCE: Data from W. Back and B. B. Hanshaw, *Journal of Hydrology*, 10 (1970):330–68.

shifting along the flow path. This is due to an increase in the Mg^{++}/Ca^{++} and the $SO_4^{=}/HCO_3^{-}$ ratios with increasing distance from the recharge area. The change in these ratios is due to the solution of gypsum ($CaSO_4 \cdot 2H_2O$) and dolomite ($Ca(Mg)CO_3$) along the flow paths. It is important to note that these reactions involve only the solution of minerals in fresh water. If this water mixes with sea water,

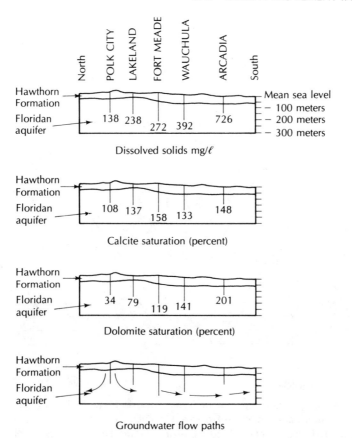

FIGURE 9.12. Hydrogeochemical cross sections from Polk City through Arcadia through the Floridan aquifer. SOURCES: Data from Back, Cherry, and Hanshaw, 1966; Back and Hanshaw, 1970; and Plummer, 1977 (References 39, 40, and 41, respectively).

water chemistry would rapidly change and be dominated by sodium and chloride, which both are minor constituents in fresh water.

Calculation of the age of groundwater from the Floridan aquifer on the basis of ^{14}C activity is complex due to the solution of dead carbon from carbonate. Complicating the dating is the fact that the flow path from Polk City to Ft. Meade is partially open to soil CO_2, and then it is closed to CO_2 from Ft. Meade to Wauchula. However, from Wauchula to Arcadia, it is again open to CO_2, the source being the oxidation of lignite from sulfate reduction (41). Several authors have dealt with the ^{14}C dating of this water (30, 40, 41) and have cited dates ranging from 20,600 to 24,100 to 36,000 years B.P. Based on the age-dating and the reaction coefficients, apparent rates of solution for several mineral species have been calculated (41).

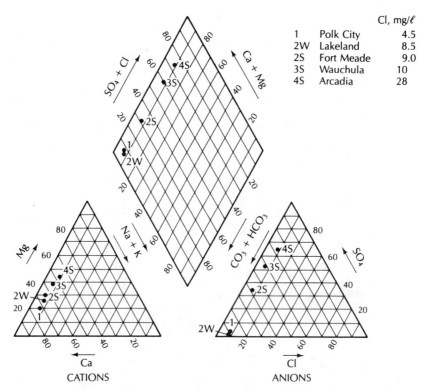

		Cl, mg/ℓ
1	Polk City	4.5
2W	Lakeland	8.5
2S	Fort Meade	9.0
3S	Wauchula	10
4S	Arcadia	28

CATIONS

ANIONS

FIGURE 9.13. Trilinear diagram of water analysis from Floridan aquifer of central Florida. Locations are shown on Figure 9.11. SOURCE: W. Back and B. B. Hanshaw, *Journal of Hydrology,* 10 (1970):330–68 (Amsterdam: North-Holland Publishing Company).

This type of study illustrates what can be accomplished in chemical geohydrology. It can be a companion to the more commonly performed hydrodynamic study of flow paths and rates. The flow velocity as determined from hydrodynamic considerations can be compared with radiocarbon estimates.

References

1. HEM, J. D. *Study and Interpretation of the Chemical Characteristics of Natural Water.* U.S. Geological Survey Water Supply Paper 1473, 1970, 363 pp.

2. BACK, W. and B. B. HANSHAW. "Chemical Geohydrology." In *Advances in Hydroscience,* vol. 2, ed. V. T. Chow. New York: Academic Press, 1965, pp. 49–109.

3. KRAUSKOPF, K. B. *Introduction to Geochemistry.* New York: McGraw-Hill Book Company, 1967, 721 pp.

4. GARRELS, R. M. and C. L. CHRIST. *Solutions, Minerals and Equilibria.* New York: Harper & Row, 1965, 450 pp.

5. HEM, J. D. and W. H. CROPPER. *Survey of the Ferrous-Ferric Chemical Equilibria and Redox Potential.* U.S. Geological Survey Water Supply Paper 1459-A, 1959, 31 pp.

6. BACK, W. and I. BARNES. *Relation of Electrochemical Potentials and Iron Content to Groundwater Flow Patterns.* U.S. Geological Survey Professional Paper 498-C, 1965, 16 pp.

7. HEM, J. D. *Restraints in Dissolved Ferrous Iron Imposed by Bicarbonate Redox Potentials, and pH.* U.S. Geological Survey Water Supply Paper 1459-B, 1960, 55 pp.

8. BASS BECKING, L. G. M., I. R. KAPLAN, and D. MOORE. "Limits of the Natural Environment in Terms of pH and Oxidation-Reduction Potential." *Journal of Geology,* 68 (1960): 243–84.

9. FETTER, C. W., JR. "Attenuation of Wastewater Elutriated through Glacial Outwash." *Ground Water,* 15 (1977):365–71.

10. WAYMAN, C. H. "Adsorption on Clay Mineral Surfaces." In *Principles and Applications of Water Chemistry,* ed. S. D. Faust and J. V. Hunter. New York: John Wiley & Sons, 1967, pp. 127–67.

11. BALLARD, R. and J. G. A. FISKELL. "Phosphorous Retention in Coastal Plain Soils: I. Relation to Soil Properties." *Soil Science Society of America, Proceedings,* 38 (1974):250–55.

12. BARROW, N. J. and T. C. SHAW. "The Slow Reactions between Soil and Anions: 2. Effect of Time and Temperature on the Decrease in Phosphate Concentration in the Soil Solution." *Soil Science,* 119 (1975):167–77.

13. BARROW, N. J. and T. C. SHAW. "The Slow Reactions between Soil and Anions: 3. The Effects of Time and Temperature on the Decrease in Isotopically Exchangeable Phosphate." *Soil Science,* 119 (1975):190–97.

14. MATTINGLY, G. E. G. "Labile Phosphate in Soils." *Soil Science,* 119 (1975): 369–75.

15. VIJAYACHANDRAN, P. K. and R. D. HARTER. "Evaluation of Phosphorus Adsorption by a Cross Section of Soil Types." *Soil Science,* 119 (1975):119–26.

16. OLSEN, S. R. and F. S. WATANABE. "A Method to Determine a Phosphorus Adsorption Maximum of Soils as Measured by the Langmuir Isotherm." *Soil Science Society of America, Proceedings,* 21 (1957):144–49.

17. SYERS, J. K., M. G. BROWMAN, G. W. SMILLIE, and R. B. COREY. "Phosphate Sorption by Soils Evaluated by the Langmuir Adsorption Equation." *Soil Science Society of America, Proceedings,* 37 (1973):358–63.

18. FITTER, A. H. and C. D. SUTTON. "The Use of the Freudlich Isotherm for Soil Phosphate Sorption Data." *Journal of Soil Science,* 26 (1975):241–46.

19. MITCHELL, J. "The Origin, Nature and Importance of Soil Organic Constituents Having Base Exchange Properties." *Journal of American Society of Agronomy,* 24 (1932):256–75.

20. KENNEDY, V. C. *Mineralogy and Cation Exchange Capacity of Sediments from Selected Streams.* U.S. Geological Survey Professional Paper 433-D, 1965, 28 pp.

345

21. WENTINK, G. R. and J. E. ETZEL. "Removal of Metal Ions by Soil." *Journal of the Water Pollution Control Federation,* 44, (1972):1561–74.

22. ELLIS, B. G. "The Soil as a Chemical Filter." In *Recycling Treated Municipal Wastewater and Sludge through Forest and Cropland,* ed. W. E. Sopper and L. T. Kardos. University Park: University of Pennsylvania Press, 1973, pp. 46–70.

23. RICHARDS, L. A., ed. *Diagnosis and Improvement of Saline and Alkali Soil.* U.S. Department of Agriculture Agricultural Handbook 60, 1954.

24. PRATT, P. F. and F. L. BLAIR. "Sodium Hazard of Bicarbonate Irrigation Waters." *Soil Science Society of America, Proceedings,* 33 (1969):880–83.

25. BOWER, C. A., G. OGATA, and J. M. TUCKER. "Sodium Hazard of Irrigation Waters as Influenced by Leaching Fraction and by Precipitation on Solution of Calcium Carbonate." *Soil Science,* 106 (1968):29–34.

26. BROWN, E., M. W. SKOUGSTAD, and M. S. FISHMAN. "Methods for Collection and Analysis of Water Samples for Dissolved Minerals and Gases." In *Techniques of Water Resources Investigations,* U.S. Geological Survey, 1970, chap. A-1.

27. FETTER, C. W., JR. "Use of Test Wells as Water Quality Predictors." *Journal American Water Works Association,* 67 (1975):516–18.

28. BORMAN, R. G. *Groundwater Hydrology and Geology Near the July 1974 Phenol Spill at Lake Beulah, Walworth County, Wisconsin.* Wisconsin Geological and Natural History Survey Special Report 4, 1975, 13 pp.

29. SMITH, C. L. and J. I. DREVER. "Controls of the Chemistry of Springs at Teels Marsh, Mineral County, Nevada." *Geochemica et Cosmochemica Acta,* 40 (1976):1081–94.

30. DE VRIES, H. "Measurement and Use of Natural Radiocarbon." In *Researches in Geochemistry,* vol. 1, ed., P. H. Abelson. New York: John Wiley & Sons, 1959, pp. 169–89.

31. WIGLEY, T. M. L. "Carbon 14 Dating of Ground-water from Closed and Open Systems." *Water Resources Research,* 11 (1975):324–28.

32. PLINES, P., D. LANGMUIR, and R. S. HARMON. "Stable Carbon Isotope Ratios and the Existence of a Gas Phase in the Evolution of Carbonate Ground Waters." *Geochemica et Cosmochemica Acta,* 38 (1974):1147–64.

33. PIPER, A. M. "A Graphic Procedure in the Geochemical Interpretation of Water Analyses." *Transactions, American Geophysical Union,* 25 (1944):914–23.

34. BACK, W. "Origin of Hydrochemical Facies in Groundwater in the Atlantic Coastal Plain." *Proceedings, International Geological Congress* (Copenhagen), I (1960):87–95.

35. BACK, W. *Hydrochemical Facies and Groundwater Flow Patterns in Northern Part of Atlantic Coastal Plain.* U.S. Geological Survey Professional Paper 498-A, 1966, 42 pp.

36. BRUCH, J. C. and R. L. STREET. "Two-dimensional Dispersion." *Journal, Sanitary Engineering Division, American Society of Civil Engineers,* 93, SA6 (1967):17–39.

37. HOOPES, J. A. and D. R. F. HARLEMAN. "Wastewater Recharge and Dispersion in Porous Media." *Journal, Hydraulics Division, American Society of Civil Engineers,* 93, HY5 (1967):51–71.

38. OGATA, A. *Theory of Dispersion in a Granular Medium.* U.S. Geological Survey Professional Paper 411-I, 1970.

39. BACK, W., R. N. CHERRY, and B. B. HANSHAW. "Chemical Equilibrium between Water and Minerals of a Carbonate Aquifer." *National Speleological Society Bulletin,* 28 (1966):119–26.

40. BACK, W. and B. B. HANSHAW. "Comparison of Chemical Hydrogeology of the Carbonate Peninsulas of Florida and Yucatan." *Journal of Hydrology,* 10 (1970):330–68.

41. PLUMMER, L. N. "Defining Reactions and Mass Transfer in Part of Floridan Aquifer." *Water Resources Research,* 13 (1977):801–12.

Quality of Water

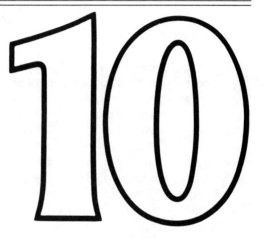

chapter

10

10.1 INTRODUCTION

The chemical reactions between water and the geological environment may be considered by some to be removed from practical concerns. However, the results of these reactions pertain directly to the quality of natural waters, the significance of which is not at all abstract. The quality of water is its suitability for a specific use. It is a consequence of the natural physical and chemical state and such alterations as may have occurred through the actions of humans. Any solute that enters the hydrologic cycle through human action is a **contaminant**; if it renders the water unfit for use, it is a **pollutant**.

Lake Michigan is an example of a body of water having very high natural quality. It is still low in **total dissolved solids (TDS)**, although the amount has been gradually increasing in the lake over the past 75 years (Figure 10.1). From 1895 to 1965, total dissolved solids increased about 20 milligrams per liter (1), with sulfate responsible for most of the increase. As basic hydrogeochemical reactions in the area have not changed, this increase can be attributed to the release of waste products by humans. Waste products are transported by increased sediment in rivers caused by land-use changes in the drainage basin; discharge of wastewater containing dissolved and suspended solids into the lake and tributary river; dumping of solid wastes into the lake; and fallout of air pollutants into the lake.

A large-volume lake, such as Michigan, can accept some amount of common salts and unreactive sediment without significant water quality degradation. However, when a lake is lacking in a mineral nutrient critical to plant growth, the addition of only a small amount of the critical nutrient can overly stimulate plant growth and result in an increased rate of eutrophication (2). In Lake Michigan, studies have shown that there is a greater concentration of phosphorous near Milwaukee than at mid-lake. There is also a greater concentration of diatoms (algae) near shore, where the phosphorous content is high (3). The source of the near-shore phosphorous is agricultural and urban runoff, as well as sewage effluent carried into the lake by the Milwaukee River.

Even very small concentrations of some toxic substances may be highly detrimental to water quality. Synthetic organic substances are widely

350

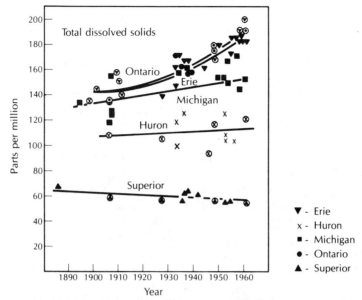

FIGURE 10.1. Concentrations of total dissolved solids in the Great Lakes. Circled points are averages of twelve or more determinations. SOURCE: A. M. Beeton, *Limnology and Oceanography*, 10 (1965):240–54.

used in industry and agriculture. One class of compounds, chlorobiphenols (PCBs), was widely used in industry in electrical and heat-transfer applications as well as in hydraulic fluids. These compounds are very stable and persist in the environment; being fat-soluble, they accumulate in food chains. In areas where PCBs were released into the environment, they are known to have accumulated in fish and other species. In the Milwaukee, Wisconsin, area, they are found in fish taken from both the Milwaukee River and near-shore Lake Michigan (4). There can be considerable amounts of pollutants carried into lakes by atmospheric pathways. Lake Superior receives an estimated PCB loading of 5 metric tons a year from dry atmospheric fallout, and an additional input of 3.1 metric tons per year associated with rain and snow (5).

Groundwater quality is also subject to degradation by chemical and biological pollutants. Typical sources of groundwater pollutants are septic tank disposal systems, landfills, sewage-treatment lagoons, land applications of agricultural chemicals, and chemical spills (6). Saline groundwater may also be drawn into a freshwater aquifer.

In this chapter, we will consider water quality standards and criteria that have been developed for the United States. Standards in other countries may vary. The U.S. Environmental Protection Agency has issued **National Interim Primary Drinking Water Standards** (7). These regulations establish a **maximum contaminant level (mcl)** for several inorganic and organic substances in public water supplies, as well as bacteriological standards. Authority to issue these standards was granted by Public Law 93-523, the Safe

351

Drinking Water Act. The standards are based on known human toxicites of compounds at a consumption level of two liters of water per day. An appropriate safety factor was included in determining each maximum contaminant level. Acceptable methods of analysis for each contaminant were also specified. In Public Law 92-500, Section 302, the United States Congress directed each state to establish water quality standards for surface-water bodies. These water quality standards specify maximum concentrations of substances for purposes of protecting aquatic life, users of surface water, and consumers of aquatic life.

The aquatic food chain is complex, and water quality criteria must ensure that important links in the chain be protected. The phenomenon of bioaccumulation of toxic materials in food chains is well known, and is reflected in water quality criteria for certain substances. Toxic effects may be short-term (24 to 96 hours) or long-term. **Bioassay** to determine toxicity are performed by exposing a number of individuals of a sensitive species to varying concentrations of a substance for a specified period of time. A concentration of a substance such that 50 percent of the test organisms die (LC_{50}) is the normal way in which bioassay results are reported, along with length of exposure. Water quality criteria are generally set at some fraction of the 96-hour LC_{50} for a sensitive species found in the local aquatic environment. The U.S. Environmental Protection Agency has published water quality criteria reflecting the "latest scientific knowledge" (8). State water quality standards closely follow the published water quality criteria.

10.2 WATER QUALITY CRITERIA FOR METALS

10.2.1 ARSENIC

Arsenic possesses the properties of both a nonmetal and a metal. It is widely distributed, the common compounds having a low solubility in water. Trivalent arsenic compounds are the most toxic to both mammals and aquatic species. Arsenic is absorbed from the gastrointestinal tract and comes to be distributed throughout the body. In addition to its toxicity, arsenic has been reported to be carcinogenic (9, 10). Arsenic in drinking water has been associated with high arsenic concentrations in hair (11) and with skin cancer (12).

Arsenic compounds, e.g., sodium arsenite, have been widely used in the past as herbicides to control both terrestrial and aquatic vegetation. Soluble arsenic salts can be harmful to irrigated crops (13).

The water quality criteria for arsenic are 50 micrograms* per liter for domestic water supplies and 100 micrograms per liter for irrigation water (8).

---

*One thousand micrograms is equal to one milligram.

10.2.2 BARIUM

Barium is an alkaline earth metal occurring in nature as insoluble salts such as barite ($BaSO_4$) and witherite ($BaCO_3$). Soluble barium salts are poisonous, with a toxic dose of 0.2 to 0.6 milligram and a fatal dose of 2.4 grams (14). Public drinking water supplies contain barium ranging from less than 1 microgram per liter to as much (rarely) as 3 milligrams per liter (15).* Because of the adverse effects of barium on the heart and blood vessels, its drinking water standard has been set at 1 mg/ℓ.

10.2.3 BERYLLIUM

This element is widely used in industry and occurs naturally in beryl ore and as an impurity in coal. Beryllium is a highly toxic compound when breathed (16), but has a demonstrably lower toxicity when ingested orally (17). As beryllium carbonate and hydroxide are almost insoluble, water that is hard and/or alkaline would have little beryllium in solution. Bioassay studies have shown that beryllium salt toxicity to several aquatic species is inversely proportional to water hardness (18, 19).

Beryllium reduces photosynthesis of terrestrial plants (20) and has been shown to reduce growth rates in several crop species, although plant toxicity in calcareous soil is less than in acid soil. This is doubtless due to formation of insoluble salts.

There are no drinking water standards for beryllium. The water quality criteria for protection of aquatic life are 11 μg/ℓ in soft fresh water and 1.1 mg/ℓ in hard fresh water (8). Irrigation water should not contain more than 100 μg/ℓ, except that up to 500 μg/ℓ is permissible for neutral-to-alkaline, fine-textured soils (8).

10.2.4 BORON

Boron has both metallic and nonmetallic properties. It occurs in nature, primarily as a sodium or calcium borate salt. It is generally found in low concentrations in natural waters and is an essential trace nutrient for plant growth. Some crops, such as citrus, are sensitive to water containing moderate levels of boron; the recommended water quality criterion is 750 μg/ℓ for long-term irrigation of sensitive crops (8).

10.2.5 CADMIUM

In the late 1960s, an environmentally related disease characterized by extreme bone pain (osteomalacia) and multiple bone fractures was diagnosed in several

---

*Hereafter, the abbreviations μg/ℓ and mg/ℓ will be used for micrograms per liter and milligrams per liter, respectively.

localities in Japan. The victims were found to have very high levels of cadmium in body tissues and bone. In separate occurrences, cadmium was found in very high concentrations in the soil (22) and also in drinking water (23). The sources of cadmium in Japan included mine tailings and metal smelters.

There is no known physiological need for cadmium, and it is toxic to almost all body systems. It is stored primarily in the kidneys and liver (24). It is also known to cross the placenta, causing injury to the fetus (25). Cadmium can induce hypertension in humans (26); chronic exposure can result in kidney disease and pulmonary edema (27), as well as osteomalacia. Symptoms of acute exposure can resemble food poisoning symptoms (28).

Cadmium in the aquatic environment in moderate concentrations is toxic to a number of species of fish, although fish in hard water can tolerate more cadmium than those in soft water. Salmonid species are less tolerant than most other fish (29, 30).

The drinking water standard for cadmium is 10 $\mu g/\ell$. For soft fresh water the water quality criteria are 0.4 $\mu g/\ell$ for sensitive species, such as the salmonids, and 4.0 $\mu g/\ell$ for less sensitive aquatic species. In hard fresh water, the criteria are 1.2 $\mu g/\ell$ and 12.0 $\mu g/\ell$, depending on the sensitivity of the species (8). The saltwater criterion of 5.0 $\mu g/\ell$ was established to protect persons who eat oysters, in which cadmium from sea water can be concentrated (8).

Industrial water pollution has released cadmium into both surface and groundwaters (31). It is used in metal plating, storage battery manufacture, and numerous other industrial processes. About 4 percent of 720 samples of rivers and reservoirs in the United States were found to contain cadmium in excess of the drinking water standard (32).

10.2.6 CHROMIUM

Chromium is a moderately abundant element in the earth's crust; however, it is not generally present in very high concentrations in natural water. In a survey of 700 lakes, streams, and reservoirs in the United States, only 11 samples had 5 $\mu g/\ell$ or more of hexavalent chromium, and none of the samples exceeded 50 $\mu g/\ell$ (32). Chromium can exist in a number of valence states, but the trivalent (+3) is more common in nature. The hexavalent state (+6) is the one in which chromium is often found in industrial applications. Chromium in the trivalent state is an essential trace nutrient in mammals. No harmful effects were found in laboratory animals ingesting moderate amounts of Cr^{+3} (8). However, hexavalent chromium is a systemic poison in addition to being corrosive (24). The toxicity of chromium to aquatic species is variable, depending upon valence state, water hardness and pH, and the species in question (33). The mcl for drinking water is 50 $\mu g/\ell$ total chromium, and the water quality criterion is 100 $\mu g/\ell$ for aquatic species (8). Hexavalent chromium of natural origin is present in the groundwater of Paradise Valley, Maricopa County, Arizona, in amounts exceeding 200 $\mu g/\ell$. The groundwater is alkaline, with a positive Eh, so that Cr^{+3} in the sediments is oxidized to soluble Cr^{+6} (98).

10.2.7   COPPER

This common metal occurs in nature as both the native metal and various copper minerals and salts. It is an essential trace element for the growth of plants and is also required in trace amounts in both vertebrate and invertebrate animals. In excessive amounts, however, copper can be toxic. It is found in relatively low concentrations in natural waters (8). Copper sulfate is sometimes used as an aquatic herbicide and as a gastropod control. The necessary levels are far greater than those found in natural water.

Some aquatic species are adversely affected by copper, the toxicity dependent upon the hardness, pH, and alkalinity of the water. Copper is more toxic in soft water and in water with low alkalinity (34, 35). The toxic level of copper in the aquatic environment varies widely with the species and the native water chemistry. The recommended water quality criterion is 0.1 of the 96-hour LC_{50} as determined by a bioassay using a sensitive resident species (8). The drinking water standard is 1.0 mg/ℓ based on a metallic taste at higher levels (7).

10.2.8   IRON

Iron is a very common element and is found in many of the rocks and soils of the earth's crust. It is also an essential trace element for both plant and animal growth. Soluble ferrous iron is present only in natural water with a low Eh. Some groundwaters, hypolimnitic water from some thermally stratified lakes, and water from some marshes may be devoid of oxygen and thus contain ferrous iron. When such water becomes oxygenated, the ferrous iron oxidizes to ferric iron and precipitates. Well-oxygenated surface waters normally contain almost no dissolved iron. In some cases, iron is found where groundwater containing ferrous iron is discharging into a lake through a spring.

Iron is toxic to some aquatic species at concentrations of 0.32 to 1.0 mg/ℓ (36, 37). Iron in drinking water imparts a metallic taste at concentrations of 1.8 mg/ℓ (38); however, soluble iron at concentrations in excess of 0.3 mg/ℓ can stain surfaces. A water quality criterion for iron of 0.3 mg/ℓ has been suggested for domestic use to avoid objectionable staining of plumbing fixtures (8). For aquatic life, a maximum iron content of 1.0 mg/ℓ is the criterion (8). Because of the stability field for ferrous iron, this criterion should be met wherever the water contains dissolved oxygen, and there is no iron pollution.

10.2.9   LEAD

Lead is a toxic metal which fulfills no known physiological requirement, and which has long been associated with occupational and environmental disease (24, 39). Lead is taken into the body via ingestion of food and fluids as well as by inhalation. It inhibits the formation of hemoglobin, leading to symptoms of anemia. Lead accumulates in the bones and soft tissue, such as the kidney, aorta, liver, and brain cortex (40). In advanced cases of lead poisoning, kidney damage is often encountered, in addition to neurological symptoms. Lead

355

poisoning is a known cause of mental retardation, cerebral palsy, and optic nerve atrophy in children (41).

Lead is predictably toxic to freshwater aquatic species. The lead toxicity appears to be greater in soft water than in hard water (33). The 96-hour LC_{50} for lead ranges from 0.1 to 482 mg/ℓ, depending upon the species in question and the water hardness (8). Because of this wide range, the suggested water quality criterion is 0.01 of the 96-hour LC_{50} value using the receiving water as a dilutant and a sensitive resident species (8). The drinking water standard is 50 μg/ℓ based on the toxic properties of lead (7).

10.2.10 MANGANESE

Another essential trace element for both animals and plants is manganese. Plants lacking sufficient manganese can exhibit chlorosis, or the leaves may fail to develop. Manganese deficiency in animals can disturb reproduction, cause bone deformities and central nervous system disorders, or retard growth. A low level of certain enzyme activities is the precursor of these conditions (24). Manganese is known to be a poison, generally related to occupational exposure to manganese dust (42).

Manganese in water is apparently hardly ever associated with toxicity in humans or livestock, but may be somewhat toxic to plants grown in acid soils (8). In concentrations above 150 μg/ℓ, it can precipitate and cause staining problems similar to those encountered with iron. To avoid such staining, a criterion of 50 μg/ℓ is suggested for domestic water supplies (8). Shellfish are known to accumulate manganese; a criterion of 100 μg/ℓ for marine water will protect consumers of shellfish from potential manganese toxicity (8).

10.2.11 MERCURY

Mercury is widely disbursed through the environment. In a survey of surface materials in the United States, mercury concentrations ranged from 10 μg/ℓ to 4.2 mg/ℓ and averaged 112 μg/ℓ (43). The mercury content of unpolluted rivers in areas of the United States with no known mercury deposits were below the detection limit of 0.1 μg/ℓ in one study (44), although water near mercury deposits and sources of mercury pollution can be 5 to 100 μg/ℓ (24).

Mercury can occur as the metal, as mercurous (Hg^+) or mercuric (Hg^{++}) salts, and in organic forms. It is naturally occurring and is widely used in industry and as an antifungal agent in agriculture (45). It is found in coal in concentrations of 0.5 to 3.3 mg/ℓ, and when the coal is burned, the mercury is released as an air pollutant which can later settle on land and water. Petroleum and asphalt also contain mercury that can be released into the environment. Mercury mining and smelting are other sources of pollution.

Mercury is toxic in both inorganic or organic form. The mercurous salts are less soluble; hence, less toxic. The most toxic forms are organic;

e.g., methyl mercury, which is extremely soluble. Inorganic mercury accumulates in the liver and kidneys, tending to damage these organs (24). The organic mercury compounds can readily pass through biologic membranes, accumulating in the brain and causing atrophy of brain cells in the cerebellum and cortex, with resulting impairment in voluntary muscle coordination, speech, hearing, and vision (24). Alkyl mercury can also penetrate the placental barrier so as to poison the fetus (46).

Elemental mercury and inorganic salts in the aquatic environment can be converted by bacterial processes into the highly toxic methyl mercury form (47). Thus, any mercury has the potential to form toxic methyl mercury, so that standards have been established for total mercury. Mercury is also biologically concentrated as it is released from the body at very low rates. Fish have been shown to concentrate mercury to levels 16,600 to 27,800 times that of the surrounding aquatic environment (48). The mercury content of certain species of fish from Lake St. Claire, on the Detroit River, increased by a factor of 100 from 1935 to 1970 (44).

The drinking water standard for total mercury is 2 $\mu g/\ell$ based on the toxic properties. In order to protect the indigenous species and the consumers of fish, the aquatic freshwater criterion is 0.05 $\mu g/\ell$. The marine criterion for water quality is 0.1 $\mu g/\ell$.

10.2.12   NICKEL

Nickel is a whitish metal with soluble salts. Apparently, it is relatively nontoxic to humans in some forms (49), but the gaseous nickel carbonyl is highly toxic (50). Airborne nickel compounds are also suspected carcinogens (24). The U.S. National Interim Primary Drinking Water Standards have not included a standard for nickel in water (8).

In the aquatic environment, nickel has been shown to be toxic to embryos of the marine hardshell clam (48-hour LC_{50} of 310 $\mu g/\ell$) (51), and to adversely affect reproduction of the freshwater crustacean, *Daphnia*, at concentrations of 95 $\mu g/\ell$ (52) and the fathead minnow at concentrations of 730 $\mu g/\ell$ (53). A water quality criterion of 0.01 of the 96-hour LC_{50} has been proposed for both freshwater and marine aquatic life (8).

10.2.13   SILVER

Silver, one of the noble metals, occurs in both the elemental form and as various salts. If it is injested, it tends to accumulate in human skin, eyes, and mucous membranes. It is nonbeneficial to humans and is potentially detrimental (8). In low concentrations, silver solutions are bactericidal and can be used to disinfect drinking water.

Studies of silver toxicity to freshwater and marine aquatic life show a wide range of toxic levels depending upon species (54). Silver has also

357

been shown to accumulate in fish tissue at levels as low as 7 $\mu g/\ell$ (55). Because of the tendency to accumulate in the body, the drinking water standard has been set at 50 $\mu g/\ell$ (7). The suggested water quality criterion is 0.01 of the 96-hour LC_{50} as determined through bioassay using a sensitive resident species.

10.2.14   ZINC

This metal is another of the trace elements necessary for human metabolism. It performs essential roles in the synthesis of protein and DNA, and is required in many enzymes. A deficiency of zinc can result in stunted growth and in poor healing of wounds (56). Zinc can be toxic, causing gastrointestinal distress if ingested in large quantities. This has been known to occur when acid food or drink that has been stored or prepared in galvanized containers is consumed (57).

Natural waters generally contain some zinc, but not in large quantities. In one study in United States waterways, 76 percent of all samples had measurable zinc with a mean value of 64 $\mu g/\ell$ and a maximum of 1.183 mg/ℓ (8). The wide use of zinc in industry has resulted in some zinc pollution of water. Community water supplies with zinc levels of 11 to 27 mg/ℓ have caused no observed harm, although some people can taste zinc at concentrations of 4 mg/ℓ (8). The suggested criterion for drinking water is 5 mg/ℓ, although there is no EPA drinking water standard for zinc.

Zinc is toxic to aquatic life. The degree of toxicity varies with pH, alkalinity, and hardness of the water. These factors affect the solubility of various zinc compounds. In a study of the fathead minnow, the toxicity decreased with increasing hardness and decreasing pH (58). The water quality criterion for zinc is 0.01 of the 96-hour LC_{50} for a sensitive species using the receiving water as a dilutant.

10.3   WATER QUALITY CRITERIA FOR INORGANIC NONMETALS

10.3.1   CHLORINE

This greenish-yellow gas is not a natural constituent of water; however, it is widely used in disinfecting sewage effluent, in controlling bacteria in cooling-water systems, and in bleaching operations. Hence, it is frequently found in surface water. When dissolved in water it dissociates into a weak acid. It also reacts with nitrogenous organic compounds to form chloramines. Both chlorine and chloramines are toxic, especially to fish.

The aquatic life in streams below chlorinated sewage outfalls is often drastically different from aquatic life below discharges of unchlorinated effluent. In one study of streams receiving chlorinated effluent, there was considerably less diversity of species, and in reaches where chlorine residuals exceeded 370 $\mu g/\ell$, fish were not present. Trout were not found in water exceeding 20 $\mu g/\ell$ (59). Salmonoid species are the most sensitive to chlorine, with concentrations as low as 4 to 6 $\mu g/\ell$ toxic to fry and 10 $\mu g/\ell$ toxic to adults (60, 61). A water quality criterion of 2 $\mu g/\ell$ of total residual chlorine will protect salmonoid fish, with 10 $\mu g/\ell$ providing protection for other, less sensitive, marine and freshwater species.

10.3.2   FLUORIDE

Fluoride is a halogen widely used in industry and naturally present in such minerals as fluorite, cryolite, and apatite. A person will typically ingest 2 to 5 milligrams per day of fluoride in food and water (62). Fluoride is stored in bones and teeth. Chronic overexposure leads to dental fluorosis or mottling of teeth. It can also cause skeletal fluorosis, in which tissue and ligaments of the joints become calcified, forming excess bony protrusions (24). Dental fluorosis usually develops only in children living in areas where the drinking water is high in fluoride, with the greatest impact in warmer climates. For this reason, the National Interim Primary Drinking Water Standards set maximum levels of fluoride dependent upon average annual maximum daily air temperatures (8):

Degrees Celsius	mcl (mg/ℓ)
12.0 and below	2.4
12.1 to 14.6	2.2
14.7 to 17.6	2.0
17.7 to 21.4	1.8
21.5 to 26.2	1.6
26.3 to 32.5	1.4

10.3.3   NITROGEN, INORGANIC

Nitrogen is a transmutable element, passing through a number of valence states as a result of biological reactions in the nitrogen cycle. The inorganic salts of the nitrogen cycle include nitrate ($NO_3^=$) and nitrite ($NO_2^=$). In addition, ammonia can exist as a gas (NH_3), which is highly soluble in water. When dissolved in water, some of the ammonia reacts with the water to form the ammonium ion (NH_4^+). Plants can incorporate both the ammonium ion and the nitrate ion from water and use them to form protein. Organically bound nitrogen, such as protein, can be broken down by bacteria to ammonia. Under aerobic conditions, certain bacteria can oxidize ammonia to nitrite, and nitrite to nitrate. Denitrification of nitrate and ammonia to gaseous elemental nitrogen can occur by bacterial action under anerobic conditions.

Nitrate nitrogen has proved to be a health hazard when it occurs in drinking water at concentrations in excess of 10 mg/ℓ as nitrogen. Nitrate is reduced to nitrite in the gastrointestinal tract. The nitrite can then enter the bloodstream, whereupon it reacts with hemoglobin, impairing the blood's ability to transport oxygen. The condition is known as methemoglobinemia, and is known only to occur in fetuses and infants under three months of age (63). Nitrite nitrogen obviously poses the same health danger, although this species is almost always in very limited concentrations in natural waters, as it is highly reactive. The National Interim Primary Drinking Water Standards set a limit of 10 mg/ℓ of nitrate as nitrogen (7).

Nitrate salts are slightly toxic to fish, but only at concentrations greater than 90 mg/ℓ (64). Nitrite is more toxic, but at levels of 5 mg/ℓ or lower it should not harm most fish species (65). Levels below 0.06 mg/ℓ are necessary, however, to protect salmonids (66).

The toxic levels of nitrate and nitrite are not likely to be encountered in any but very highly polluted waters. On the other hand, ammonia may be found in moderate concentrations, especially in wastewater effluent. Ammonia is toxic to fish, the toxicity dependent upon the pH, temperature, and ionic strength of the water. The toxic agent is the un-ionized ammonia (67). Carp can tolerate as much as 5 mg/ℓ NH_3, but trout succumb at 0.2 mg/ℓ (68). To protect sensitive species, a criterion of 0.02 mg/ℓ of un-ionized ammonia has been established (8). At low pH and temperature, only a very small percent of the ammonia is in this form, with the percentage increasing with temperature and pH. In neutral waters at 5° C, a total ammonia concentration of 16 mg/ℓ will have 0.02 mg/ℓ of un-ionized ammonia. At 20° C and pH 7.0, this amount of un-ionized ammonia would be present in 5.1 mg/ℓ of total ammonia. At pH 8.0 and 5° C, the permissible total ammonia is 1.6 mg/ℓ, and at pH 8.0 and temperature 20° C, it is 0.52.

Nitrogen is also an element that may limit growth of aquatic plants if it is in very low concentrations in lakes and ponds. If surface waters are overfertile, excessive growth of weeds and algae may occur and contribute to eutrophication. Nutrient-budget models for lakes have been developed which incorporate nutrient loading as a function of surface area of the lake, nutrient retention in the lake, and outflow (69, 70). In general, springtime concentrations of inorganic nitrogen greater than 0.3 mg/ℓ may be excessive. Other sources of nitrogen include certain blue-green algae which can fix atmospheric nitrogen and lightening discharges which convert elemental nitrogen to inorganic forms.

10.3.4 PHOSPHORUS

Phosphorus in the organic and inorganic phosphate forms is probably the key nutrient in controlling eutrophication of surface waters. It is an essential micronutrient for plant growth. The natural distribution of phosphorus is rather limited; it is commonly found only in one mineral, apatite. Runoff waters from areas of phosphate rock may contain high amounts of phosphate; otherwise, its usual source is organic pollution and phosphate detergent. An unusual case of phosphate pollution occurs in river sediments in Oshkosh, Wisconsin. The sediments from the site of a former match factory are so high in phosphorus that they reportedly ignite spontaneously when air-dried.

Phosphorus models for predicting lake eutrophication include parameters for the loading factor in terms of grams of phosphorus per square meter of lake surface, and mean depth (71), flushing time (72), and stratification state (73). No simple statement can cover all possible conditions. However, total phosphorus levels of 0.01 mg/ℓ or greater are potentially eutrophic, as are loading rates of 0.1 to 1.0 grams of total phosphorus per square meter of surface area per year.

10.3.5   SELENIUM

Selenium is a necessary micronutrient for plants and animals, occurring as a structural element in some proteins; but it is toxic in quantities not much greater than those required. It occurs in most soils and can accumulate in plants (74). It can be added to soil by use of superphosphate fertilizer and from air pollution from burning coal and paper (75). High selenium content of food and drinking water can cause selenosis (76). This disease is characterized by symptoms of mental depression, pallor, weakness, nervousness, dizziness, liver and kidney damage, and a garliclike body odor (24). There is not as great a safety factor between "normal" ingestive levels of selenium and the toxic level as there is for arsenic, bromine, cobalt, chromium, or mercury. Normal ingestion of selenium is 0.2 milligram per day, and the toxic level is 5 milligrams per day (77). The National Interim Primary Drinking Water Standard for selenium is 0.01 mg/ℓ.

10.3.6   SULFATE AND SULFIDE

Sulfur occurs in water as the inorganic sulfate salt as well as the dissolved gas, hydrogen sulfide. Sulfate is not a noxious substance, although high-sulfate water may have a laxative effect. Hydrogen sulfide, with its characteristic foul odor, is highly toxic and highly soluble in water. It comes from the anerobic decomposition of organic matter and can be found in both ground and surface waters. Hydrogen sulfide can also form from sulfide wastes disposed in water. Because of the odor, it renders water unpalatable; hence, no drinking water standard has been set. Due to its toxicity to aquatic life (78), a water quality criterion of 2 μg/ℓ of undissociated hydrogen sulfide has been established (8). The proportions of sulfide and undissociated hydrogen sulfide present is a function of the pH.

361

10.4 WATER QUALITY CRITERIA FOR ORGANIC COMPOUNDS

The variety of organic compounds used in modern industry is staggering. More than two million organic chemicals are known, and among them are some 1500 suspected carcinogens (79). Many of these carcinogens are organic and have the potential to exist in trace concentrations in water. For the vast majority of these, the safe concentration level necessary to avoid carcinogenic or sublethal toxic effects is unknown.

In a study of 5500 samplings of environmental waters, 1296 different organic compounds were found (79). Since it is difficult to extract organic substances from water and then anaiyze them, it can be assumed that many more compounds are present at concentrations below the detection limits. By one estimate, only 10 to 20 percent of all organic matter in water is analyzed (79). New organic chemicals are being developed every year, and some portion of them will find their way into ground and surface waters through manufacturing processes, use, and disposal. In the United States the Toxic Substances Control Act (Public Law 94-469) allows the Environmental Protection Agency to regulate substances that "may present an unreasonable risk of injury to health or the environment" (80). This legislation is useful in helping to protect United States surface waters from organic contamination. There is also concern that chlorination of drinking water supplies could create carcinogenic chloro-organic compounds. It would be prudent to treat all surface-water supplies used for drinking with activated carbon to remove trace organics.

The National Interim Primary Drinking Water Standards set maximum contaminant levels for only a few of the large number of organic contaminants that could prove to be dangerous. New chemicals will be added to the list with additional research, and some may be dropped if their use is eliminated.

10.4.1 CHLORINATED HYDROCARBONS

These compounds are used as pesticides, and the class includes a large number of chemicals. The use of some (e.g., DDT, heptachlor, epoxide, and chlordane) has been eliminated or greatly restricted due to adverse environmental effects. These substances are also toxic to mammals, fish, and invertebrates (81–84). Table 10.1 lists maximum contaminant levels and water quality criteria for selected chlorinated hydrocarbons.

TABLE 10.1. Drinking water standards and contaminant criteria for four chlorinated hydrocarbons.

Compound	Drinking Water Standard (7) ($\mu g/\ell$)	Criteria (8)	
		Fresh Water ($\mu g/\ell$)	Marine Water ($\mu g/\ell$)
Endrin	0.2	0.004	0.004
Lindane	4.0	0.01	0.004
Methoxychlor	100.0	0.03	0.03
Toxaphene	5.0	0.005	0.005

10.4.2 CHLOROPHENOXYS

These compounds are used as herbicides in both home and commercial applications. They are also biologically active against mammals (85). The maximum contaminant level for 2,4-D is 100 $\mu g/\ell$; for 2,4,5-TP (Silvex), it is 10 $\mu g/\ell$ (7). Some chlorophenoxys contain an impurity, dioxin, which is highly carcinogenic.

10.4.3 HALOMETHANES

Halomethane compounds pose a risk due to cancer. However, the threshold concentrations for carcinogenic compounds at low levels of long-term exposure are controversial (90). Halomethanes include chloroform, bromodichloromethane, dibromochloromethane, bromoform, 1,2-dichlomethane, and carbon tetrachloride. These compounds may form from the chlorination of drinking water containing aromatic hydrocarbons. A survey of 80 United States municipal water supplies found one or more of these carcinogens present in 17 of the water systems. Only 2 of the 17 had a groundwater source (99). A target level of 100 $\mu g/\ell$ of trihalomethanes in drinking water has been proposed by the U.S. Environmental Protection Agency (90).

ADDITIONAL WATER QUALITY CRITERIA 10.5

The total amount of solids dissolved in water determines whether it is fresh or saline. Drinking water is recommended to have no more than 500 mg/ℓ of dissolved solids (86), although water with up to 1000 mg/ℓ is often used if no other source is available. Water with less than 1000 mg/ℓ of total dissolved solids is fresh; from 1000 to 10,000 it is brackish; from 10,000 to 100,000 it is saline; and if the concentration of TDS exceeds 100,000 mg/ℓ, the water is a brine (87).

The concentration of total dissolved solids is sometimes indirectly measured by the **specific electrical conductance** of the water. The primary reason for this is that portable, reliable, inexpensive meters are available. There is no simple relation, however, between total dissolved solids and specific electrical conductivity. The electrolytic capacity of a natural water is a function of the number of moles of each ion and the electrical charge of the ion. A mole of Ca^{++} could transport twice as many electrons in an electrical current as could a mole of Na^+.

The most common anion in many natural waters is chloride. The recommended maximum amount of chloride in drinking water is 250 mg/ℓ; at higher levels the chloride would impart a salty taste (86). The most common cation found with high chloride is sodium. For some individuals on a low-sodium diet, the sodium content of drinking water must be included in computing the daily sodium intake.

The amount of calcium, magnesium, and iron contribute to the **hardness** of water. These cations react with soap to form a precipitate, thus reducing the cleansing action of the soap. Detergents are less affected by hardness than are soaps. The hardness of water is expressed as that due to an equivalent amount of calcium carbonate. Soft water has less than 50 mg/ℓ hardness, moderately soft water 50 to 100 mg/ℓ hardness, hard water 100 to 200 mg/ℓ hardness, and very hard water more than 200 mg/ℓ hardness. Hardness is removed by exchanging the calcium, magnesium, and iron for sodium using an exchange resin bed. Home water softeners can contribute to sodium and chloride pollution of groundwater in unsewered areas (88).

In surface water, the most critical dissolved gas is oxygen. It is obviously necessary to support fish and invertebrate life. Dissolved oxygen levels close to saturation are necessary to support a wide spectrum of aquatic life. Should the dissolved oxygen content drop too low, anerobic processes can occur. The minimum concentration of dissolved oxygen necessary for a viable fish population is 5.0 mg/ℓ (89). This level should also protect invertebrates in the water. Salmonids require higher dissolved oxygen concentrations, especially in warmer waters.

The presence of fine silt and clay in water imparts a cloudiness called **turbidity.** Because bacteria may cling to these fine particles, there is a National Interim Primary Drinking Water Standard of 1 turbidity unit for surface-water supplies. A turbidity unit relates to the amount of light the sample absorbs, as measured on a special instrument.

PATHOGENIC ORGANISMS 10.6

Water can be a transmitting medium for a variety of disease-causing organisms, some of which may be parasitic. Disease-causing organisms typically come from human waste, as do some intestinal and many other parasites. Drinking infected water or skin contact with it can transmit the disease. Drinking contaminated well or surface water has been known to transmit typhoid fever, cholera, bacillary dysentary, and paratyphoid fever. All of these are caused by a specific bacterium. The diseases may also be transmitted by consumption of shellfish from contaminated water. Polio and infectious hepatitis are caused by viruses sometimes transmitted by water, as is amoebic dysentary, which is due to the proterozoan *Entamoeba histolytica*.

A particularly debilitating disease is caused by blood flukes, which belong to the genus *Schistosoma*. They live part of their life cycle in human blood vessels. Eggs are discharged with urine or feces. After the eggs hatch, the organism must come into contact with a specific species of snail. In the snail, which is an intermediate host, the fluke forms a new stage which is discharged into water. In this stage, it enters a human, either through the skin or via drinking water. Within the human host, the life cycle of the fluke is completed.

Pathogenic organisms in drinking water can be removed by filtration and/or disinfection with chlorine or iodine. Shellfish from areas where sewage is discharged are unfit for consumption. Water contact should be avoided in areas of parasitic infection.

Between 1971 and 1977, there were 192 outbreaks of waterborne disease in the United States, a country of high health standards. More than 36,000 persons were affected (90). Ninety-seven percent of the outbreaks involved pathogens, and about half involved the use of untreated or inadequately treated groundwater.

GROUNDWATER CONTAMINATION 10.7

The large volume of solid and liquid waste produced by a modern technological society presents the problem of disposal. In the past, raw wastewater was pumped into rivers and solids were dumped into swamps and marshes. The emphasis on clean water and wetland preservation has changed that pattern of disposal. Paradoxically, this has resulted in increased problems of soil and groundwater contamination. Sewage treatment results in the formation of semisolid sludges containing sediment, organic waste, bacteria, metals, etc. Some sludge may be used as soil conditioner and fertilizer; however, those sludges with high concentrations of metals cannot be applied to soil systems safely. Fly ash from air-pollution control equipment is a large-volume waste product that may contain sulfur, mercury, or other potential contaminants. Industrial wastes must be disposed of, as well. The solution is usually a land-disposal site where the waste is buried. This may lead to a groundwater contamination problem.

Groundwater pollution is of great concern; it may take many years for contaminated water to be flushed from a system because of the slow rate of movement. However, groundwater contamination is usually a local rather than a regional problem. Groundwater contamination from a point source would be unlikely to extend more than a few thousand meters. But if there are many point sources, such as in large suburban developments with septic tanks used for waste disposal, area-wide groundwater contamination may occur.

Polluted water moving through soil and rock may be attenuated to some degree. Attenuation processes include dilution and dispersion, mechanical filtration, chemical reactions, volitization, adsorption, biological assimilation, membrane filtration, and radioactive decay (91). However, even when such processes have been operative, groundwater has been contaminated in many places. Table 10.2 lists chemical and biological contaminants which have been documented as groundwater contaminants. There are doubtless other chemicals which could be added to this listing.

TABLE 10.2. Chemicals and organisms known to have caused groundwater contamination (various sources)

Metals	Nonmetals	Organics	Organisms
aluminum	acids	alcohol	Giardia lamblia
arsenic	ammonia	aldrin	Salmonella sp.
barium	boron	BOD	Shigella sp.
cadmium	chloride	chlordane	typhoid
chromium	cyanide	DDT	Yersinin
copper	fluoride	detergents	enterocolitica
iron	nitrate	ethyl acrylate	viral hepatitis
lead	phosphate	gasoline	
lithium	radium	hydroquinone	
manganese	selenium	lindane	
mercury	sulfate	paramethylanimo-	
molybdenum	various radio-	phenol	
nickel	active isotopes	PBB	
silver		PCB	
uranium			
zinc			

10.7.1   SEPTIC TANKS AND CESSPOOLS

The disposal of domestic wastewater is accomplished in many areas through use of septic tanks and drain tile fields. Anerobic decomposition of wastes takes place in the septic tank. The liquid waste is carried to a drain tile field where it seeps through the ground to the water table. An analysis of the typical septic tank effluent is given in Table 10.3. Bacteria can be mechanically filtered by fine-grained soils; in coarser soils, an organic crust can form in the drain tile field, which also serves to filter out bacteria. Phosphorus is removed by adsorption on fine-grained soils and by precipitation with calcium and iron.

TABLE 10.3.   Effluent quality from six septic tanks*

Site	Avg. Flow (gpd)	BOD (mg/ℓ)	COD (mg/ℓ) (unfiltered)	COD (mg/ℓ) (filtered)	TSS (mg/ℓ)	Fecal Coliforms (no./mℓ)	Fecal Strep (no./mℓ)	Total N (mg/ℓ)	Ammonia N (mg/ℓ)	Nitrate-Nitrogen (mg/ℓ)	Total P (mg/ℓ)	Ortho P (mg/ℓ)
A	75	131	325	249	69	2907	2.7	50.5	34.1	0.68	12.3	10.8
B	125	176	361	323	44	4127	39.7	57.8	42.5	0.46	14.1	13.6
C	245	272	542	386	68	27,931	1387	76.3	45.6	0.60	31.4	14.0
D	315	127	291	217	52	11,113	184	40.2	33.2	0.35	11.0	10.1
E	860†	120	294	245	51	2310	20.7	31.6	20.1	0.16	11.1	10.5
F	150	122	337	281	48	3246	25.3	56.7	38.3	0.83	11.6	10.5

SOURCE: R. J. Otis, W. C. Boyle, and D. K. Sauer, Small-Scale Waste Management Program, University of Wisconsin-Madison, 1973.

*All values are means.

†Includes 340-gpd sewer flow and 520 gpd from foundation drain.

Infectious disease has been traced to groundwater contaminated by septic tanks. In Polk County, Arkansas, in 1971, an outbreak of viral hepatitis was traced to a well that was contaminated by seepage from a septic tank 30 meters away (92). In 1972, typhoid in Yakima, Washington, was attributed to well water from driven well points. A typhoid carrier was living nearby, and the contaminated water traveled through "extremely pervious gravel" 64 meters from a drain field to the well (92).

Regional groundwater contamination from septic tank discharge has also been documented. On Long Island, New York, there are many densely populated areas with no sewer systems. The surface deposits are coarse glacial sand and gravels. The groundwater has been contaminated by nitrate and ammonia and, to a lesser extent, by detergents (93, 94). The background levels of ammonia-nitrogen and nitrate nitrogen are less than 1 mg/ℓ, but due to septic tank discharges, both may exceed 20 mg/ℓ as nitrogen. Groundwater must then be pretreated to remove nitrate, as the drinking water criterion of 10 mg/ℓ is exceeded.

10.7.2 LANDFILLS

Land disposal by burial is a common means of disposal of municipal refuse, ashes, garbage, and sludge from municipal and industrial wastewater treatment. Some landfill sites are used for disposal of toxic chemicals and low-level radioactive waste material. A growing awareness of the environmental dangers associated with refuse disposal in old gravel pits and quarries has led many states to require careful site selection based on sound hydrogeological study. It has been shown that water from recharge or moving groundwater can leach chemicals from buried solid waste. Table 10.4 contains analyses of leachate from a sanitary landfill at DuPage, Illinois. A summary of leachate analyses from a number of landfills in Illinois is presented in Table 10.5. The analyses indicate that landfill leachates can contain extremely high concentrations of inorganic and organic materials.

TABLE 10.4. Chemical analysis of landfill leachate at DuPage, Illinois

Na	748.	mg/ℓ
K	501.	mg/ℓ
Ca	46.8	mg/ℓ
Mg	233.	mg/ℓ
Cu	<0.1	mg/ℓ
Zn	18.8	mg/ℓ
Pb	4.46	mg/ℓ
Cd	1.95	mg/ℓ
Ni	0.3	mg/ℓ
Hg	0.0008	mg/ℓ
Cr	<0.10	mg/ℓ
Fe	4.2	mg/ℓ
Mn	<0.1	mg/ℓ
Al	<0.1	mg/ℓ
NH_4	862.	mg/ℓ
As	0.11	mg/ℓ
B	29.9	mg/ℓ
Si	14.9	mg/ℓ
Cl	3484.	mg/ℓ
SO_4	<0.01	mg/ℓ
PO_4	<0.1	mg/ℓ
COD	1340.	mg/ℓ
Organic acids	333.	mg/ℓ
Carbonyls as acetophenone	57.6	mg/ℓ
Carbohydrates as dextrose	12.	mg/ℓ
pH	6.9	
Eh	+7 mv	
Conductivity	10.20 mmhos/cm	

SOURCE: K. Cartwright, R. A. Griffin, and R. H. Gilkeson, *Ground Water*, 15 (1977):294–305.

368

TABLE 10.5. Mean and range of values of chemical constituents of 45 landfill leachates from Illinois

Parameter	Observed Range (mg/ℓ)			Observed Extremes (mg/ℓ)	
	N	$\bar{Y}$	S	High	Low
Calcium (Ca)	32	686	553	2300	100
Magnesium (Mg)	32	284	258	1102	68
Sodium (Na)	31	380	259	1100	25
Potassium (K)	31	204	172	740	2.4
Ammonia (NH₃-N)	42	167	172	670	1.23
Iron (Fe)	44	402	473	1920	13.5
Chloride (Cl)	45	616	382	1820	75
Sulfate (SO₄)	36	379	458	2000	0
Hardness	39	2093	1794	6500	430
COD	44	7171	6792	26000	28

SOURCE: T. P. Clark, *Ground Water*, 13 (1975):321–31.

NOTE: N = number of samples, $\bar{Y}$ = mean of samples, S = standard deviation of samples.

The volume of leachate produced is a function of the amount of water percolating through the refuse. Land disposal of solid waste is more likely to produce large volumes of leachate in humid climates than in arid climates. The vadose zone of arid regions may receive no recharge of precipitation. Under such conditions, solid waste disposal would not be likely to result in groundwater contamination (91, 95).

In order to minimize leachate production in humid regions, disposal above the seasonally high water table is generally practiced. The final landfill cover is often compacted soil of a high clay content, which retards infiltration. The clay cover is a more important factor in retarding leachate formation than the location above the water table (96). Water accumulates in most landfills, which raises the water table to form a mound at the landfill site. Under these conditions, only infiltrating water can move through the refuse, even if the base is below the water table.

Soils of low natural permeability (10^{-6} centimeter per second or less) are the most desirable for landfill locations. They result in small amounts of infiltration and slow groundwater movement. This reduces the volume of leachate produced and retards the movement from the area. Natural attenuation sites are located in clay with a high ion-exchange capacity (96). Adsorption of organics, ion-exchange removal of metals, and dilution of total solids are some of the factors that can attenuate leachate.

Leachate can also be collected from the site and then treated by conventional means for wastewater. Collection sites are often used where permeable sediments and bedrock are present. The site is lined with compacted clay, asphalt, plastic sheeting, or similar material. Drain tile above the liner collects the leachate, which drains into a holding tank. Some sites have been built with a redundant drain tile system under the liner in case the liner should leak (97).

Groundwater contamination from improperly located and designed landfills is a continuing problem. Many old dumps and landfills were located in geologically unsuitable sites. Due to the slow rate of groundwater movement, contaminant plumes will grow for years after abandonment of such sites. To add to the problem, the growing volumes of solid wastes, the lack of geologically suitable sites, and/or inadequate regulations on landfill siting may result in contamination from sites that are new. Land disposal of solid waste can be accomplished without groundwater contamination, but only if proper geologic siting and engineering designs are followed.

10.7.3 MINING

The extraction and processing of metallic ore and coal has been the source of both surface and groundwater contamination. Groundwater moving through mineralized rock may have a higher concentration of certain heavy metals than groundwater moving through nonmineral districts (100). Mining and milling expose overburden and waste rock to oxidation. Oxidation of pyrite, a nearly ubiquitous mineral, produces sulfuric acid. In the Appalachian region of the eastern United States, 6000 tons of sulfuric acid are produced daily in this manner (101). This results in highly acid water draining from spoil banks and tailings deposits; hence, the shallow groundwater and surface water of such regions is acidic. The lower pH of the water draining through mineralized spoils and tailings can also leach heavy metals (102, 103), as well as soluble calcium, magnesium, sodium, and sulfate (104). Uranium mining and milling operations can release radioactive uranium, radium, radon, and thorium through wastewater from the mine and milling operations. These can be discharged into surface waters or seep into groundwater from tailings ponds (105).

References

1. BEETON, A. M. "Eutrophication of the St. Lawrence Great Lakes." Limnology and Oceanography, 10 (1965):240–54.

2. BEETON, A. M. Indices of Great Lakes Eutrophication. Great Lakes Research Division Publication No. 15, University of Michigan, 1966, 8 pp.

3. HOLLAND, R. E. and A. M. BEETON. "Significance to Eutrophication of Spatial Differences in Nutrients and Diatoms in Lake Michigan." Limnology and Oceanography, 17 (1972):88–96.

4. VEITH, G. and G. F. LEE. "PCBs in Fish from the Milwaukee Region." Proceedings of the 14th Conference on Great Lakes Research, International Association for Great Lakes Research, 1971, pp. 157–69.

5. EISENREICH, S. As cited in "The Invisible Snowfall." Environmental Science and Technology, 13 (1979):1337–39.

6. BALLENTINE, R. K., S. R. REZNEK, and C. W. HALL. Subsurface Pollution Problems in the United States. U.S. Environmental Protection Agency Technical Studies Report TS-00-72-02, 1972, 24 pp.

7. United States Environmental Protection Agency. "National Interim Primary Drinking Water Regulations." *Federal Register,* 40, no. 248 (December 24, 1975):59566–88.

8. United States Environmental Protection Agency. *Quality Criteria for Water.* Washington, D.C.: U.S. Government Printing Office, 1976, 256 pp.

9. HUEPNER, W. C. "Environmental Carcinogenesis in Man and Animals." *Annals of the New York Academy of Sciences,* 108 (1963):963–91 and appendices.

10. NOVEY, H. and S. H. MARTELL. "Asthma, Arsenic and Cancer." *Journal of Allergy,* 44 (1969):315–19.

11. GOLDSMITH, J. R. et al. "Evaluation of Health Implications of Elevated Arsenic in Well Water." *Water Research,* 6 (1972):1133–36.

12. TSANG, W. P. et al. "Prevalence of Skin Cancer in an Endemic Area of Chronic Arsenicism in Taiwan." *Journal of the National Cancer Institute,* 40 (1968):453–63.

13. RASMUSSEN, G. K. and W. H. HENRY. "Effects of Arsenic on the Growth of Pineapple and Orange Seedlings in Sand and Solution Nutrient Cultures." *Citrus Industry,* 46 (1965):22–23.

14. LYDTIN, H. et al. "Effect of an Adrenergic Beta-receptor Blocking Agent on the Cardiac Rhythm through Experimental Barium Poisoning." *Arzneimittel-forschung,* 17 (1967):1456–59.

15. LITTLE, A. D. "Inorganic Chemical Pollution of Fresh Water." In *Water Quality Data Book,* vol. 2. U.S. Environmental Protection Agency, 18010 DPV, 1971, pp. 24–26.

16. NISHIMURA, M. "Clinical and Experimental Studies on Acute Beryllium Disease." *Nagoya Journal of Medical Science,* 28 (1966):17–44.

17. REEVES, A. "Absorption of Beryllium from the Gastrointestinal Tract." *Archives of Environmental Health,* 11 (1965):209–14.

18. SLONIM, C. B. and A. R. SLONIM. "Effect of Water Hardness on the Tolerance of the Guppy to Beryllium Sulfate." *Bulletin of Environmental Contamination and Toxicology,* 10 (1973):195–301.

19. SLONIM, A. R. and E. E. RAY. "Acute Toxicity of Beryllium Sulfate to Salamander Larvae (*Anbystonia* spp.)." *Bulletin of Environmental Contamination and Toxicology,* 13 (1975):307–12.

20. BALLARD, E. G. and G. W. BUTLER. "Mineral Nutrition of Plants." *Annual Review of Plant Physiology,* 17 (1966):77–112.

21. ROMNEY, E. M. and J. D. CHILDRESS. "Effect of Beryllium in Plants and Soil." *Soil Science,* 100 (1965):210–17.

22. EMMERSON, B. T. "Ouch-Ouch Disease: The Osteomalacia of Cadmium Nephropathy." *Annals of Internal Medicine,* 73 (1970):854–55.

23. KOBAYSHI, J. *Relation between the "Itai-Itai" Disease and the Pollution of River Water by Cadmium from a Mine.* Fifth International Water Pollution Research Conference, I-15, 1970, pp. 1–7.

24. WALDBOTT, G. L. *Health Effects of Environmental Pollutants.* St. Louis: C. V. Mosby Company, 1973, 316 pp.

371

25. CARPENTER, S. J. "Teratogenic Effect of Cadmium and Its Inhibition by Zinc." *Nature*, 216 (1967):1123.

26. THIND, G. S. "Role of Cadmium in Human and Experimental Hypertension." *Journal of the Air Pollution Control Association*, 22 (1972):267–70.

27. KENDREY, G. and F. J. C. ROE. "Cadmium Toxicology," *Lancet*, 1 (1969):1206–7.

28. FRANT, S. and I. KLEEMAN. "Cadmium 'Food Poisoning.'" *Journal of the American Medical Association*, 117 (1941):86–89.

29. PICKERING, O. H. and M. H. GAST. "Acute and Chronic Toxicity of Cadmium to the Fathead Minnow (*Pimephales promelas*)." *Journal of Fisheries Research Board, Canada*, 29 (1972):1099–1106.

30. EATON, J. G. "Cadmium Toxicity to the Bluegill (*Lepomis macrochirus Rafinesque*)." *Transactions of the American Fisheries Society*, 103 (1974):729–35.

31. MC CAULL, J. "Building a Shorter Life." *Environment*, 13, no. 7 (1971):3–14, 38–41.

32. DURUM, W. H. et al. *Reconnaissance of Selected Minor Elements in Surface Waters of the United States, October 1970.* U.S. Geological Survey Circular 643, 1971.

33. PICKERING, Q. H. and C. HENDERSON. "The Acute Toxicity of Some Heavy Metals to Different Species of Warm Water Fishes." *International Journal of Air-Water Pollution*, 10 (1960):453–63.

34. JONES, J. R. E. *Fish and River Pollution.* London: Butterworth, 1964.

35. MOUNT, D. I. and C. E. STEPHAN. "Chronic Toxicity of Copper to the Fathead Minnow (*Pimephales promelas*) in Soft Water." *Journal of Fisheries Research Board, Canada*, 26 (1969):2449–57.

36. WARNICK, S. L. and H. L. BELL. "The Acute Toxicity of Some Heavy Metals to Different Species of Aquatic Insects." *Journal of the Water Pollution Control Federation*, 41 (1969):280–84.

37. DOUDOROFF, P. and M. KATZ. "Critical Review of Literature on the Toxicity of Industrial Wastes and Their Components to Fish. II. The Metals, as Salts." *Sewage and Industrial Wastes*, 25 (1953):802–39.

38. COHEN, J. M. et al. "Taste Threshold Concentrations of Metals in Drinking Water." *Journal American Water Works Association*, 52 (1960):660–70.

39. SHUKLA, S. S. and H. V. LE LAND. "Heavy Metals: A Review of Lead." *Journal of the Water Pollution Control Federation*, 45 (1973):1319–31.

40. BARRY, P. S. I. and D. B. MOSSMAN. "Lead Concentrations in Human Tissues." *British Journal of Industrial Medicine*, 27 (1970):339–51.

41. PERLSTEIN, N. A. and R. ATTALA. "Neurologic Sequelae of Plumbism in Children." *Clinical Pediatrics*, 5 (1966):292–98.

42. ROSENSTOCK, H. H., D. G. SIMONS, and J. S. MEYER. "Chronic Manganism, Neurologic and Laboratory Studies During Treatment with Leva-dopa." *Journal of the American Medical Association*, 217 (1971):1354–56.

43. SHACKLETTE, H. T. et al. *Mercury in the Environment—Surficial Materials of the Conterminous United States.* U.S. Geological Survey Circular 644, 1971, 5 pp.

44. WERSHAW, R. L. "Mercury in the Environment." U.S. Geological Survey Professional Paper 713, 1970.

45. GRANT, N. "Mercury in Man." *Environment,* 13, no. 4 (1971):2–15.

46. SUZUKI, T., T. MIYAMA, and H. KATSUNUMA, "Comparison of Mercury Contents in Maternal Blood, Umbilical Cord Blood and Placental Tissues." *Bulletin of Environmental Contamination and Toxicology,* 5 (1970):502–8.

47. WOOD, J. M. "Biological Cycles for Toxic Elements in the Environment." *Science,* 183 (1974):1049–52.

48. MC KIM, J. M. et al. "Long-term Effects of Methylmercuric Chloride on Three Generations of Brook Trout (*Salvelinus fontinalis*): Toxicity, Accumulation, Distribution, and Elimination." *Journal of Fisheries Research Board, Canada,* 33 (1976):2726–39.

49. SCHROEDER, H. A. et al. "Abnormal Trace Elements in Man—Nickel." *Journal of Chronic Diseases,* 15 (1962):51–65.

50. BRIEF, R. S. et al. "Metal Carbonyls in the Petroleum Industry." *Archives of Environmental Health,* 23 (1971):373–84.

51. CALABRESE, A. and D. A. NELSON. "Inhibition of Embryonic Development of the Hard Clam, *Mercenaria mercenaria,* by Heavy Metals." *Bulletin of Environmental Contamination and Toxicology,* 11 (1974):92–97.

52. BIESINGER, K. E. and G. M. CHRISTENSEN. "Effects of Various Metals on Survival, Growth and Reproduction of *Daphnia magna.*" *Journal of Fisheries Research Board, Canada,* 29 (1972):1691–1700.

53. PICKERING, Q. H. "Chronic Toxicity of Nickel to the Fathead Minnow." *Journal of the Water Pollution Control Federation,* 46 (1974):760–65.

54. TERHAAR, C. J. et al. "Toxicity of Photographic Processing Chemicals to Fish." *Photographic Science and Engineering,* 16 (1972):370–77.

55. COLEMAN, R. L. and J. E. CLEARLY. "Silver Toxicity and Accumulation in Largemouth Bass and Bluegill." *Bulletin of Environmental Contamination and Toxicology,* 12, no. 1 (1974):53–61.

56. PRASAD, A. S. "Zinc: Human Nutrition and Metabolic Effects." *Annals of Internal Medicine,* 73 (1970):631–35.

57. BRAUN, M. A. et al. "Food Poisoning Involving Zinc Contamination." *Archives of Environmental Health,* 8 (1964):657–60.

58. MOUNT, D. I. "The Effect of Total Hardness and pH on Acute Toxicity of Zinc to Fish." *Air-Water Pollution,* 10 (1966):49–56.

59. TSAI, C. "Water Quality and Fish Life below Sewage Outfalls." *Transactions of the American Fisheries Society,* 102 (1973):281–92.

60. DANDY, J. W. T. "Activity Response to Chlorine in Brook Trout, *Salvelinus fontinalis.*" *Canadian Journal of Zoology,* 50 (1972):405–10.

61. BRUNGS, W. A. "Effects of Residual Chlorine on Aquatic Life." *Journal of the Water Pollution Control Federation,* 45 (1973):2180–93.

62. SPENCER, H. et al. "Availability of Fluoride from Fish Protein Concentrate and from Sodium Fluoride in Man." *Journal of Nutrition,* 100 (1970): 1415–24.

373

63. VIGIL, J. et al. *Nitrates in Municipal Water Supplies Cause Methemoglobinemia in Infants.* Public Health Report 80, 1965, pp. 1119–21.

64. KNEPP, G. L. and G. F. ARKIN. "Ammonia Toxicity Levels and Nitrate Tolerance of Channel Catfish." *The Progressive Fish Culturist,* 35 (1973):221–24.

65. MC COY, E. F. *Role of Bacteria in the Nitrogen Cycle in Lakes.* Water Pollution Control Research Series, U.S. Environmental Protection Agency (EP 2.10: 16010 EHR 03/72), 1972.

66. RUSSO, R. C. et al. "Acute Toxicity of Nitrite to Rainbow Trout (*Salmo gairdneri*)." *Journal of Fisheries Research Board, Canada,* 31 (1974):1653–55.

67. DOWNING, K. M. and J. C. MERKENS. "The Influence of Dissolved Oxygen Concentration on the Toxicity of Un-ionized Ammonia to Rainbow Trout (*Salmo gairdneri Richardson*)." *Annals of Applied Biology,* 43 (1955):243–46.

68. BALL, I. R. "The Relative Susceptibilities of Some Species of Freshwater Fish to Poisons—I. Ammonia." *Water Research,* 1 (1967):767–75.

69. VOLLENWEIDER, R. A. "Möglichkeiter und Grenzen elementarer Modelle der Stoffbilanz von Seen." *Archiv fuer Hydrobiologie,* 66 (1969):1–36.

70. DILLON, P. J. "A Critical Review of Vollenweider's Nutrient Budget Model and Other Related Models." *Water Resources Bulletin,* 10 (1974):969–89.

71. VOLLENWEIDER, R. A. *Water Management Research.* OECD Paris, DAS/CSI/ 68.27, 1968, 183 pp.

72. DILLON, P. J. "The Phosphorus Budget of Cameron Lake, Ontario: The Importance of Flushing Rate to the Degree of Eutrophy of Lakes." *Limnology and Oceanography,* 20 (1975):28–39.

73. IMBODEN, D. M. "Phosphorus Model of Lake Eutrophication." *Limnology and Oceanography,* 19 (1974):297–304.

74. LAKIN, H. W. "Selenium Accumulation in Soil and Its Absorption by Plants and Animals." *Bulletin, Geological Society of America,* 83 (1972):181–89.

75. JOHNSON, H. "Determination of Selenium in Solid Waste." *Environmental Science and Technology,* 4 (1970):850–53.

76. ROSENFELD, I. and O. A. BEATH. *Selenium.* New York: Academic Press, 1964.

77. COPELAND, R. "Selenium: The Unknown Pollutant." *Limnos,* 4, no. 3 (1970):7–9.

78. SMITH, L. L. and D. M. OSEID. "Effects of Hydrogen Sulfide on Fish Eggs and Fry." *Water Research,* 6 (1972):711–20.

79. DONALDSON, W. T. "Trace Organics in Water." *Environmental Science and Technology,* 11 (1977):348–51.

80. "TOSCA: Paves the Way for Controlling Toxic Substances." *Environmental Science and Technology,* 11 (1977):28–29.

81. TREON, J. F. et al. "Toxicity of Endrin for Laboratory Animals." *Journal of Agricultural and Food Chemistry,* 3 (1955):842–48.

82. POST, G. and T. R. SCHROEDER. "The Toxicity of Four Insecticides to Four Salmonid Species." *Bulletin of Environmental Contamination and Toxicology,* 6 (1971):144–55.

83. JENSEN. L. D. and A. R. GAUFIN. "Acute and Long-Term Effects of Organic Insecticides on Two Species of Stonefly Naiads." *Journal of the Water Pollution Control Federation,* 38 (1966):1273–86.

84. MACEK, K. J. and W. A. MC ALLISTER. "Insecticide Susceptibility of Some Common Fish Family Representatives." *Transactions of the American Fisheries Society,* 99 (1970):20–27.

85. SEABURY, J. H. "Toxicity of 2,4 Dichlorophenoxyacetic Acid for Man and Dog." *Archives of Environmental Health,* 7 (1963):202–9.

86. U.S. Department of Health, Education and Welfare. *Drinking Water Standards.* Public Health Bulletin 956, 1962.

87. GORRELL, H. A. "Classification of Formation Waters Based on Sodium Chloride Content." *Bulletin, American Association of Petroleum Geologists,* 42 (1953):2513.

88. HOFFMAN, J. I. and C. W. FETTER, JR. "Water Softener Salt: A Major Source of Groundwater Contamination." Geological Society of America, *Abstracts with Programs,* 10, 7 (1978):423.

89. EVERETT, G. V. "Rainbow Trout, *Salmo gairdneri (Rich.),* Fishery of Lake Titicaca." *Journal of Fish Biology,* 5 (1973):429–40.

90. CRAUN, G. F. "Waterborne Disease—A Status Report Emphasizing Outbreaks in Groundwater Systems." *Ground Water,* 17 (1979):183–91.

91. RUNNELLS, D. D. "Wastewater in the Vadose Zone of Arid Regions: Geochemical Interactions." *Ground Water,* 14 (1976):374–84.

92. MC CABE, L. J. and G. F. CRAUN. "Status of Waterborne Disease in the U.S. and Canada." *Journal American Water Works Association,* 67 (1975):95–98.

93. SMITH, S. D. and D. H. MYOTT. "Effect of Cesspool Discharge on Ground-Water Quality on Long Island, New York." *Journal American Water Works Association,* 67 (1975):456–58.

94. FETTER, C. W., JR. "Water Quality and Pollution, South Fork of Long Island, New York." *Water Resources Bulletin,* 10 (1974):779–88.

95. MANN, J. F., JR. "Wastewater in the Vadose Zone of Arid Regions: Hydrologic Interactions." *Ground Water* 14 (1976):367–75.

96. CLARK, T. P. "Survey of Ground-Water Protection Methods for Illinois Landfills." *Ground Water* 13 (1975):321–31.

97. GIDDINGS, M. T., JR. "The Lycoming County, Pennsylvania, Sanitary Landfill: State of the Art in Groundwater Protection." *Ground Water* 15 (1977):5–14.

98. ROBERTSON, F. N. "Hexavalent Chromium in the Groundwater in Paradise Valley, Arizona." *Ground Water* 13 (1975):516–27.

99. SYMONS, J. M. et al. "National Organics Reconnaissance Survey for Halogenated Organics." *Journal American Water Works Association,* 67 (1975):634–47.

100. KLUSMAN, R. W. and K. W. EDWARDS. "Toxic Metals in Groundwater of the Front Range, Colorado." *Ground Water,* 15 (1977):160–69.

101. AHMAD, M. U. "Coal Mining and Its Effect on Water Quality." *American Water Resources Association Proceedings No. 18,* 1974, pp. 138–48.

375

102. NORBECK, P. N., L. L. MINK, and R. E. WILLIAMS. "Ground Water Leaching of Jig Tailing Deposits in the Coeur d'Alene District of Northern Idaho." *American Water Resources Association Proceedings No. 18*, 1974, pp. 149–57.

103. RALSTON, D. R. and A. G. MORILLA. "Ground-Water Movement through an Abandoned Tailings Pile." *American Water Resources Association Proceedings No. 18*, 1974, pp. 174–83.

104. MC WHORTER, D. B., R. K. SKOGERBOE, and G. V. SKOGERBOE. "Potential of Mine and Mill Spoils for Water Quality Degradation." *American Water Resources Association Proceedings No. 18*, 1974, pp. 123–37.

105. KAUFMAN, R. F., G. G. EADIE, and C. R. RUSSELL. "Effects of Uranium Mining and Milling on Ground Water in the Grants Mineral Belt, New Mexico." *Ground Water*, 14 (1976):296–308.

Groundwater Development and Management

chapter 11

11.1 INTRODUCTION

The area of water resources development and management is so broad that we will make no attempt to cover all aspects in this book. Instead we will focus on groundwater, since it is this aspect of the general topic to which hydrogeology most closely relates. However, as groundwater is not isolated from surface water, a study of groundwater development necessarily encompasses many aspects of surface-water flow.

Surface water occurs in readily discernible drainage basins. The boundaries are topographic and may be easily delineated on a topographic map. The water conveniently flows in the direction in which the land surface is sloping. Moreover, surface water does not cross topographic divides (except, perhaps, during floods) and the locations of the drainage divides are fixed. Groundwater, on the other hand, occurs in aquifers that are hidden from view. The boundaries of an aquifer are physical: it can crop out, abut an impermeable rock unit, grade into a lower permeability deposit, or thin and disappear. At a given location, the land surface may be underlain by several aquifers. Each aquifer may have different chemical makeup and different hydraulic potential; each may be recharged in a different location and flow in a different direction. Moreover, groundwater divides do not necessarily coincide with surface-water divides. Clearly, the development and management of groundwater is more complicated than that of surface water, simply on the basis of the mode of occurrence.

11.2 DYNAMIC EQUILIBRIUM IN NATURAL AQUIFERS

Under natural conditions, an aquifer is usually in a state of **dynamic equilibrium** (1). A volume of water recharges the aquifer and an equal volume is discharged. The potentiometric surface is steady and the amount of water in storage in the aquifer is a constant. The aquifer transmits the water from the recharge areas to the discharge zones. The maximum amount of water any section of the aquifer can transmit is a function of the transmissivity and the maximum gradient of the potentiometric surface. If the water table is close to the surface of an unconfined aquifer, the aquifer is full and is transmitting the maximum amount of water. If, however, the water table is far below the surface, the aquifer is not transmitting water at full capacity.

The amount of water that recharges an unconfined aquifer is determined by three factors: (a) the amount of precipitation that is not lost by evapotranspiration and runoff, and is thus available for recharge; (b) the vertical hydraulic conductivity of surficial deposits and other strata in the recharge area of the aquifer, which determines the volume of recharged water capable of moving downward to the aquifer; and (c) the transmissivity of the aquifer and potentiometric gradient, which determine how much water can move away from the recharge area. Should an aquifer be transmitting the maximum volume of water, it is more than likely that some potential recharge is being rejected in the recharge area. This is often the case in humid areas. Should the water table be low, indicating that the aquifer is not flowing at full capacity, there is probably either a lack of potential recharge or low vertical hydraulic conductivity in the recharge area retarding downward movement. Aquifers in arid regions typically have deep water tables in the recharge areas, indicating a deficiency in the amount of potential recharge.

Recharge to confined aquifers can occur in places in which the confining layer is absent. Under such conditions, the three factors affecting unconfined aquifer recharge are controlling. If there is a hydraulic gradient across the confining layer in a direction that promotes flow into the aquifer, then recharge can occur across the confining layer. In this case, the vertical hydraulic conductivity of the confining layer, the thickness of the confining layer, and the head difference across it control the amount of recharge. Recharge to a confined aquifer may come from both downflow from a higher aquifer or upflow from a lower aquifer.

When a well begins to pump water from an aquifer, the water is withdrawn from storage around the well and from vertical leakage (1). As the cone of depression grows, an increasingly larger portion of the aquifer will be contributing water from storage. The amount of water discharging naturally from the aquifer will remain at the predevelopment rate until the pumping cone reaches the recharge or discharge area. When the pumping cone reaches a discharge area, the potentiometric gradient toward the discharge area is lowered, and the amount of natural discharge proportionally reduced. If the pumping cone reaches the recharge area of an aquifer, it may induce additional recharge of water that was previously rejected. It is even possible for a section of the aquifer to change from a discharge area to a recharge area. For example, drawdown near a river may eliminate groundwater discharge to the river and induce infiltration from the river into the aquifer, reversing the prior direction of flow.

In any event, the pumping cone will continue to grow until it has sufficiently reduced natural discharge or increased recharge to balance the volume of water removed by pumping. With this occurrence, a new condition of dynamic equilibrium is reached (1). However, it is important to note that in

379

order for this to happen a pumping cone must form. This means the potentiometric surface in parts of the aquifer must be lowered. The resultant cone depression is usually a complex surface resulting from the withdrawals of hundreds of wells. Should the sum of the remaining natural discharge and the pumping withdrawals exceed the available recharge, the cone of depression will not stabilize and water levels will continue to fall.

The rate at which the cone of depression spreads is a function of the storativity of a confined aquifer or the specific yield of a water-table aquifer. As storativity values are 100 to 1000 times smaller than specific yields, the cone of depression will spread 100 to 1000 times faster in an artesian aquifer than in a water-table aquifer. It can take many years for the cone of depression to influence recharge or discharge areas sufficiently for an aquifer to regain dynamic equilibrium.

CASE STUDY: DEEP SANDSTONE AQUIFER OF NORTHEASTERN ILLINOIS

The deep sandstone aquifer beneath northeastern Illinois is confined by the Maquoketa Shale, a leaky layer. It has an outcrop area that receives direct recharge west of the limit of the shale. Under predevelopment conditions, the potentiometric surface stood at or above land level and had a very gentle slope toward Lake Michigan of about 3×10^{-4} meter per meter. Flow through the aquifer from the recharge area was about 3400 cubic meters (0.9 million gallons) per day (2). Natural discharge was by slow upward leakage through the shale into Silurian limestone beneath Lake Michigan. Most of the groundwater recharge in the recharge area of the regional aquifer was circulating in local flow systems. It could not move as a part of regional flow due to the low carrying capacity of the aquifer caused by the gentle slope of the potentiometric surface. As groundwater development began, the well discharge soon exceeded the natural discharge and the pumping cone rapidly grew. As the hydraulic gradient steepened, increasingly greater amounts of water were drawn into the regional flow system. By 1958, an estimated 72,000 cubic meters (19 million gallons) per day were moving into the pumping cone (2). At that time, pumpage from the aquifer was 163,000 cubic meters (43 million gallons) per day. The difference between pumpage and recharge was coming from storage as the cone of depression grew. By 1975, groundwater pumpage was 553,000 cubic meters (150 million gallons) per day, and the hydraulic gradient was 4×10^{-3} meter per meter, with water levels falling several meters per year.

If pumpage from the deep sandstone aquifer in the Chicago region had been limited to 174,000 cubic meters (46 million gallons) per day, the cone of depression would have eventually stabilized. The period to reach a stable configuration would have been about 50 years from the time pumpage first reached 174,000 cubic meters per day (3). However, so long as the rate of deep-well pumpage exceeds 174,000 cubic meters per day, the cone of depression will continue to grow (4).

380

GROUNDWATER BUDGETS 11.3

Some knowledge of the amount of natural recharge to an aquifer is mandatory in a groundwater development program. There are several methods available to make such an estimate. In Section 3.7, a method was described for estimating annual groundwater recharge from baseflow-recession curves. This method is useful for areas in which the groundwatershed and the river basin coincide.

A second method is to determine the flow in the aquifer across a vertical plane at the boundary of the recharge area and the discharge area (5). If there is sufficient knowledge of the potential field of the aquifer, the area(s) of recharge and discharge can be determined. The volume of steady flow from recharge areas to discharge areas is determined using either Darcy's law for a confined aquifer or the Dupuit equation for an unconfined aquifer. The flow from the recharge areas is equal to the volume of recharge for aquifers in dynamic equilibrium.

A **water budget** for the recharge area of an aquifer is a very useful means of determining groundwater recharge. An advantage of the water-budget method is that the aquifer does not have to be in dynamic equilibrium in order to use it. Many of the parameters used for a hydrologic budget are measured directly: precipitation, streamflow, transported water, and reservoir evaporation. Groundwater inflow, outflow, and change in storage are computed from the hydraulic aquifer characteristics and measured potentiometric data. The amount of evapotranspiration could be measured in lysimeters, but it is more typically computed using an appropriate formula such as the Thornthwaite method (Section 2.3). Basin-wide evapotranspiration may also be determined by a water-budget analysis as follows:

$$
\begin{aligned}
\text{Evapotranspiration} = &\ (\text{precipitation} + \text{surface-water inflow} \qquad \textbf{(11-1)} \\
&\ + \text{imported water} + \text{groundwater inflow}) \\
&- (\text{surface-water outflow} + \text{groundwater} \\
&\quad \text{outflow} + \text{reservoir evaporation} \\
&\quad + \text{exported water}) \\
&\pm \text{changes in surface-water storage} \\
&\pm \text{changes in groundwater storage}
\end{aligned}
$$

The natural recharge to an undeveloped aquifer may be determined by a water-budget analysis of the recharge area:

$$
\begin{aligned}
\text{Groundwater recharge} = &\ (\text{precipitation} + \text{surface-water inflow} \qquad \textbf{(11-2)} \\
&\ + \text{imported water} + \text{groundwater inflow}) \\
&- (\text{evapotranspiration} + \text{reservoir} \\
&\quad \text{evaporation} + \text{surface-water outflow} \\
&\quad + \text{exported water} + \text{groundwater outflow}) \\
&\pm \text{changes in surface-water storage}
\end{aligned}
$$

381

Equation (11-2) can account for groundwater recharge not only from precipitation, but also from losing streams, irrigation water, unlined canals, and so forth. Its usefulness may be limited, however, if evapotranspiration cannot be determined. Detailed knowledge of all the factors is necessary if the computation of recharge is to be accurate.

If the land surface of the recharge area is developed for agriculture, industry, urban growth, and so forth, additional computations of recharge may be necessary. Water used for many purposes may be recharged to the groundwater reservoir. Excess irrigation water is one example. On Long Island, New York, all groundwater pumped for cooling or air conditioning must be returned to the aquifer from which it was removed. Water used for domestic purposes is often recharged by septic tank drain fields or cesspools. The increasing emphasis on the use of land systems* for municipal wastewater treatment means that treated sewage effluent which formerly flowed into rivers may now be recharging aquifer systems. The amount of recharge from such sources can be determined by a supplemental water-budget analysis:

(11-2a)

Additional
groundwater recharge = (industrial use + municipal use
+ domestic use + irrigation use)

− (cooling-water evaporation
+ irrigation-water evapotranspiration
+ water exported in products
+ sewage discharge into surface waters)

The determination of the additional groundwater recharge is often more complicated than an analysis of the basic amount of recharge. Water-use records of dozens or hundreds of individual water users and sewage dischargers must be collected. These records may range from excellent to nonexistent. Many of the factors must be estimated. For example, the owners of private wells for home use will almost never know how much water they pump. An accurate accounting of this type involves long and tedious inventory analysis.

Should an attempt be made to balance the additions to the water supply of an area with the depletions, the result should be accurate to within the accuracy of measurement or estimation of the various parameters. Each parameter will have an accuracy of estimation dependent upon how precisely the measurement can be made. Measurement of streamflow might be accurate to ±5 percent for a good measurement. If total streamflow is 30 cubic meters per second, then there is a variability of 30 cubic meters per second × 0.05,

---

*Land systems of treatment include, among other techniques, spraying wastewater as irrigation water, recharging it through seepage basins, and allowing it to flow across the land surface. Biological, chemical, and physical processes act to remove pollutants and purify the water.

or ± 1.5 cubic meters per second. The overall accuracy of the estimate can be determined by taking a weighted average of the individual variability.

EXAMPLE PROBLEM

A water-budget analysis of a watershed indicates an estimated total outflow of 90 cubic meters per second. The accuracy of estimation of each of the individual components is known. What is the accuracy of the estimate of the total discharge?

Factor	Estimated Flow	Accuracy of Estimate	Variability of Factor
Evapotranspiration	60 m³/sec	±25%	±15 m³/sec
Surface ourflow	20 m³/sec	± 5%	± 1 m³/sec
Groundwater outflow	5 m³/sec	±10%	± 0.5 m³/sec
Exported water in canal	5 m³/sec	± 2%	± 0.1 m³/sec
TOTAL	90 m³/sec		±16.6 m³/sec

The sum of the variability of each factor is ± 16.6 cubic meters per second; therefore, the accuracy of the total flow is 16.6/90 or ± 18 percent.

The amount of water available for use from an aquifer is not the natural recharge. It is the increase in recharge or leakage from adjacent strata induced by development, along with the reduction in discharge. As water levels fall to accommodate the development, there will also be some water available from storage. A water-budget analysis is helpful in determining the amount of natural recharge and discharge. Further hydrologic analysis is then necessary to evaluate the effects that pumping will have on these figures.

MANAGEMENT POTENTIAL OF AQUIFERS 11.4

Aquifers can play many roles in the overall development of the water resources of an area. Some of the functions have been recognized for many years; others have been recognized only recently. The most obvious use of an aquifer is to supply water to wells — the **supply function**. One of the more vexing problems in groundwater management has been to determine how much water an aquifer can supply. This problem will be discussed in Section 11.5.

Aquifers also transmit water from one location to another. This has been called a **pipeline function** (6). Many communities are dependent upon groundwater sources that are recharged elsewhere. In this case, the aquifer acts as an aqueduct. However, when the user community does not have zoning and land-use control, as, for example, when the aquifer is transmitting water some distance, there may be difficulty in protecting the recharge area of the aquifer from overdevelopment or contamination (7). Aquifers are not as efficient in carrying large volumes of water as are surface canals. However, surface canals can lose large amounts of water by evaporation; furthermore, they require capital for construction.

Groundwater can be mined in the same manner as minerals. Whenever groundwater is withdrawn at a rate greater than replenishment, mining is occurring. In some aquifers, the rate of replenishment is so low that it is almost nonexistent. For example, the average annual recharge to the Ogallala aquifer of the southern High Plains of New Mexico and Texas is 3.8 centimeters per year (8). Based on the average rate of withdrawal for the 1951–1960 period, the groundwater mine will be exhausted in 100 years. That pumping rate is twenty-eight times the natural recharge. Property owners of land in the High Plains are permitted an income-tax depletion allowance to compensate for the loss of property value due to a falling water table (9).

The unsaturated zone overlying an aquifer can act as a waste-treatment system. This has been called the **"filter-plant" function** of aquifers (7). However, the unsaturated zone can do much more than act as a physical filter to remove bacteria and viruses. It is also effective in removing phosphorus and heavy metals (10). Passage of water through the saturated zone can also improve the quality. Degradation of native water quality can occur if the treatment potential of soil systems is exceeded.

Groundwater can also have an **energy-source function**. The groundwater heat pump is a viable alternative to conventional heat pumps in some localities (11). As the thermal energy of the aquifer is removed by a heat pump in colder climates, it is another type of mining. In Wisconsin, the thermal impact is very small (12). In states further south, where heating and air conditioning demands are more equal, the net impact on groundwater temperatures is even less.

Groundwater reservoirs sometimes also have a **storage function**. This is not true for an undeveloped aquifer in dynamic equilibrium if the recharge zone is rejecting potential recharge. However, if the groundwater reservoir has unused storage capacity, it can effectively store water from wet periods for use during times of drought (13–16). Aquifers with available storage capacity may either be those which are not filled by the natural recharge to them, or those in which the potentiometric surface has been lowered by pumping. Water put into storage could be from natural sources, especially extreme precipitation or flood events (13). Increasingly, treated wastewater effluent is being used to replenish aquifers (17).

When groundwater stored in aquifers can be used to replace surface-water reservoirs, a number of benefits accrue (18). The expense of surface-water reservoirs is circumvented, as there are no capital costs involved. There are no evaporative losses from groundwater storage, nor are there any infiltration losses. Surface-water reservoirs sometimes create great ecological disruption with the destruction of riverine and floodplain environments. Productive farmland may also be lost beneath surface reservoirs. Such disruption may be avoided with groundwater storage; furthermore, there is no worry over dam safety. On the other side of the coin, a person cannot waterski in a groundwater reservoir!

Aquifers are also used for the storage of natural gas. The aquifer must be confined, and a structural or stratigraphic feature, such as an anticline, is required to hold the gas in place. Wells are drilled through the structure and the gas is pumped into the aquifer under pressure. It displaces water and forms a bubble in the aquifer. Fresh water could also be stored in saltwater aquifers, as the former is less dense and would float as a bubble in the saline water.

PARADOX OF SAFE YIELD

It is a natural inclination of scientists to compare and classify phenomena in quantitative terms. Thus, it is to be expected that hydrogeologists have attempted to define the amount of water which could be developed from a groundwater reservoir. The term **safe yield** was apparently used in this regard as early as 1915 (19). At that time, safe yield was regarded as the amount of water which could be pumped "regularly and permanently without dangerous depletion of the storage reserve." Later, other factors that need to be considered were added, such as economics of groundwater development (20), protection of the quality of the existing store of groundwater (21), and protection of existing local rights and potential environmental degradation (22). Synonyms for safe yield appear in the literature, including "potential sustained yield" (2), "permissive sustained yield" (24), and "maximum basin yield" (25). A composite definition, based on the ideas of many authors, could be expressed as follows: Safe yield is the amount of naturally occurring groundwater that can be withdrawn from an aquifer on a sustained basis, economically and legally, without impairing the native groundwater quality or creating an undesirable effect such as environmental damage.

The concept of groundwater withdrawals causing environmental damage warrants more than a mention. Many surface-water systems are dependent upon natural groundwater discharge. It has been shown by model studies that groundwater development may reduce streamflow and, as a consequence, lower lake levels and dry wetlands (26). As these may be environmentally sensitive areas, the danger of environmental harm is real. Likewise, groundwater withdrawals have been linked to subsidence of the land surface (27). This has resulted in land-surface cracking and damage to structures, highways, pipelines, dams, and tunnels. The gradients of irrigation canals have been changed—even reversed—and low areas have become flooded by sea water. In a broader sense, environmental impacts include ecological, economical, social, cultural, and political values (28).

Many authorities are uneasy with the concept of safe yield. For some, the term is too vague (29). Obviously, the amount of groundwater that can be produced will vary under varied patterns of pumping and development. In addition, the question of what would constitute an undesirable result to be avoided is open to debate (30). The abandonment of the term safe yield has been proposed on the grounds that it does not take into account the interrelationship of groundwater and surface water and may preclude the development of the storage functions of an aquifer (6).

However, in spite of the reservations of many hydrogeologists with regard to the concept of safe yield and its implications, the basic concept must be applied whenever the use of an aquifer is planned and managed. Groundwater management programs obviously imply that water must be pumped from the ground (31). If there is no evaluation of the hydrologic and environmental impacts of various withdrawal programs, it is possible that uncontrolled withdrawals will exceed prudent levels.

A single value for the safe yield of an aquifer cannot be provided in the same sense as a quantity such as mean annual precipitation. Safe-yield values are based on a number of constraints; such values must be determined by a team of professionals, in the same manner that an environmental impact statement is prepared. Economists, engineers, engineering geologists, plant and wildlife ecologists, and lawyers might all participate with the hydrogeologist in preparing a safe-yield determination for an aquifer or a groundwater basin. The safe-yield evaluation should include a statement of the legal and economic constraints that were considered, as well as the limiting values of environmental damage that were considered. Indeed, such a study should provide a series of safe-yield values and the different factors that applied to each determination. This is obviously not a simple matter. Computer models of groundwater flow systems are ideal tools for estimating the series of values. All of the hydraulic factors can be evaluated. R. A. Freeze has shown how a computer model can compute a "maximum basin yield" (25).

The safe yield of an aquifer system is only one facet of a groundwater management program. Artificial augmentation of precipitation or recharge could increase the amount of water that can be withdrawn on a sustained basis. The use of groundwater reservoirs for cyclic storage means that in drought years it is necessary, and desirable, to pump water on a temporary basis far in excess of the safe yield. Under these conditions, the groundwater supply would replace surface-water supplies that might be critically low or be used to irrigate crops normally watered by rainfall. In wet years, the groundwater reservoir would be replenished by above-average recharge and pumping at rates below the safe yield.

The underlying principle of groundwater development is that by withdrawing water from an aquifer, some of the natural discharge may be made available for use (31).

386

LEGAL CONSTRAINTS
TO GROUNDWATER DEVELOPMENT 11.6

One of the basic considerations in evaluating safe yield is the legal constraints on groundwater development. Likewise, the legal basis for groundwater management must be established. As each state in the United States has its own body of groundwater law, it is not possible to provide a complete discussion of this subject. However, general principles can be elucidated and some specific cases illustrated. Because of the interaction of surface water and groundwater, an overview of surface-water law is also necessary. A **water right**, as defined by law, is not legal title to the water, but the legal right to use it in a manner dictated by state law.

11.6.1 FEDERAL ASPECTS OF WATER LAW

Surface-water law is a combination of both state and federal regulations, agreements, and law. For example, the amount of water that may be diverted from Lake Michigan at Chicago was fixed by the U.S. Supreme Court following a legal case that began in 1925 and concluded in 1967 (32). The appropriation of surface water in the Colorado River Basin has been made on the basis of a concatenation of events which included the Colorado River Compact (1922), the Boulder Canyon Project Act (1928), a treaty with Mexico (1944), the Upper Colorado River Compact (1948), the Colorado River Storage Project Act (1956), a Supreme Court decision, *Arizona* v. *California* (1963), and the Colorado River Basin Project Act (1968) (33).

No overt interstate conflicts have yet arisen over groundwater claims, even though there are a number of potential cases (34). An example of an interstate groundwater conflict is the deep sandstone aquifer of northeastern Illinois and southeastern Wisconsin. Pumping from urban areas in both Chicago and Milwaukee have caused lowering of groundwater levels in the adjacent state. In 1973, at the stateline, drawdown due to Illinois pumpage had lowered the potentiometric surface 52 meters (171 feet), while drawdown due to Wisconsin pumpage was 29 meters (94 feet). Due to the greater drawdown of the Chicago pumping cone, about 35,200 cubic meters per day (9.3 million gallons per day) of groundwater was flowing from Wisconsin into Illinois (3).

A matter of potential litigation in federal aspects of groundwater law is the **Winters Doctrine**. In a historic 1908 Supreme Court decision, it was held that when Indian reservations were established, the federal government also reserved the water necessary to make the land productive (35). States could not appropriate such water to other users. This included water on non-reservation lands that had been opened to settlement. The Court held that the amount of water reserved for Indian use was that sufficient to irrigate all the potentially arable reservation land. While *Winters* v. *United States* was specifically applied to surface water, the Supreme Court has also recently applied the doctrine to groundwater (36). Water resources management in the United States will be complicated for many years by the need to meet Indian water rights, especially in the West (37, 38).

11.6.2 STATE SURFACE-WATER LAW

In the eastern United States, the **riparian doctrine** of ownership of surface-water rights is recognized (39). This concept holds that the riparian landowner; that is, the property owner adjacent to a surface-water body, has the first right to withdraw and use the water. This right is often controlled by the state in the sense that application to the state for a permit to withdraw is necessary. Ownership of the water right is held with ownership of the land, and all riparian owners have equal rights to the water. Water withdrawals are limited to "reasonable" use in comparison with other riparians. As all riparians have equal rights to use reasonable amounts of water—even new users—it is fortunate that there is a large amount of surface water available in the eastern United States. As eastern riparian owners generally return most of the water to the stream, albeit sometimes more polluted, there have been no crises due to lack of water. Water problems in these areas typically involve quality rather than quantity of water.

In the seventeen western United States, a different doctrine, **prior appropriation**, is followed for surface water. The right to use water is separate from other property rights. The water-right holder does not necessarily need to be riparian, and riparian property owners may have no water rights. The first person to divert water from a surface course has the primary water right, and it passes to successive owners. Junior users have lower rights and in time of drought may not receive any water, even though senior users always get their full share. However, some states have established a priority of use, with domestic use receiving top priority, irrigation receiving second priority, and industrial and commercial receiving lowest priority. Most states have some limitation on the transfer of appropriative rights, sometimes going so far as to link the water right to a specific use and piece of land (39). Some states provide for the forfeiture of a water right if it is not exercised for a given time period.

11.6.3 STATE GROUNDWATER LAW

There are several different types of state law governing groundwater use, with a number of variations of the basic concepts. The right of absolute ownership of the water under a property holder's land is known as the **English Rule**. In 1843, an English court held that a landowner could pump groundwater at any rate, even if an adjoining property owner were harmed (40). The idea of absolute ownership was transported across the Atlantic and planted in the United States (41). In 1903, a Wisconsin court ruled in *Huber* v. *Merkel* that a property owner could pump groundwater, even with malicious intent to harm a neighbor, the water being put to no good use (42). However, this ruling was overturned in 1974 with the opinion of Wisconsin Justice Landry that "the Huber case and its misconceived progeny can no longer be relied upon as conferring to the owner of land an absolute right to use with impunity all of the water that can be pumped from the subsoil underneath" (43). Although Wisconsin courts now hold English Rule in disfavor, it is still the law of the land in Texas and in many other states (44).

Even before the 1903 *Huber* v. *Merkel* decision—as early as 1862—an American state court had ruled that a landowner had the right to use only a reasonable amount of groundwater. The rights of adjacent property owners were also recognized (45). This is known as the **American Rule**, and, as evidenced by the Wisconsin case, a state can change from the English Rule to the American Rule by judicial action. As case law moves with glacially slow progress, it is unusual for a state to shift from one rule to the other by court action.

Some western states have groundwater laws dictating that waters within the state boundaries belong to the public. Individuals can establish water rights to groundwater by use; the first to draw from an aquifer has established a senior right. New Mexico is a state with this doctrine for both ground and surface water (44). However, the state engineer of New Mexico can regulate the withdrawal of groundwater from "declared" underground water basins.

One area of conflict arises when surface water is fed by discharge of groundwater. Development of the groundwater reservoir typically depletes the baseflow of the stream. Holders of surface-water rights may find that the allocated surface water is diminishing in volume as groundwater withdrawals increase. On the South Platte River Basin of Colorado, the stream depletion due to groundwater withdrawals is about equal to the consumptive use of the groundwater pumped from the basin for irrigation (46). In some western states, the surface-water right can be usurped from the holder by groundwater pumpage. In the case of Colorado, this conflict is being resolved by integrating the claims of both surface-water and groundwater users with regard to the same basin (47). In some groundwater basins of New Mexico, would-be groundwater developers must acquire sufficient surface-water rights to compensate for the reduced flow from the stream (44).

In California, a groundwater system based on correlative rights is followed. The right to use groundwater belongs to the owner of the overlying land, so long as the use is reasonable. Water in excess of that used by the overlying owners can be allocated by appropriation. Thus, there are two systems of ownership of water rights. The determination of available water is based on the average recharge to the aquifer, termed the safe yield (although it is not the safe yield as defined in Section 11.4). If withdrawals are less than the average annual recharge, the excess can be apportioned. If an **overdraft** (defined as pumping in excess of the average annual recharge) exists for at least five years, the groundwater is apportioned among all users in amounts proportional to their individual pumping rates, with the total pumping set at the average annual recharge. This is the **doctrine of mutual prescription**, which was first adjudicated in the case of the Raymond Basin (48).

If one wishes to artificially recharge and store water in an aquifer, can someone else pump out that water? Can a public agency use the aquifer under a property holder's land for storage of imported water? These are questions central to the use of aquifers as storage reservoirs, and they must be answered if cyclic storage is to become an important procedure in water management. Two California cases dealing with aquifer storage have apparently settled these questions. They also may form a basis for resolution of these problems in other states (49).

The first California case involved the rights of the landowner to exploit an aquifer. Alameda County has a water district formed to protect the aquifer, especially from saltwater intrusion. As a conservation practice, local and imported water is injected into the aquifer to maintain the seaward hydraulic gradient. A sand and gravel pit was operating and, by 1969, the sand-mining operation extended 80 feet into the aquifer. The injected water was flooding the sand pit, requiring the mining company to pump large amounts of water from it. The mining company sued the water district to enjoin it from recharging the aquifer. The mining company lost the case, as the court held that public agencies can store water in an aquifer, up to the historic groundwater level, even if it decreases the usefulness of land to the property holders (50).

In the second case, a public entity injected water for underground storage with the intention of withdrawing it at a later time. This right to recapture injected waters was recognized by a court ruling in 1943 (51) and then reaffirmed in 1975 (52). In both instances, the aquifers in question were in the Los Angeles area, where several cities were competing for use of groundwater that comes from both natural sources and artificial recharge.

11.7 ARTIFICIAL RECHARGE

Resource management sometimes is interpreted as limiting the development and use of a resource in order to conserve it. Groundwater management has a somewhat broader scope, in that artificial recharge can be used to expand the amount of available water. **Water spreading** has been used in the western United States for decades to capture additional runoff. Much of the recharge of alluvial basin-fill aquifers comes from stream-bed infiltration during the wet season. In order to increase the amount of infiltration, stream water is diverted onto empty land below the mouth of the canyons carrying the streams (21).

Surface spreading is feasible given the following circumstances: the upper soil layers are permeable, the water table is not close to the surface, the land is relatively flat, and the aquifer to be recharged has a transmissivity great enough to carry the water away from the spreading area. It is extensively practiced simply by placing low dams across the ephemeral streams draining from the mountains. The dams act to flood the land along the stream channel.

Surface spreading tends to raise the water table over a rather extensive area. The same is true of other diffused types of artificial recharge. Irrigation is also a form of artificial recharge. Because of problems of salt accumulation in the soil, the amount of applied irrigation water is in excess of the needs of plants for evapotranspiration. The unevaporated water percolates through the unsaturated zone and recharges the water table. When spray irrigation is used as a means of wastewater treatment, the amount of water applied is far in excess of the plant requirement (53). The result may be a substantial increase in the elevation of the water table (54). Similar diffused sources of artificially recharged water include septic tank drain fields. Wastewater from these sources percolates through the unsaturated zone and recharges the water table (55, 56).

Recharge basins are frequently used to recharge unconfined aquifers, especially where land costs are high. Basins are advantageous in that a substantial hydraulic head can be maintained in order to increase the infiltration rate. They are inexpensive to construct and operate. On Long Island, New York, basins are used to recharge storm-water runoff from roadways and parking lots (57, 58). They are also used in Peoria, Illinois (59, 60) and Fresno, California (61) to recharge aquifers with river water. The infiltration capacity through coarse river gravels at Peoria is very high, with rates of 6.1 to 30.5 meters per day being typical annual averages.

Recharge basins concentrate a large volume of infiltrating water in a small area. As a result, a groundwater mound forms beneath the basin. As the recharge starts, the mound begins to grow; when the recharge ceases, the mound decays as the water spreads through the aquifer. The growth and decay can be described mathematically (62, 63). Digital-computer models have been used to evaluate the impact of recharge basins on the water table (64, 65). It is important to maintain the unsaturated zone beneath the recharge basins in order to help maintain high infiltration rates, while still providing water-quality improvements (10, 17).

The infiltration capacity of recharge basins is initially high and then declines as recharge progresses. This is due to surface clogging by fine sediments (66) and biological growths in the upper few centimeters of the soil (67). It has been found that the operation of recharge basins with alternating flooding and drying-out periods will maintain the best infiltration rates (17). Fine surface sediments may occasionally need to be removed mechanically (60).

If the intent of a management program is to recharge a confined aquifer, then **recharge wells** must be used. The design of a well for artificial recharge is similar to that of a supply well. The principal difference is that water flows out of the recharge well under either a gravity head or a head maintained by an injection pump.

Injection wells for artificial recharge are prone to clogging. A large amount of water is being pushed through a small volume of aquifer near the well face. Clogging can be due to a number of factors; for example,

1. air entrainment caused by aeration of water falling into the well;
2. filtration of suspended sediment and organic matter;
3. development of bacterial growths in the aquifer;
4. formation of precipitates due to geochemical reactions between the recharge water and native groundwater;
5. swelling of clay colloids in the aquifer;
6. dispersal of clay particles due to ion exchange between recharging water and aquifer materials;
7. precipitation of iron from native groundwater due to recharge of water with a pH and Eh in the range of ferric iron;
8. growth of iron bacteria; or
9. mechanical compaction of aquifer materials due to high injection pressures.

Frequent maintanance of injection wells may be necessary. This typically consists of pumping water out of the wells to redevelop the flow capacities. Chlorination may also be used to reduce biological growths (17). Wisconsin is one state that does not permit the use of injection wells for any purpose. The reason for this prohibition is to prevent well disposal of wastes which may result in groundwater contamination.

11.8 PROTECTION OF WATER QUALITY IN AQUIFERS

One of the most important aspects of groundwater management is the protection of the water quality in an aquifer. There are a number of artificial sources of potential groundwater contamination. Pollution from such sources as septic tanks, sanitary landfills, land-treatment systems for municipal wastewater, waste injection wells, toxic chemical disposal sites, cemetaries, mine tailings, acid mine drainage, water softener regeneration salt, highway deicing salt, oil-field brines, agricultural chemicals and fertilizers, and accidental oil, gasoline, and chemical spills have been well documented in scientific literature (69). An incident such as the burial in Michigan of slaughtered cattle which had been given feed contaminated by PBB, a chemical fire retardant, points out that the earth is typically thought of as the most convenient repository for material that society does not know how to handle.

Groundwater contamination can also occur when water of poor quality is drawn into a well field that originally had been developed in high-quality water. The best example of this is saltwater intrusion in coastal areas. Heavy pumping of coastal aquifers can cause a landward migration of the interface separating fresh and salty groundwater.

One critical aspect of preventing groundwater pollution is the identification of the recharge areas of aquifers. In such areas, protection of the aquifer is vital. Recharge areas should be zoned as water-quality conservation areas, with close control of potential sources of contamination. The hydrogeology of sites for such facilities as sanitary landfills and land-treatment systems should be intensively studied to ensure that the condition of local soils and the rate and direction of local groundwater flow preclude possible aquifer contamination. Hazardous and toxic waste storage and disposal should be barred from aquifer recharge areas. Sanitary sewers for the collection of domestic wastewater are preferable to septic tanks in aquifer recharge areas. Under most conditions, land areas underlain by extensive confining layers or situated in the discharge areas of unconfined aquifers are preferable for uses that might contaminate the groundwater. Agricultural practices in recharge areas should also be regulated. Overapplication of chemicals and fertilizers should be discouraged, and controlled collection and disposal of animal waste encouraged.

One of the most difficult aspects of aquifer protection is the control of abandoned wells. Many states require that abandoned wells be backfilled. This regulation can be enforced for a newly drilled water, oil, or gas well that proves to be unproductive. However, in the case of wells that fall into disuse and casual abandonment, the owners might not be aware of the danger they can pose to groundwater quality. Likewise, improper well construction can cause shallow groundwater or surface water to migrate downward into the aquifer. Abandoned wells may also be used for the disposal of liquid or solid waste, an obvious threat to water quality. Education is more efficient than regulation in preventing contamination of this type.

Saltwater encroachment can be prevented by regulating the spacing and withdrawal rates of wells. The objective is to avoid establishing a hydraulic gradient that slopes from the zone of contaminated water toward the well field. If the aquifer contains fresh water overlying salty water, pumping rates must be low to avoid drawing up salty water from below, a phenomenon known as **upconing** (Figure 11.1). Saline-water aquifers underlie up to two-thirds of the land area of the conterminous United States, so the problems of water encroachment and upconing are not limited to coastal areas (70).

Control of seawater intrusion in coastal areas has been practiced in a number of localities. In coastal areas of southern California, where there is a confined aquifer containing salt water and fresh water, a pressure-ridge system has been used (71). A line of injection wells parallel to the coast injects water into the aquifer. The result is a ridge in the potentiometric surface (Figure 11.2). Water levels behind the barrier can be drawn down below sea level with no fear of saltwater encroachment. Similar barriers could be used in uncon-

393

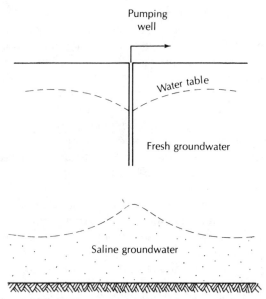

FIGURE 11.1. Upconing of brackish water caused by a well pumping from an overlying freshwater zone.

fined aquifers using artificial recharge from wells, pits, or trenches. Artificial recharge in the area of pumping-well fields could also be used to maintain the elevations of the potentiometric surface above sea level (Figure 11.3). A row of pumping wells could be installed parallel to the coast in either a confined or an unconfined aquifer. They would create a trough in the potentiometric surface lower than either sea level or the well-field areas behind the trough. The trough wells would pump salty water, which would not be suitable for most uses. However, wells behind the trough would pump fresh water (Figure 11.4).

11.9 GROUNDWATER MINING AND CYCLIC STORAGE

From the previous sections it is apparent that in order to develop the groundwater reservoir a cone of depression must be created. This means that water is removed from storage or is, in a sense, mined. Likewise, in order to utilize many aquifers for cyclic storage, the water levels must be drawn down in order to create storage space. Again, some water may have to be mined in order to prepare the aquifer to perform an essential service (72). Groundwater management programs with an objective to maintain natural equilibrium water levels may not be as efficient as other management plans.

The most dramatic need for limited groundwater mining is to provide supplemental water for use when precipitation and surface supplies are

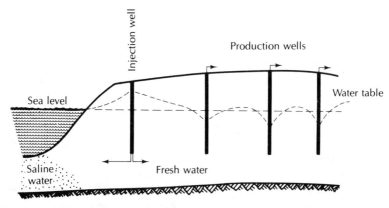

FIGURE 11.2. Use of injection wells to form a pressure ridge to prevent saltwater intrusion in an unconfined coastal aquifer.

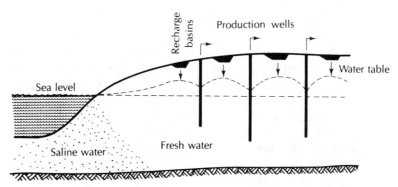

FIGURE 11.3. Use of artificial recharge in the area of production wells in an unconfined coastal aquifer. The artificially recharged water maintains the water table above sea level to prevent saltwater intrusion.

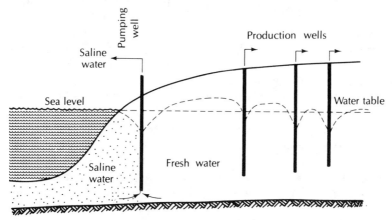

FIGURE 11.4. Use of pumping wells at the coastline to form a trench in the water table, which acts as a barrier to further saltwater encroachment.

limited by drought. Normal rates of industrial and agricultural productivity can be maintained by the use of supplemental groundwater (15). Water managers may find it difficult to convince themselves that the best time to deplete the groundwater reservoir artificially is when the amount of natural recharge is lowest. Groundwater levels will be falling at a seemingly alarming rate. However, the drawdown will create space for storage when nature provides an especially wet year (13). Water managers must be ready to capture this excess by artificial recharge to supplement the natural infiltration. It has been suggested that major recharge events that can result in a rapid rise in groundwater levels can be expected every 10 to 15 years (13). If water levels have not been depleted by "overpumping" during drought, much of the potential recharge may be lost due to lack of aquifer storage space. This concept of **cyclic storage** is not applicable to aquifers to which the amount of possible recharge is limited by their capacity to transport and store groundwater rather than by the amount of water available to them.

In some arid lands, considerable groundwater is known to exist. Furthermore, there may be rather large hydraulic gradients, which could mean that this groundwater is moving through the aquifers. This interpretation is at variance with the extremely low rates of recharge known to occur at present. Model studies have suggested that the gradients are "fossil" (73); i.e., decayed remnants of higher groundwater gradients of 10,000 years ago, when pluvial periods during the late Pleistocene provided large amounts of recharge. Management programs, however, cannot be based on these extremely long cycles between recharge events. For practical purposes, such aquifers must be considered to be unreplenished, and all groundwater pumpage to be simply mining when recharge events are millennia apart.

The mining of groundwater is occurring in many localities. The Ogallala aquifer of the southern High Plains of Texas and New Mexico is an example of a mined aquifer in arid regions, and the deep sandstone aquifer of northeastern Illinois is a humid-region example. Mining the deep sandstone aquifer appears to be an integral part of the area-wide development of the groundwater and surface-water resources of northeastern Illinois (74, 75). Water levels will be drawn down by overdrafts until a "critical level" is reached. Groundwater pumping will then be curtailed to the extent that no further drawdown will occur.

Many areas of the world are underlain by aquifers containing saline groundwater. Although desalinization is technically feasible, it is not economic except in some very limited applications. In water-short areas of the western United States, there is increasing competition between energy development and agricultural interests for the available water (76). Some of this water is used for power-plant cooling, with subsequent evaporational losses. It has been suggested that saline groundwater resources could be mined to provide the necessary cooling water (14). In this approach, it would be necessary to ensure that the available saline groundwater supply is sufficient for the design life of the power plant. Furthermore, the problem of the disposal of brine or salt would have to be dealt with. Saline groundwater could also serve as a source of emergency cooling water for nuclear reactors.

CONJUNCTIVE USE OF GROUND AND SURFACE WATER   11.10

It is obvious that in stream-aquifer systems it is counterproductive to consider surface-water management and groundwater management as separate actions. When water can flow from the stream to the aquifer, or vice versa, it is tantamount to having two separate policies or plans for the same water. No one would open a joint checking account with a total stranger. However, the legal separation of groundwater and surface water, or the administration of groundwater and surface-water management by separate agencies, could have a similar result: a rapid depletion of the total resources due to overuse.

As is the usual case, the need for management of groundwater and surface water as elements of an interrelated system is most critical when demand exceeds supply. The greatest application of such management in the United States has come in the arid west. Overpumping of aquifers near a river could deplete the surface resources, while development of a losing river could reduce the normal recharge to an aquifer system. On a regional basis, the incremental drawdown caused by any one pumper is small. It is the cumulative effect of many pumpers which can cause a depletion of the total resource. The individual user has no economic incentive to reduce pumping (77). The situation is exacerbated when groundwater pumping results in a reduction in the flow of surface waters. Groundwater users may feel social pressure if a neighbor's well is adversely affected; however, this pressure is not experienced when reduction in availability of surface water affects an unknown user many miles away. Individual litigation among tens or hundreds of water users along a stream-aquifer system would present a legal nightmare. Some type of institutional control offers the most workable management approach.

One management plan might be a total ban on groundwater withdrawals during such times that the flow of the river falls below a specified value. This, however, would not prevent the streamflow from falling even more, as the response time of an aquifer is quite long (78). In order to fully protect the rights of senior surface-water users, groundwater pumping might be prohibited altogether. While this would serve one purpose, it would also eliminate the use of the groundwater reservoir for storage of excess water. In addition, groundwater development usually provides water supplies in excess of streamflow alone.

The most extreme case occurs when a river fed by mountain runoff flows across an aquifer system that receives no recharge from precipitation. Under such conditions, the river is the only source of water for both groundwater and surface-water users. Even in this extreme case, the river and aquifer system can be managed conjunctively to maximum benefit. If the water is used for irrigation, there is a demand only part of the year. The remainder of the year, the streamflow passes downstream. The maximum benefit would accrue if some of the water flowing in the nongrowing season could be stored for use when it is needed. This can be accomplished by taking advantage of the delayed response of the aquifer system (78). Wells located away from the river and pumped during the irrigation season would draw water from storage at that

397

time, and then, when the cone of depression reaches the river, infiltration would be induced mostly during the nongrowing season (Figure 11.5). Wells located closer to the river would be pumped only near the end of the irrigation season.

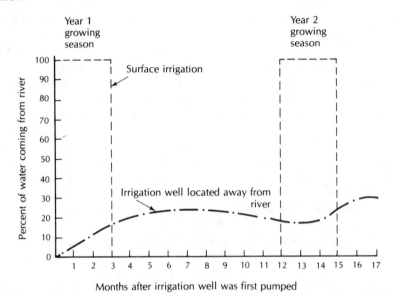

FIGURE 11.5. Comparison of withdrawal of irrigation water from a surface stream with induced infiltration to an irrigation well. The irrigation well delivers the same amount of water during the irrigation season as the surface withdrawal. SOURCE: R. A. Young and J. D. Bredehoeft, *Water Resources Research,* 8, no. 3 (1972):533-56.

While the benefits of conjunctive groundwater and surface-water use are obvious, the implementation of management is not easy. The legal framework for each state is different. In New Mexico, groundwater users in designated groundwater basins fall under the jurisdiction of the New Mexico state engineer. New appropriations of groundwater are made only on the condition that the groundwater user acquire sufficient surface-water rights to cover the impact of the groundwater pumping on surface-water flow (44, 79). In the state of Colorado, the Water Right Determination and Administration Act of 1969 is enforced by the state engineer. In part, this act reads ". . . it shall be the policy of this state to integrate the appropriation, use and administration of underground water tributary to a stream with the use of surface water in such a way as to maximize the beneficial use of all the waters of this state." However, the Colorado state engineer does not have unlimited license to maximize available water, as the Act also states that "no reduction of any lawful diversion because of the operation of the priority system shall be permitted unless such reduction would increase the amount of water available to and required by water rights having senior priorities." A management plan which would increase the total amount of water but which would result in junior users receiving more water at the expense of senior users is apparently not what the Colorado legislature intended.

It has been pointed out that, in most cases, a number of different management plans might be possible (80). For example, an applicant for groundwater might be allocated a large amount of water, but it would be available only during the nongrowing season. A total volume not nearly as great might be allocated if it were to be pumped during the irrigation season. Both plans could protect the senior water rights equally.

Conjunctive groundwater and surface-water management lends itself to the use of computer models (78-83). The hydrologic interaction between the stream-aquifer system can be described by partial differential equations, subsequently solved by numerical methods (Chapter 12). Techniques of operations theory can be used to maximize the amount of water available while minimizing the cost of development. The models can be established so that the holders of senior water rights are protected and water is allocated to the most beneficial use.

TRENDS IN WATER RESOURCES MANAGEMENT   11.11

Water is increasingly held to be a public rather than a private resource. In states where the right to use water is held to be a property right, it is protected by the U.S. Constitution. However, even in England, where the English Rule of absolute ownership of groundwater originated, the Water Resources Act of 1963 changed concepts of ownership of water. Withdrawal of surface water and groundwater in England and Wales is subject to approval of a permit application (84).

In arid states of the United States West, competition for water between public and private users is increasing—especially as population shifts from rural to urban areas (85). Municipalities are able to acquire water rights in the same manner as other entities: they may find an unallocated source or purchase a water right from its owner. However, they are also able to garner water rights by means not available to individuals or corporations: the exercise of the power of eminent domain. Through legal action, a municipal entity may acquire private property, including water rights (86). The use must generally be for a public purpose, and the owner must be compensated for the loss of property. As with any case of public condemnation of private property, the question of what constitutes "adequate compensation" is full of thorns. In Colorado, the compensation is set as "fair market value." This can be established on the basis of prior voluntary sales of water rights. Compensation can also include the loss of value to land that is separated from a water right necessary to put it to a given use (86).

Traditionally, water rights have been considered in connection with off-stream purposes such as irrigation or municipal use. However, there are a number of in-stream uses that are important for public purposes. This includes the maintenance of water quality and fish and wildlife habitat. In

399

Montana, there are increasing demands for water for energy development. In order to protect the ecological integrity of the Yellowstone River, the Montana Fish and Game Commission and the Montana State Water Quality Bureau have formally requested the appropriation of large amounts of water for in-stream flows (87). This is a further example of the supersession of private needs by public in the allocation of a scarce resource.

References

1. THEIS, C. V. "The Significance and Nature of the Cone of Depression in Ground-water Bodies." *Economic Geology,* 38 (1938):889–902.

2. SUTER, M. et al. *Preliminary Report on Ground-water Resources of the Chicago Region, Illinois.* Cooperative Ground-water Report 1, Illinois State Water Survey and Illinois State Geological Survey, 1959, 89 pp.

3. FETTER, C. W., JR. and H. YOUNG. Unpublished results of a computer model study of the Deep Sandstone Regional Aquifer, 1978; and Harley Young, *Digital-Computer Model of the Sandstone Aquifer in Southeastern Wisconsin.* Southeast Wisconsin Regional Planning Commission Technical Report 16, 1976, 42 pp.

4. SASMAN, R. T. et al. *Water Level Decline and Pumpage in Deep Wells in the Chicago Region, 1971-75.* Illinois State Water Survey Circular 125, 1977, 35 pp.

5. WALTON, W. C. *Leaky Artesian Aquifer Conditions in Illinois.* Illinois State Water Survey Report of Investigation 39, 1960, 27 pp.

6. KAZMANN, R. G. "'Safe Yield' in Ground-water Development, Reality or Illusion?" *Proceedings of the American Society of Civil Engineers,* 82, IR3 (1956):1103:1–1103:12.

7. HORDON, R. M. "Water Supply as a Limiting Factor in Developing Communities; Endogenous vs. Exogenous Sources." *Water Resources Bulletin,* 13 (1977):933–39.

8. BRUTSAERT, W., G. W. GROSS, and R. M. MC GEHEE. "C.E. Jacob's Study on the Prospective and Hypothetical Future of the Mining of the Ground Water Deposited under the Southern High Plains of Texas and New Mexico." *Ground Water,* 13 (1975):492–505.

9. SELLERS, J. H. "Tax Implications of Ground-water Depletion." *Ground Water,* 11, no. 4 (1973):27–35.

10. FETTER, C. W., JR. "Attenuation of Waste-water Elutriated through Glacial Outwash." *Ground Water,* 15 (1977):365–71.

11. GASS, T. E. and J. LEHR. "Ground-water Energy and the Ground-water Heat Pump." *Water Well Journal,* 31, no. 4 (1977):42–47.

12. ANDREWS, C. B. "The Impact of the Use of Heat Pumps on Ground-water Temperatures." *Ground Water,* 16 (1978):437–43.

13. AMBROGGI, R. P. "Underground Reservoirs to Control the Water Cycle." *Scientific American,* 236, no. 5 (1977):21–27.

14. GREYDANUS, H. W. "Management Aspects of Cyclic Storage of Water in Aquifer Systems." *Water Resources Bulletin,* 14 (1978):477–83.

15. LEHR, J. H. "Ground Water: Nature's Investment Banking System." *Ground Water,* 16 (1978):142–43.

16. THOMAS, H. E. "Cyclic Storage, Where Are You Now?" *Ground Water,* 16 (1978):12–17.

17. FETTER, C. W., JR. and R. G. HOLZMACHER. "Ground-water Recharge with Treated Waste-water." *Journal of Water Pollution Control Federation,* 46 (1974):260–70.

18. HELWIG, O. J. "Regional Ground-water Management." *Ground Water,* 16 (1978):318–21.

19. LEE, C. H. "The Determination of Safe Yield of Underground Reservoirs of the Closed Basin Type." *Transactions, American Society of Civil Engineers,* 78 (1915):148–51.

20. MEINZER, O. E. *Outline of Ground-water Hydrology, with Definitions.* U.S. Geological Survey Water Supply Paper 494, 1923.

21. CONKLING, H. "Utilization of Ground-water Storage in Stream System Development." *Transactions, American Society of Civil Engineers,* 3 (1946):275–305.

22. BANKS, H. O. "Utilization of Underground Storage Reservoirs." *Transactions, American Society of Civil Engineers,* 118 (1953):220–34.

23. FETTER, C. W., JR. "The Concept of Safe Groundwater Yield in Coastal Aquifers." *Water Resources Bulletin,* 8 (1972):1173–76.

24. American Society of Civil Engineers. *Groundwater Basin Management.* Manual of Engineering Practices 40, 1961, 160 pp.

25. FREEZE, R. A. "Three-Dimensional, Transient, Saturated-Unsaturated Flow in a Groundwater Basin." *Water Resources Research,* 7 (1971):347–66.

26. COLLINS, M. A. "Ground-Surface Water Interactions in the Long Island Aquifer System. *Water Resources Bulletin,* 6 (1972):1253–58.

27. BOUWER, H. "Land Subsidence and Cracking Due to Ground-water Depletion." *Ground Water,* 15 (1977):358–64.

28. FETTER, C. W., JR. "Hydrogeology of the South Fork of Long Island, New York: Reply." *Bulletin, Geological Society of America,* 88 (1977):896.

29. THOMAS, H. E. *Conservation of Ground Water.* New York: McGraw-Hill Book Company, 1951, 327 pp.

30. ANDERSON, M. P. and C. A. BERKEBILE. "Hydrogeology of the South Fork of Long Island, New York: Discussion." *Bulletin, Geological Society of America,* 88 (1977):895.

31. PETERS, H. J., "Ground-water Management." *Water Resources Bulletin,* 8 (1972):188–97.

32. Wisconsin et al. v. Illinois et al., 388 US 426 (1967).

33. JACOBY, G. C., JR., G. D. WEATHERFORD, and J. W. WEGNER. "Law, Hydrology and Surface Water Supply in the Upper Colorado River Basin." *Water Resources Bulletin,* 12 (1976):973–84.

34. BITTINGER, M. W. "Survey of Interstate and International Aquifer Problems." *Ground Water,* 10, no. 2 (1972):44–54.

35. Winters v. United States, 207 US 564 (1908).

36. Cappaert v. United States, 426 US 128 (1976).

37. FOSTER, K. E. "The Winters Doctrine: Historical Perspective and Future Applications of Reserved Water Rights in Arizona." *Ground Water,* 16 (1978):186–91.

38. BERRY, M. P. "The Importance of Perceptions in the Determination of Indian Water Rights." *Water Resources Bulletin,* 10 (1974):137–43.

39. HIRSHLEIFER, J., J. C. DE HAVEN, and J. W. MILLIMAN. *Water Supply.* Chicago: University of Chicago Press, 1960, 378 pp.

40. Acton v. Blundell, 12 M&W. 324, 354 (1843).

41. Frazier v. Brown, 12 Ohio St. 294 (1861).

42. Huber v. Merkel, 117 Wis. 355, 94 N.W. 354 (1903).

43. Wisconsin v. Michaels Pipeline Constructors, Inc., 63 Wis. 2d 278 (1974).

44. THOMAS, H. E. "Water-Management Problems Related to Ground-water Rights in the Southwest." *Water Resources Bulletin,* 8 (1972):110–17.

45. Basset v. Salisbury Manufacturing Co., Inc., 43 N.H. 569, 82 Am. Dec. 179 (1862).

46. DANIELSON, J. A. and A. R. QAZI. "Stream Depletion by Wells in the South Platte Basin—Colorado." *Water Resources Bulletin,* 8 (1972):359–66.

47. PEAK, W. "Institutionalized Inefficiency: The Unfortunate Structure of Colorado's Water Resources Management System." *Water Resources Bulletin,* 13 (1977):551–62.

48. Pasadena v. Alhambra, 33 Calif. (2d) 908, 207 Pac. (2d) 17 (1949).

49. GLEASON, V. E. "The Legalization of Ground Water Storage." *Water Resources Bulletin,* 14 (1978):532–41.

50. Niles Sand and Gravel Co., Inc., v. Alameda County Water District, 37 Calif. App. 3d 924 (1974): Cert. Denied 419 US 869.

51. City of Los Angeles v. Glendale, 23 C2d 68 (1943).

52. City of Los Angeles v. City of San Fernando, 14 Calif. 3d 199 (1975).

53. KARDOS, L. T. "Waste-water Renovation by the Land—A Living Filter." In *Agriculture and the Quality of Our Environment,* ed. N.C. Brady. American Association for the Advancement of Science Publication 85, 1967, pp. 241–50.

54. PARIZEK, R. R. and E. A. MEYERS. "Recharge of Groundwater from Renovated Sewage Effluent by Spray Irrigation." *American Water Resources Association, Proceedings of the Fourth Conference,* 1967, pp. 426–43.

55. DUDLEY, J. G. and D. A. STEPHENSON. *Nutrient Enrichment of Ground-water from Septic Tank Disposal Systems.* Inland Lake Renewal and Shoreland Management Demonstration Project Report, University of Wisconsin—Madison, 1973, 131 pp.

56. BEATTY, M. T. and J. BOUMA. "Application of Soil Surveys to Selection of Sites for On-Site Disposal of Liquid Household Wastes." *Geoderma*, 10 (1973):113–22.

57. SEABURN, G. E. *Preliminary Results of Hydrologic Studies at Two Recharge Basins on Long Island, New York.* U.S. Geological Survey Professional Paper 627-C, 1970, pp. 1–17.

58. SEABURN, G. E. *Preliminary Analysis of Rate of Movement of Storm Runoff through the Zone of Aeration beneath a Recharge Basin on Long Island, New York.* U.S. Geological Survey Professional Paper 700-B, 1970, pp. 196–98.

59. SMITH, H. F. *Artificial Recharge and Its Potential in Illinois.* International Association of Scientific Hydrology Publication No. 72 (Haifa), 1967, pp. 136–42.

60. HARMESON, R. H., R. L. THOMAS, and R. L. EVANS. "Coarse Media Filtration for Artificial Recharge." *Journal American Water Works Association,* 60 (1968):1396–1403.

61. NIGHTINGALE, H. I. and W. C. BIANCHI. "Ground-water Recharge for Urban Use: Leaky Acres Project." *Ground Water*, 11, no. 6 (1973):36–43.

62. HANTUSH, M. S. "Growth and Decay of Ground-water Mounds in Response to Uniform Percolation." *Water Resources Research,* 3 (1967):227–34.

63. SINGH, R. "Prediction of Mound Geometry under Recharge Basins." *Water Resources Research,* 12 (1976):775–80.

64. BIANCHI, W. C. and HASKELL, E. E., JR. "Field Observations Compared with Dupuit-Forcheimer Theory for Mound Heights under a Recharge Basin." *Water Resources Research,* 4 (1968):1049–57.

65. HUNT, B. W. "Vertical Recharge of Unconfined Aquifer." *Journal of Hydraulics Division, American Society of Civil Engineers,* 97 (1971):1017–30.

66. BEHNKE, J. J. "Clogging in Surface Spreading Operations for Artificial Ground-water Recharge." *Water Resources Research,* 5 (1969):870–76.

67. MORAVCOUA, V., L. MASINOVA, and V. BERNATOVA. "Biological and Bacteriological Evaluation of Pilot Plant Artificial Recharge Experiments." *Water Research,* 2 (1968):265–76.

68. SNIEGOCKI, R. T. and R. F. BROWN. "Clogging in Recharge Wells, Causes and Cures." *Proceedings of the Artificial Ground-water Recharge Conference, The Water Resources Association* (England), 1970, pp. 337–57.

69. BOUWER, H. *Ground-water Hydrology.* New York: McGraw-Hill Book Company, 1978, pp. 396–455.

70. BRUINGTON, A. E. "Saltwater Intrusion into Aquifers." *Water Resources Bulletin,* 8 (1972):150–60.

71. BRUINGTON, A. E. and F. D. SEARES. "Operating a Sea Water Barrier Project." *Journal, Irrigation and Drainage Division, American Society of Civil Engineers,* 91 (1965):117–40.

72. RALSTON, D. R. "Administration of Ground-water as Both a Renewable and Non-renewable Resource." *Water Resources Bulletin,* 9 (1973):908–17.

73. LLOYD, J. W. and M. H. FARAG. "Fossil Ground-water Gradients in Arid Regional Sedimentary Basins." *Ground Water,* 16 (1978):388–93.

403

74. SCHICHT, R. J., J. R. ADAMS, and J. B. STALL. *Water Resources Availability, Quality, and Cost in Northeastern Illinois.* Illinois State Water Survey Report of Investigation 83, 1976, 90 pp.

75. SCHICHT, R. J. and J. R. ADAMS. *Effects of Proposed 1980 and 1985 Lake Water Allocations in the Deep Sandstone Aquifer in Northeastern Illinois.* Illinois State Water Survey Contract Report for Illinois Division of Water Resources, 1977.

76. PLOTKIN, S. E., H. GOLD, and I. L. WHITE. "Water and Energy in the Western Coal Lands." *Water Resources Bulletin,* 15 (1979):94–107.

77. BREDEHOEFT, J. D. and R. A. YOUNG. "The Temporal Allocation of Ground-water—A Simulation Approach." *Water Resources Research,* 6 (1970):3–21.

78. YOUNG, R. A. and J. D. BREDEHOEFT. "Digital Computer Simulation for Solving Management Problems of Conjunctive Ground-water and Surface-water Systems." *Water Resources Research,* 8 (1972):533–56.

79. BRUTSAERT, W. F. and T. G. GEBHARD, JR. "Conjunctive Availability of Surface and Ground-water in the Albuquerque Area, New Mexico: A Modelling Approach." *Ground Water,* 13 (1975):345–53.

80. MOREL-SEYTOUX, H. J. "A Simple Case of Conjunctive Surface–Ground-water Management." *Ground Water,* 13 (1975):506–15.

81. MADDOCK, T., III. "The Operation of a Stream-Aquifer System under Stochastic Demands." *Water Resources Research,* 10 (1974):1–10.

82. HAIMES, Y. Y. and Y. C. DREIZIN. "Management of Ground-water and Surface-water via Decomposition." *Water Resources Research,* 13 (1977):69–77.

83. FLORES, E. Z., A. L. GUTJAHR, and L. W. GELHAR. "A Stochastic Model of the Operation of a Stream-Aquifer System." *Water Resources Research,* 14 (1978):30–38.

84. TRELEASE, F. J. "New Water Laws for Old and New Countries." In *Contemporary Developments in Water Law,* ed. C. W. Johnson and S. H. Lewis. University of Texas Center for Research in Water Resources, Symposium No. 4, 1970, pp. 40–54.

85. ANDERSEN, R. L. and N. I. WENGERT. "Developing Competition for Water in the Urbanizing Areas of Colorado." *Water Resources Bulletin,* 13 (1977):769–73.

86. RADOSEVICH, G. E. and M. B. SABEY. "Water Rights, Eminent Domain, and the Public Trust." *Water Resources Bulletin,* 13 (1977):747–57.

87. THOMAS, J. L. and D. KLARICH. "Montana's Experience in Reserving Yellowstone River Water for Instream Beneficial Uses—Legal Framework." *Water Resources Bulletin,* 15 (1979):60–74.

Field and Computer Methodology

chapter

12.1 INTRODUCTION

The day is past when the only activity of the hydrogeologist was to locate and design a water well. Today, hydrogeologists are involved in many phases of resource management, including environmental impact analysis, as integral members of a multidisciplinary team. Hydrogeological studies are necessary and generally required by regulatory agencies for site studies prior to construction of such projects as sanitary landfills, land-treatment systems for wastewater, surface mines, power plants, artificial-recharge lagoons, nuclear-waste repositories, dams, and reservoirs.

In this chapter, we will introduce a number of techniques that can be applied to both the exploration for groundwater supplies and to various aspects of environmental hydrogeology. This includes the use of aerial photographs and other remote sensing data, as well as both surface and borehole geophysical methods. The methods of site evaluation will be examined.

One of the most powerful tools for the evaluation and management of groundwater resources is the digital-computer model. Both flow of groundwater and mass transport of contaminants can be modeled. These models are also a valuable adjunct to the site study in environmental hydrogeology. The use of numerical methods and computer models will be discussed.

12.2 FRACTURE-TRACE ANALYSIS

One technique that has been gaining acceptance among hydrogeologists is the use of **fracture-trace analysis**. As we discussed in Chapter 7, groundwater is known to be concentrated in fracture zones found in many different rock types. The fracture traces are located by study of linear features on aerial or satellite photographs. On air photos, natural linear features consist of tonal variation in soils, alignment of vegetative patterns, straight stream segments or valleys, aligned surface depressions, gaps in ridges, or other features showing a linear orientation (1). Some linear features may be visible on the ground; for instance, surface sags or straight stream segments. Others, such as variation in soil tone, or alignment or height of vegetation of a certain type, may not be noticeable except on aerial photographs (1). Many of these natural linear features consist of interrupted segments, which may be of different types. For instance, a straight stream segment in a flood plain may align with a row of trees in a nearby woods. Natural linear features from 300 meters to around 1500 meters in length are fracture traces. Those greater than 1500 meters are termed **lineaments** (1). Some lineaments are up to 150 kilometers long (2).

Fracture traces are surface expressions of joints, zones of joint concentration, or faults (3). It is generally believed that the joint sets tend to be nearly perpendicular (2). They are known to extend to a depth of 1000 meters at one Arizona location (3), although this is probably far deeper than is typical. Mapped fracture traces have been traced to cliffs where the fracture zone can be seen in cross section (2, 3). Under these observed conditions, the fracture zones dip at approximately 87 to 89 degrees. Figure 12.1 shows an exposure of a fracture in a cliff in central Pennsylvania.

FIGURE 12.1. Cross section of a zone of fracture concentration revealed by a fracture trace near Spring Creek in Centre County, Pennsylvania. Photo courtesy of R. R. Parizek.

These fracture zones are less resistant to erosion than rock, which is less fractured. Hence, valley and stream segments tend to run along them. They may be zones of groundwater drainage, so that soils over them have a deeper water table or are not as moist as soils in surrounding areas. The soil color or vegetation may appear to be different from that of surrounding soils. If they are zones of concentrated groundwater discharge, there may be a line of springs or seeps. Fracture traces in carbonate rocks are typically areas of solution. Aligned sinkholes or surface sags are typical surface expressions.

Fracture traces may be related to regional tectonic activity. They tend to be oriented at a constant angle to the regional structural trend; however, the orientation appears to be independent of local folds (3). Lineaments are known to cut across rocks of many ages and cross folds and faults (2). They have been observed to be parallel to the major joint sets in flat-lying or gently dipping strata, but this is not the case if the strata are steeply dipping. If surface areas are separated by major faults, the individual fault blocks may have fracture traces of different orientation (2). The majority of fracture traces in an area appear to be grouped into two subparallel sets that are approximately perpendicular. Streams developed in rocks where fracture control is evident have been described as having a "stair-step" pattern (4). In the area of central Pennsylvania shown in Figure 12.2, the valley development follows fracture traces. Most fractures are generally N-S or E-W, with lesser numbers running NW-SE and SW-NE.

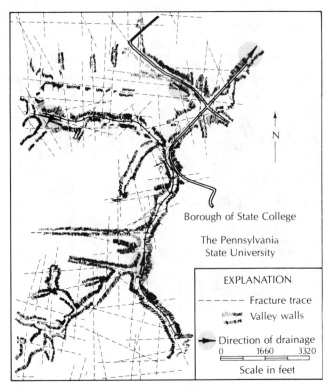

FIGURE 12.2. Valley development in an area of folded carbonate rocks in central Pennsylvania. The valleys tend to follow fracture traces. SOURCE: R. R. Parizek, *Hydrogeologic Framework of Folded and Faulted Carbonates—Influence of Structure,* Mineral Conservation Series Circular 82, College of Earth and Mineral Sciences, Pennsylvania State University, 1971, pp. 28–65.

Statistical studies of wells in carbonate terrain have shown that those located on fracture traces, either intentionally or accidentally, have a greater yield than those not on fracture traces (5). Figure 12.3 illustrates that the productivity of fracture-trace wells is significantly above that of other wells, not on fracture traces. The use of fracture-trace analysis has resulted in a very successful location of well fields in areas where random location has yielded very erratic results (6). The greatest yields come from wells located at the intersection of two fracture traces. Caliper logs of wells on fracture traces in carbonate-rock terrain showed many more cavernous openings and enlarged bedding planes than logs of those wells drilled in interfracture areas (Figure 12.4).

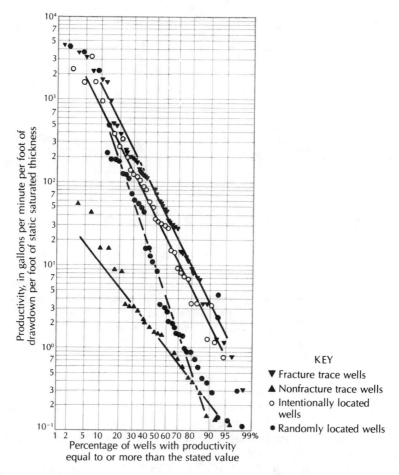

FIGURE 12.3. Production-frequency graph for water wells grouped according to whether or not they fall on a fracture trace. SOURCE: S. H. Siddiqui and R. R. Parizek, *Water Resources Research, 7* (1971):1295–1312.

Many hydrogeologists are successfully using fracture-trace analysis to locate high-yield wells. The technique has been applied to carbonate-rock terrain (6) but is also applicable to most other rock types (2). It is

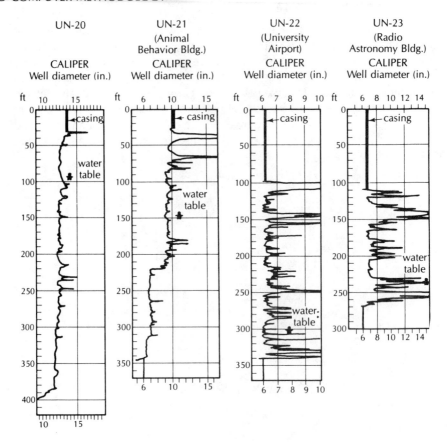

FIGURE 12.4. Caliper logs of wells in an area of carbonate rocks in central Pennsylvania. Wells UN-20 and UN-21 were drilled in interfracture areas; Wells UN-22 and UN-23 were located on fracture traces. SOURCE: L. H. Lattman and R. R. Parizek, *Journal of Hydrology* (Elsevier Scientific Publishing Company) 2 (1964):73–91. Used with permission.

reportedly usable even if the bedrock is mantled by up to 50 meters of glacial drift (7). Fracture-trace analysis is also widely used in selecting sites for sanitary landfills. Naturally, the landfill locations are most suitable if they fall in interfracture areas. Other uses include analysis of foundation and dam sites, evaluation of potential water problems in mines and tunnels, and control of mine drainage (2).

In the identification of fracture traces on aerial photographs, a low-magnification stereoscope is generally used (1). Possible fracture traces are indicated by drawing directly on the photograph. One problem in identification is the confusion of linear features of human origin (fences, cowpaths, roads, power lines, plow and harvest patterns, etc.) with natural linear features. There is also a tendency to map fracture traces at oblique angles to regular grid systems on the photographs. As section lines almost always appear on air photos, especially in cultivated areas, there is a tendency to preferentially map

410

NW-SE and NE-SW features as fracture traces. Following the stereoscopic mapping, the photos should be checked without use of the stereoscope to see if any other features are noticed. A typical air-photo size for fracture-trace analysis is 1:20,000. Linear features that show up in more than one expression, and those crossing roads or fields, are more likely to represent fracture traces. In Figure 12.5, subtle fracture traces are indicated in an area of farmland in central Pennsylvania.

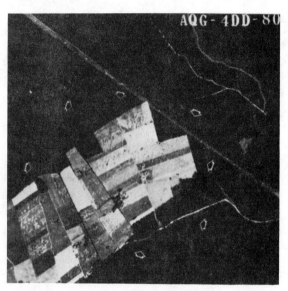

FIGURE 12.5. Aerial photograph of an area with 5 to 80 meters of transported sediments overlying folded and faulted dolomite, limestone, and sandy dolomite, Centre County, Pennsylvania. The line of each fracture trace is indicated by arrows at both ends. Photo courtesy of R. R. Parizek.

Following the mapping of linear features on air photos, it is necessary to make a field check. Some mapped features will usually turn out to be due to man. The more inexperienced the geologist, the more likely it is that this will occur. If a suspected fracture trace has a surface expression, it will be easier to locate in the field. Those without obvious surface expression must be located by virtue of their spatial relation to individual trees or buildings which are visible on the photographs and can be identified on the ground. In urbanizing areas, it may be possible to use older photographs taken before extensive development to map fracture traces. This makes the field location of the fracture traces even more difficult.

The ready availability of satellite photographs has made the mapping of lineaments a feasible part of fracture-trace analysis (8). Aerial imagery from the LANDSAT satellite can be used to locate major lineaments. These may be 100 kilometers or longer. Figure 12.6 is a LANDSAT image of central Pennsylvania in which two major lineaments are indicated. One of them has a series of water gaps, as a river crosses the ridges. Sulfide mineraliza-

411

tion is also known to occur along this lineament. The relationship of lineaments and groundwater is expected to be similar to that of fracture traces (8).

FIGURE 12.6. LANDSAT image of a section of the Valley and Ridge Province and the Appalachian Plateau Province in Pennsylvania. Two major lineaments are shown. Sulfide mineralization is concentrated along the Tyrone-Mt. Union lineament with known locations indicated by the black dots. There are also several water gaps along this feature, where the river has cut across ridges. Photo courtesy of R. R. Parizek.

12.3 SURFICIAL METHODS OF GEOPHYSICAL INVESTIGATIONS

Geophysical surveys have been used by the mining and petroleum industries for many decades. Groundwater geologists soon discovered the usefulness of these surveys in exploring the shallow subsurface (within a few hundred meters) where groundwater supplies are usually found (9, 10, 11). A number of different techniques are used, the most common of which are direct-current resistivity, seismic refraction, and gravity and magnetics methods. Seismic reflection is less widely used, although it is the preferred method in petroleum exploration.

Geophysical methods may be used to determine indirectly the extent and nature of the geologic materials beneath the surface. The thickness of unconsolidated surficial materials, the depth to the water table, the location of subsurface faults, and the depth of the basement rocks can all be determined. In some instances, the location, thickness, and extent of subsurface bodies, such as gravel deposits or clay layers, can also be evaluated. The correlation of geophysical data with well logs or test-boring data is generally more reliable than either type of information used by itself. As with all hydrogeological investigations, a careful definition of the problem and determination of the best type of information to solve the problem should be made before geophysical work is done. The geophysical survey should then be planned to yield the greatest amount of useful data for the budgeted cost.

12.3.1 DIRECT-CURRENT ELECTRICAL RESISTIVITY

Of the several electrical geophysical methods, **direct-current electrical resistivity** has found the greatest application to hydrogeology (12). A commutated direct current or a current of very low frequency (less than 1 cycle per second) is generated in the field or provided by storage batteries. It is introduced into the ground by means of two metal electrodes. If the soil is dry, water may be needed around the electrodes to establish a good connection. The voltage in the ground is measured between two other metal electrodes, also driven into the ground. By knowing that current flowing through the ground and the potential differences or voltage between two electrodes, it is possible to compute the resistivity of the earth materials between the electrodes. The resistivity of earth materials varies widely, from 10^{-6} ohm-meter for graphite to 10^{12} ohm-meters for quartzite. Dry materials have a higher resistivity than similar wet materials, as moisture increases the ability to conduct electricity. Gravel has a higher resistivity than silt or clay under similar moisture conditions, as the electrically charged surfaces of the fine particles are better conductors.

Electrical resistivity, R, is equal to the expression

$$R = \frac{A}{L} \frac{\Delta V}{I} \qquad (12\text{-}1)$$

where

A is the cross-sectional area

L is the length of the flow path

ΔV is the voltage drop

I is the electrical current

Electrical resistivity is measured in units of ohm-meters or ohm-feet. The four electrodes used can be designated as follows:

A is the positive-current electrode

B is the negative-current electrode

$\left.\begin{array}{l} M \\ N \end{array}\right\}$ are the potential electrodes

If $\overline{XY}$ indicates the distance between Electrode X and Electrode Y, Equation (12-1) can be expressed as (12)

413

$$\overline{R} = \left(\frac{2\pi}{\dfrac{1}{\overline{AM}} - \dfrac{1}{\overline{BM}} - \dfrac{1}{\overline{AN}} + \dfrac{1}{\overline{BN}}} \right) \frac{\Delta V}{I} \tag{12-2}$$

As earth materials are almost never homogeneous and electrically isotropic, the resistivity found by Equation (12-2) is an apparent resistivity, $\overline{R}$.

There are two electrode configurations in common usage. The **Wenner array** consists of the four electrodes spaced equal distances apart in a straight line: $\overline{AM} = \overline{MN} = \overline{NB} = a$. A current electrode is on each end (Figure 12.7A). In using the Wenner array, the apparent resistivity, $\overline{R}$, may be found from the expression

$$\overline{R} = 2\pi a \frac{\Delta V}{I} \tag{12-3}$$

which is solved from Equation (12-2).

The second configuration is the **Schlumberger array**. It is a linear array, with potential electrodes placed close together (Figure 12.7B). Typically, $\overline{AB}$ is set equal to or greater than five times the value of $\overline{MN}$. The apparent resistivity is given by

$$\overline{R} = \pi \frac{(\overline{AB}/2)^2 - (\overline{MN}/2)^2}{\overline{MN}} \frac{\Delta V}{I} \tag{12-4}$$

Geophysical instruments are available to measure the value of ΔV for a known I. The appropriate formula for the electrode array is used to compute the apparent resistivity.

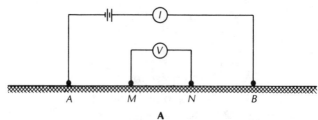

A

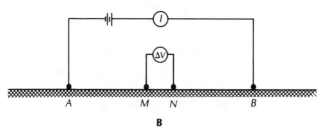

B

FIGURE 12.7. **A.** Wenner electrode array; **B.** Schlumberger electrode array. These are the two most widely used configurations for electrical resistivity surveys.

Resistivity surveys are made in two fashions. An **electrical sounding** will reveal the variations of apparent resistivity with depth. **Horizontal profiling** is used to determine lateral variations in resistivity. When the electrode spacing is expanded in making an electrical sounding, the distance between the potential electrodes and the current electrodes increases. This means that the current will travel progressively deeper through the ground, and will measure apparent resistivity to greater depths. Either the Wenner or the Schlumberger array may be used; however, the latter is more convenient for electrical sounding. This is because, for each incremental measurement, only the outer current electrodes must be moved every time. The inner electrodes are spread only occasionally. In the Wenner array, all four electrodes must be moved for each incremental measurement. The sounding is begun with the electrodes close together. After each reading, the electrodes are repositioned with a, or $\overline{AB}/2$, increased and a new measurement made. The apparent resistivity is plotted as a function of electrode spacing on logarithmic paper. For a number of reasons (12), the Schlumberger array is superior to the Wenner array for electrical soundings. There is, however, a set of theoretical type curves of Wenner apparent resistivity for two-, three-, and four-layer earth models (13). This could be helpful in interpreting the results of a Wenner-array electrical sounding.

For a homogeneous earth, there is a definite relationship between the electrode spacing and the percent of the current that penetrates to a given depth. For a nonhomogeneous and layered earth, the exact relationship cannot be easily determined. It is safe to assume that the greater the electrode spacing, the deeper the stratum influencing the apparent-resistivity curve. There are a number of possible earth models which could produce a given curve. In Figure 12.8, there are three possible theoretical interpretations ex-

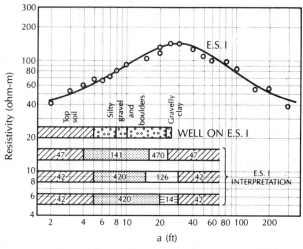

FIGURE 12.8. Wenner electrical-sounding curve of apparent resistivity as a function of electrode spacing. Three possible interpretations and a test boring are also given. SOURCE: A. A. R. Zohdy, *Ground Water*, 3, no. 3 (1965):41–48.

pressed as resistivity and a test-boring log. The rise in apparent resistivity indicates a shallow zone of high resistivity. The test boring shows this to be a layer of silty gravel and boulders from 5 to 23 feet. It should be noted that the apparent-resistivity curve peaks at 30 feet. An interpretation that the layer of maximum resistivity lay at 30 feet would have been wrong.

In horizontal profiling, the electrode spacing is kept at a constant value. The electrodes are moved in a grid pattern over the land surface. The apparent resistivity of each point on the grid is marked on a map and isoresistivity contours are drawn. Figure 12.9 shows an apparent-resistivity map based on a large number of horizontal resistivity measurements. An area of buried stream-channel gravels is delineated where the apparent resistivity exceeds 80 ohm-meters (14). A geologic cross section based on test borings and electrical resistivity soundings is shown in Figure 12.10. The location of the cross section is indicated in Figure 12.9 as line AB.

Geoelectrical methods are useful in groundwater studies for such purposes as defining buried stream channels and areas of saline versus fresh

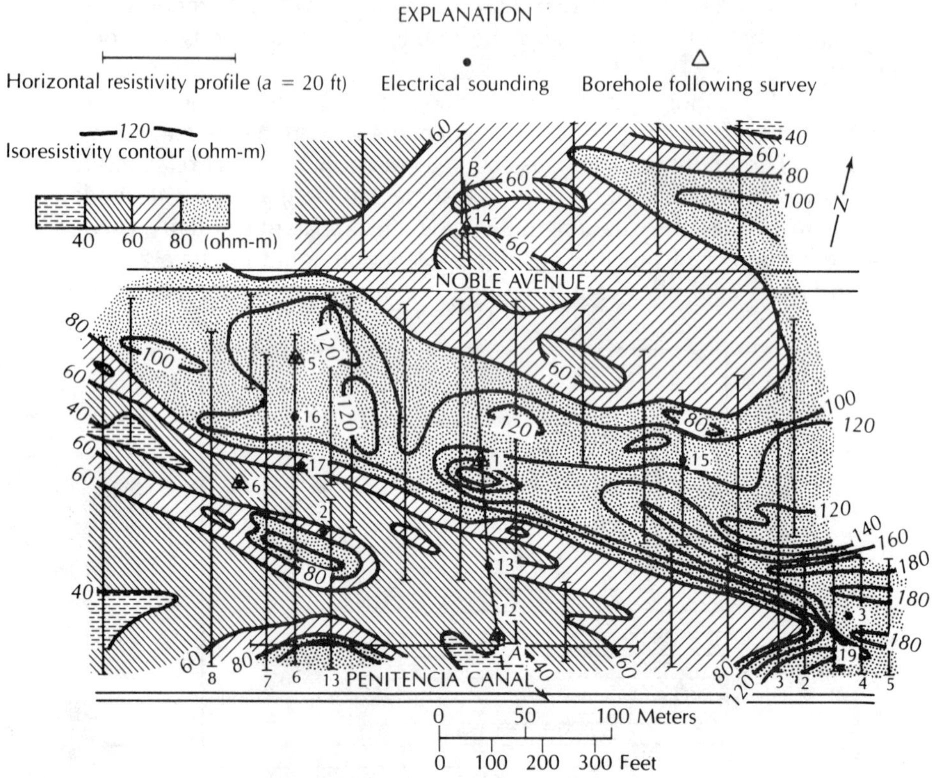

FIGURE 12.9. Apparent-resistivity map of Penjtencia, California. Locations of resistivity profiles, soundings, and boreholes are shown. The location of Borehole and Electrical Sounding 1 is in the center of the high-resistivity zone. SOURCE: A. A. R. Zohdy, *Ground Water*, 3, no. 3 (1965):41–48.

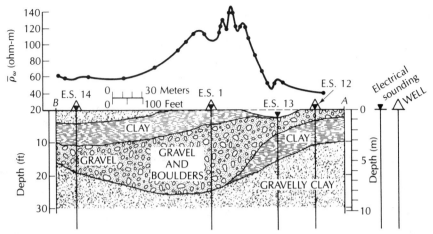

FIGURE 12.10. Geologic cross section based on test borings and electrical soundings. The position of the cross section is shown as line *AB* on Figure 12.9. SOURCE: A. A. R. Zohdy, *Ground Water*, 3, no. 3 (1965):41–48.

groundwater. Saline water has a much lower resistivity, as it is a better electrical conductor. Interpretation for such cases is relatively simple and may be done qualitatively. Layers of very low resistivity, such as clay, can also be discerned on sounding curves. It is often impossible to pick out the water table on an electrical sounding (12), although it is frequently attempted. In many respects, detailed analyses of resistivity data are best left to the experienced geophysicist.

12.3.2 SEISMIC METHODS

Seismic methods using artificially created seismic waves traveling through the ground are quite commonly employed in hydrogeology. These methods are useful in determining depth to bedrock, slope of the bedrock, depth to water table and, in some cases, the general lithology. Applied seismology has been highly developed in the petroleum industry, where the **seismic reflection method** is used almost exclusively. Structural and formational boundaries can be indicated to great depth.

Hydrogeological studies often involve finding the thickness of unconsolidated material overlying bedrock. For this purpose, the **seismic refraction method** is superior. The loose material transmits seismic waves more slowly than consolidated bedrock. By studying the arrival times of seismic waves at various distances from the energy source, the depth to bedrock can be determined.

The energy source can be a small explosive charge set in a shallow drill hole. One or two sticks of dynamite is sufficient for depths to bedrock in excess of 30 to 50 meters. Of course, explosives should be handled only by persons trained and licensed to do so. A judgment of how large a charge to use must be made in each case. For shallower work, 5 to 15 meters, a sledge

417

hammer struck on a steel plate lying on the ground may be a sufficient energy source. The seismic wave is detected by geophones placed in the earth in a line extending away from the energy source. A seismograph records the travel time for the wave to go from the energy source to the geophone. The more sophisticated seismographs are multichannel units with a number of geophones attached.

Figure 12.11 illustrates the travel paths of compressive seismic waves traveling through a two-layer earth. The seismic velocity in the lower layer is greater than that in the upper layer. As the energy travels faster in the lower layer, the wave passing through it gets ahead of the wave in the upper layer. At the boundary between the two layers, part of the energy is refracted back upward from the lower layer boundary to the surface.

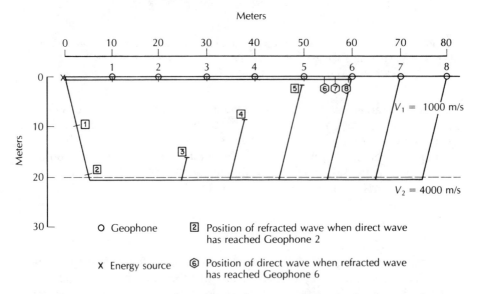

FIGURE 12.11. Travel paths of a refracted seismic wave and a direct wave. The direct wave will reach the first five geophones first, but for the more distant geophones the first arrival is from a refracted wave.

The angle of refraction of each wave front is called the **critical angle,** i_c, and is equal to the arc sin of the ratio of the velocities of the two layers:

$$i_c = \sin^{-1}\frac{V_1}{V_2} \tag{12-5}$$

Figure 12.12 illustrates a wave front and the path of the refracted energy that travels along the lower layer boundary. A direct wave in the upper layer is also shown.

418

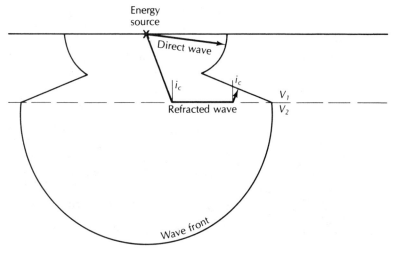

FIGURE 12.12. A seismic wave front at a given time after a charge is detonated.

EXAMPLE PROBLEM

Determine the critical angle, i_c, when $V_1 = 1000$ meters per second and $V_2 = 4000$ meters per second.

$$i_c = \sin^{-1} V_1/V_2 = \text{arc sin } 0.25 = 14.5°$$

If V_2 is less than V_1, the wave will be refracted downward and no energy will be directed upward. Thus, the refraction method will show higher velocity layers but not lower velocity layers that are overlain by a high-velocity layer.

Energy can travel directly through the upper layer from the source to the geophone. This is the shortest distance, but the waves do not travel as fast as those traveling along the top of the lower layer. The latter must go further, but they do so with a higher velocity. In Figure 12.11, the position of waves traveling to each geophone is indicated. Geophones 1 through 5 first receive waves that have traveled through only the upper layer. The sixth and succeeding geophones measure arrival times of refracted waves that have gone through the high-velocity layer, as well. The position of the trailing wave front at each time the leading front reaches each geophone is indicated in the figure.

419

A graph is made of the arrival time of the first wave to reach the geophone versus the distance from the energy source to the geophone. This is known as a **travel-time** or **time-distance curve**. Figure 12.13 shows the time-distance curve for the shot in Figure 12.11. The reciprocal of the slope of each straight-line segment is the apparent velocity in the layer through which the first arriving wave passed. The slope of the first segment is 10 milliseconds per 10 meters, so that the reciprocal is 10 meters per 10 milliseconds, or 1000 meters per second.

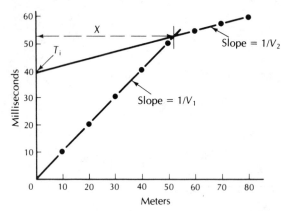

FIGURE 12.13. Arrival time-distance diagram for a two-layered seismic problem.

The projection of the second line segment backward to the time-axis ($X = 0$) yields a value known as the **intercept time**, T_i. This value can be determined graphically, as shown in Figure 12.13. T_i is 39 milliseconds and X is 52 meters. The depth to the lower layer, Z, is found from the equation (15)

$$Z = \frac{T_i}{2} \frac{V_1 V_2}{\sqrt{V_2^2 - V_1^2}} \qquad \textbf{(12-6)}$$

The depth to the lower layer can also be found from the equation (15)

$$Z = \frac{X}{2} \sqrt{\frac{V_2 - V_1}{V_1 + V_2}} \qquad \textbf{(12-7)}$$

where X is the distance from the shot to the point at which the direct wave and the refracted wave arrive simultaneously. This is shown on Figure 12.13 as the X-axis distance where the two line segments cross.

EXAMPLE PROBLEM

Find the value Z from Figure 12-13.

From the slope of each line segment, $V_1 = 1000$ meters per second and $V_2 = 4000$ meters per second. T_i is 39 milliseconds and X is 52 meters.

$$Z = \frac{T_i}{2} \frac{V_1 V_2}{\sqrt{V_2^2 - V_1^2}}$$

$$= \frac{0.039}{2} \times \frac{1000 \times 4000}{\sqrt{4000^2 - 1000^2}}$$

$$= 20 \text{ m}$$

Also,

$$Z = \frac{X}{2} \sqrt{\frac{V_2 - V_1}{V_1 + V_2}}$$

$$= \frac{52}{2} \sqrt{\frac{4000 - 1000}{4000 + 1000}}$$

$$= 20 \text{ m}$$

A more typical case in hydrogeology is a three-layer earth, the top layer being unsaturated, unconsolidated material. In the next layer, below the water table, the unconsolidated deposits are saturated, which yields a higher seismic velocity. The third layer is then bedrock. Under such conditions, the seismic method can be used to find the water table. However, similar velocities are possible from either saturated sand or unsaturated glacial till. Similar seismic refraction patterns could be obtained from a water table in a uniform sand deposit or a layer of unsaturated sand overlying unsaturated glacial till. This illustrates the point that geophysics is best interpreted in light of other data.

The three-layer seismic refraction case with $V_1 < V_2 < V_3$ is shown in Figure 12.14. The first arriving waves show three line segments. The reciprocal of the slope of each line is the seismic velocity of the respective layers. The intercept time for each of the two deeper layers is the projection of the line segment back to the time-axis. Indicated on the figure is the distance, X_1, from the shot to the point at which waves from Layers 1 and 2 arrive simultaneously and the distance, X_2, to the point at which waves from Layers 2 and 3 arrive simultaneously. The depth, Z_1, of Layer 1 is found from the values of V_1 and V_2 and either T_{i1} or X_1 using Equation (12-6) or (12-7). The thickness of the second layer, Z_2, is found using (15):

$$Z_2 = \frac{1}{2} \left((T_{i2} - 2Z_1) \frac{\sqrt{V_3^2 - V_1^2}}{V_3 V_1} \right) \left(\frac{V_2 V_3}{\sqrt{V_3^2 - V_2^2}} \right) \qquad \textbf{(12-8)}$$

The value of Z_1 must be computed before computing the value of Z_2.

The velocities computed from the reciprocals of the slope are called **apparent velocities**. If the lower layer is horizontal, they represent the actual velocity. However, if the lower layer is sloping, the arrival time for a shot measured downslope will be different from one measured upslope. Seismic lines are routinely run with a shot at either end, so that dipping beds can be

421

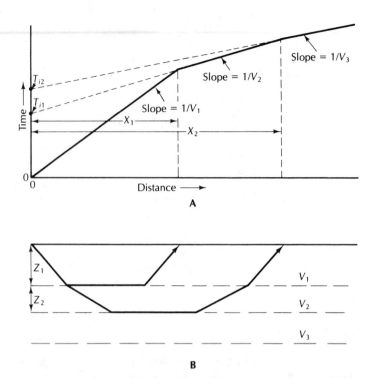

FIGURE 12.14. **A.** Diagram of arrival time versus distance for a three-layered seismic problem; **B.** Wave path for a three-layered seismic problem.

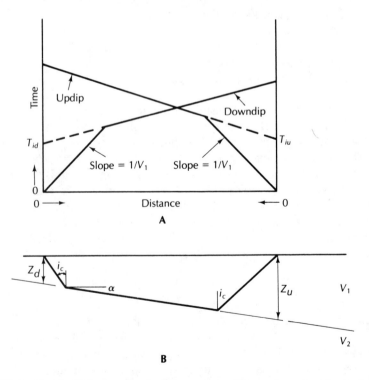

FIGURE 12.15. **A.** Diagram of arrival time versus distance for a two-layered seismic problem with a sloping lower layer; **B.** Wave path for the preceding problem.

determined. Time-distance curves for a dipping stratum are shown in Figure 12.15, with travel times measured from shots at either end of the line. The upper layer is unaffected by the dip of the lower bed, so that the reciprocal of the slope of the first line segment is V_1. In order to find the values of V_2 and the depth to the bedrock at the updip end of the line, Z_u, as well as at the downdip end, Z_d, a complex series of computations must be made (15).

The slope of the second line segment of the downdip line is m_d, and the slope of the second line segment of the updip line is m_u. The value of the angle of refraction, i_c, is found from

$$i_c = \tfrac{1}{2}(\sin^{-1}V_1 m_d + \sin^{-1}V_1 m_u) \tag{12-9}$$

The value of V_2 is given by

$$V_2 = \frac{V_1}{\sin i_c} \tag{12-10}$$

The angle of slope of the dipping layer, α, is found from

$$\alpha = \tfrac{1}{2}(\sin^{-1}V_1 m_d - \sin^{-1}V_1 m_u) \tag{12-11}$$

Finally, the depths to the lower layer at either end of the shot line are found from the expressions

$$Z_u = \frac{V_1 T_{iu}}{\cos \alpha \; 2 \cos i_c} \tag{12-12}$$

$$Z_d = \frac{V_1 T_{id}}{\cos \alpha \; 2 \cos i_c} \tag{12-13}$$

If there are more than two dipping layers, then expressions of greater complexity must be used (16).

EXAMPLE PROBLEM

A seismic survey yielded data for a dipping two-layer case in which the following values were obtained:

$$V_1 = 1570 \text{ m/sec}$$
$$m_u = 1.67 \times 10^{-4} \text{ sec/m}$$
$$m_d = 1.54 \times 10^{-4} \text{ sec/m}$$
$$T_{iu} = 0.046 \text{ sec}$$
$$T_{id} = 0.050 \text{ sec}$$

Compute V_2, Z_u, and Z_d.

$$i_c = \tfrac{1}{2}(\sin^{-1}V_1 m_d + \sin^{-1}V_1 m_u)$$
$$= \tfrac{1}{2}(\sin^{-1} 1570 \times 1.54 \times 10^{-4} + \sin^{-1} 1570 \times 1.67 \times 10^{-4})$$
$$= \tfrac{1}{2}(13.99 + 15.20)$$
$$= 14.6°$$

$$V_2 = V_1/\sin i_c$$
$$= 1570/\sin 14.6$$
$$= 6230 \text{ m/sec}$$

$$\alpha = \tfrac{1}{2}(\sin^{-1}V_1 m_d - \sin^{-1}V_1 m_u)$$
$$= \tfrac{1}{2}(\sin^{-1} 1570 \times 1.54 \times 10^{-4} - \sin^{-1} 1570 \times 1.67 \times 10^{-4})$$
$$= \tfrac{1}{2}(13.99 - 15.20)$$
$$= -0.6°$$

$$Z_u = \frac{V_1 T_{iu}}{\cos \alpha\, 2 \cos i_c}$$
$$= \frac{1570 \times 0.046}{2 \times \cos 14.6 \times \cos -0.6}$$
$$= 37.3 \text{ m}$$

$$Z_d = \frac{V_1 T_{id}}{\cos \alpha\, 2 \cos i_c}$$
$$= \frac{1570 \times 0.050}{2 \times \cos 14.6 \times \cos -0.6}$$
$$= 40.6 \text{ m}$$

The cases given in this section are but a few of the many possible cases that might be encountered. Figure 12.16 shows schematic travel-time curves for a number of nonhomogeneous earth models. Should the hydrogeologist suspect that the situation is more than a simple two- or three-layer model with homogeneous beds, an experienced geophysicist should interpret the field data.

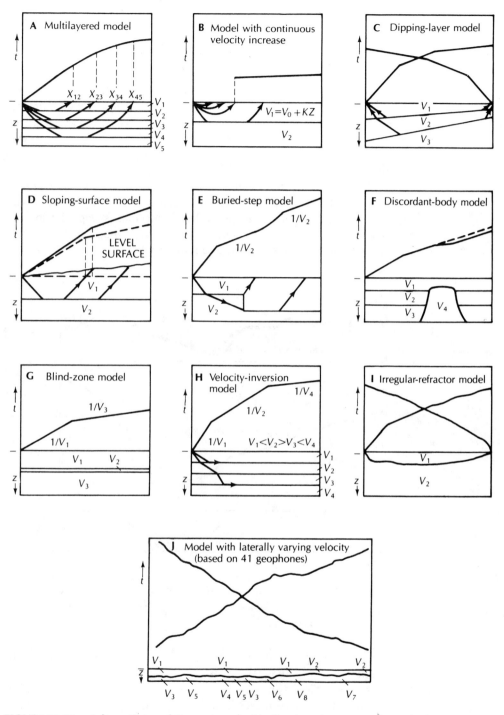

FIGURE 12.16. Schematic travel-time curves for idealized nonhomogeneous geologic models. SOURCE: G. P. Eaton, "Seismology," in *Techniques of Water-Resources Investigations*. U.S. Geological Survey, 1974, Book 2, Chap. D1, pp. 67–84.

12.3.3 GRAVITY AND MAGNETIC METHODS

Measurement of the gravimetric and magnetic fields of the earth are standard geophysical methods used to study the structure and composition of the earth. To the extent that the basic geology influences the hydrogeology, these methods are quite useful in groundwater studies. The collection of data can be relatively simple if ground stations are used. However, the reduction and correction of the data are fairly complex (12, 15). Airborne geomagnetic surveys obviously involve unique problems.

Both gravity and magnetics surveys can be used to delineate the area of unconsolidated basin fill or buried stream-channel aquifers. In Figure 12.17, a Cenozoic basin in Antelope Valley, California, is depicted in cross section. The presence of the basin and the area of the deepest part are shown on both an aeromagnetic profile and a gravity profile.

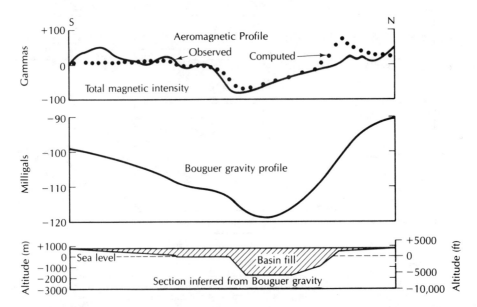

FIGURE 12.17. Gravity and aeromagnetic profiles across a basin-fill aquifer. The aeromagnetic profile was flown at 150 meters. SOURCE: D. R. Mabey "Magnetic Methods," in *Techniques of Water-Resources Investigations,* U.S. Geological Survey, 1974, Book 2, Chap. D1, pp 85–115.

Magnetic anomalies are caused by distortions of the earth's magnetic field created by magnetic materials in the crust. Magnetic anomalies indicate the type of rocks in a very general way. In hydrogeologic studies, magnetic anomalies can be useful in indicating the depth to magnetic basement rocks. Sedimentary rocks are typically nonmagnetic; hence, they do not affect the magnetic field. Some magnetic rocks, such as basalt flows, can be important aquifers. Magnetic surveys might be useful in tracing basalt flows in areas of nonmagnetic rocks.

The mass of the rocks beneath a point on the earth will affect the local value of the acceleration of gravity. To be useful, the measured values must be referenced to a common datum—usually, mean sea level. A free-air correction is made to compensate for elevation differences. To correct for the gravitational attraction of the rock that lies between the gravity station and sea level, a **Bouguer correction** is made. Corrections must also be made for tidal effects, latitude, and terrain. After the measured gravity data are corrected, the result is a **Bouguer anomaly value**, which can be mapped with gravity contours drawn. Such a map could help to define the extent of a buried bedrock valley if there were a density difference between the sediments and the bedrock.

It should be emphasized that there are many possible earth models that would result in the same gravitational or magnetic anomaly. There is no unique solution to any set of geophysical data, and the person interpreting the data must keep this in mind.

GEOPHYSICAL WELL LOGGING 12.4

Direct access to the subsurface is gained wherever there is a well or test boring. When a well is drilled, a record may be made of the geologic formations encountered. The reliability of a **lithologic well log** depends on the method of drilling and sample recovery as well as the knowledge and skills of the person making the log. There are also many existing wells for which there is no available record of the subsurface geology.

Geophysical well logging offers a great deal in the way of practical applications to hydrogeology. Borehole geophysical methods were developed primarily in the petroleum industry, and virtually all oil and gas wells are routinely logged when drilled. In the water-well industry, the use of geophysical logging is generally restricted to either research projects or high-capacity municipal and industrial wells. The cost of well logging is not justified by the marginal benefits gained for small-yield domestic wells.

Borehole geophysical data have a number of uses (see Table 12.1). The log of a well can indicate the areas of high porosity and permeability which would produce the most water. Zones of an aquifer with high-salinity water can be identified. It can be detected whether water is flowing through the well under natural gradients from one aquifer to another. The magnitude and direction of flow through the borehole can also be determined. If a number of wells are logged over an area, the logs can be used for stratigraphic correlation. The lithology of the rocks penetrated by the wells can be identified, especially if some core samples are available for baseline comparison. Regional ground-water flow patterns might be identified from such characteristics as fluid temperature. Nuclear well-logging techniques can be used in cased wells. This is the only way to get subsurface data under such conditions. Geophysical logs give a permanent record, based on repeatable measurements. Thus, data collected for one purpose are available for other, unanticipated, uses in the future.

TABLE 12.1. Summary of log applications

Required Information on the Properties of Rocks, Fluid, Wells, or the Groundwater System.	Widely Available Logging Techniques Which Might Be Utilized.
Lithology and stratigraphic correlation of aquifers and associated rocks.	Electric, sonic, or caliper logs made in open holes. Nuclear logs made in open or cased holes.
Total porosity or bulk density.	Calibrated sonic logs in open holes, calibrated neutron or gamma-gamma logs in open or cased holes.
Effective porosity or true resistivity.	Calibrated long-normal resistivity logs.
Clay or shale content.	Gamma logs.
Permeability.	No direct measurement by logging. May be related to porosity, injectivity, sonic amplitude.
Secondary permeability—fractures, solution openings.	Caliper, sonic, or borehole televiewer or television logs.
Specific yield of unconfined aquifers.	Calibrated neutron logs.
Grain size.	Possible relation to formation factor derived from electric logs.
Location of water level or saturated zones.	Electric, temperature or fluid conductivity in open hole or inside casing. Neutron or gamma-gamma logs in open hole or outside casing.
Moisture content.	Calibrated neutron logs.
Infiltration.	Time-interval neutron logs under special circumstances or radioactive tracers.
Direction, velocity, and path of groundwater flow.	Single-well tracer techniques—point dilution and single-well pulse. Multiwell tracer techniques.
Dispersion, dilution, and movement of waste.	Fluid conductivity and temperature logs, gamma logs for some radioactive wastes, fluid sampler.
Source and movement of water in a well.	Injectivity profile. Flowmeter or tracer logging during pumping or injection. Temperature logs.
Chemical and physical characteristics of water, including salinity, temperature, density, and viscosity.	Calibrated fluid conductivity and temperature in the well. Neutron chloride logging outside casing. Multielectrode resistivity.
Determining construction of existing wells, diameter and position of casing, perforations, screens.	Gamma-gamma, caliper, collar, and perforation locator, borehole television.
Guide to screen setting.	All logs providing data on the lithology, water-bearing characteristics, and correlation and thickness of aquifers.
Cementing.	Caliper, temperature, gamma-gamma. Acoustic for cement bond.
Casing corrosion.	Under some conditions caliper, or collar locator.
Casing leaks and (or) plugged screen.	Tracer and flowmeter.

SOURCE: W. S. Keys and L. M. MacCary, "Application of Borehole Geophysics to Water-Resources Investigations," in *Techniques of Water-Resources Investigations,* U.S. Geological Survey, 1971, Book 2, Chap. E1.

Because of the large number of borehole techniques that are applicable to water wells (17–23), a discussion of only the more common methods will be included in this section, with emphasis on qualitative rather than quantitative interpretation of geophysical logs. Generally, a suite of geophysical logs is made, rather than only a single type. The methods tend to be complimentary; one may confirm another. Likewise, certain interpretations are made on the basis of two or more logs. Figure 12.18 consists of six different geophysical logs made on the same borehole, along with a lithologic log. It can readily be seen that the logs deflect with the changes in lithology.

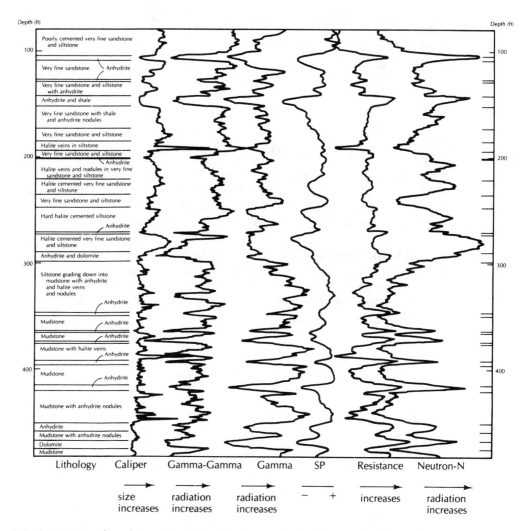

FIGURE 12.18.   The relationship of six different geophysical logs to the lithology of a well in the upper Brazos River Basin, Texas. SOURCE: W. S. Keys and L. M. MacCary, in *Techniques of Water-Resources Investigations*, U.S. Geological Survey, 1971, Book 2, Chap. E1.

429

Well logs are usually made as pen-and-ink strip charts. This gives a continuous record, which is most useful. Some instruments give point readings of various values at discrete depths. A probe is lowered into the borehole on a cable. The cable, which contains powerlines from the surface to the probe, supports the weight of the probe and transmits signals from the probe to the recorder at the surface. The probe contains the necessary electronics, energy or nuclear sources, and detectors.

12.4.1   CALIPER LOGS

A **caliper log** is used to measure the diameter of an uncased borehole in bedrock units. It can also be used to find the casing depth. The minimum hole diameter is, of course, the size of the drill bit.* The hole may be enlarged by caving of the formations into the hole or by solution of minerals by the drill water. It also may be enlarged if the drill bit is rotated at a depth while the downward pressure is removed. Another use for caliper logs is to indicate solution-enlarged bedding planes and joints in carbonate aquifers.

12.4.2   TEMPERATURE LOGS

A **temperature log** is a continuous vertical record of the temperature of the fluid in the borehole. This may or may not be indicative of the temperature of the fluid in the rocks opposite the borehole fluid. In a recently drilled well, the borehole fluid may be well mixed. After the well has had an opportunity to reach environmental equilibrium, the temperature log can reveal zones of differing temperature in the well. There will be a component of the geothermal gradient present; however, water in different aquifers may be at discrete temperatures, which may be detected on the log. Thermal logging has been used to trace the movement of water previously injected into an aquifer (23). The recharged water had a daily thermal variation of up to 17° C, and the diural fluctuation was traced in a series of observation wells (Figure 12.19).

12.4.3   SINGLE-POINT RESISTANCE

Electrical resistance can be measured in a borehole by a number of different methods. The simplest case is the **single-point resistance**, in which a single electrode is lowered into the borehole on an insulated cable. The other electrode is at the ground surface. As the electrode is lowered into the borehole, the resistance of the earth between the two electrodes is measured.

The single-point electrode is measuring the resistance of all of the rocks between the electrodes. Most of the variation in resistance is due to the

---

*When a hole is drilled, the drilling rig turns a pipe (drill stem). The rock at the bottom of the hole is broken up by the drill bit, which cuts it into pieces. Drilling fluids circulating through the hole and drill stem bring the broken rock to the surface.

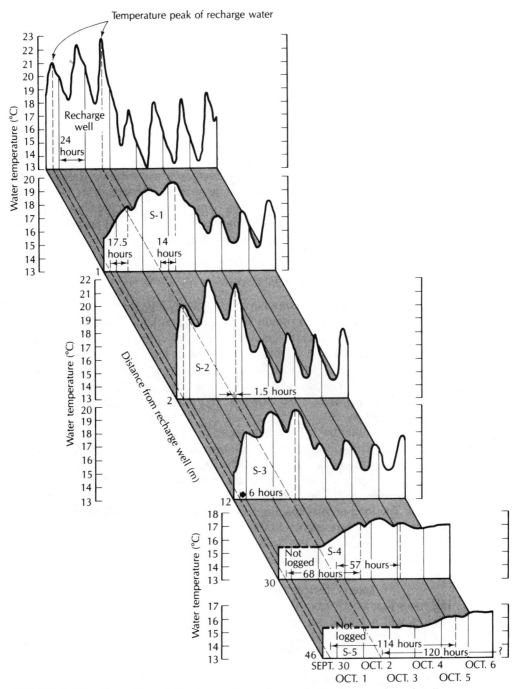

FIGURE 12.19. Fluctuations in the temperature of groundwater in a number of wells near an injection well for artificial recharge. All temperatures are measured at a depth of 49 meters. The diural variation in the temperature of the recharge water dies out by 30 meters. SOURCE: W. S. Keys and R. F. Brown, *Ground Water*, 16 (1978):32–48.

431

changes in the conductivity of the borehole fluid, and a small volume of rock around the borehole near the downhole electrode. If the borehole fluid is homogeneous, the variation in resistance will be due to the lithologic variations near the borehole.

Lithologies with a high resistance include sand, sand and gravel, sandstone, and lignite. Clay and shale have the lowest resistance. Increasing salinity will cause a decrease in resistance. If the borehole is enlarged (for instance, by a fracture), the resistance will also decrease. If the caliper log indicates a cavity, and the resistance log decreases, the decrease is due to the hole enlargement. Should the borehole be straight, a decrease in resistance might be due to either a shale layer or, perhaps, a sandstone containing a brine. However, other logs can distinguish brine from fresh water and sandstone from shale. Figure 12.20 shows a single-point resistance log.

12.4.4 RESISTIVITY

Earth **resistivity** may be measured in a borehole by lowering two current electrodes and measuring the resistivity between two additional electrodes. Resistivity is measured in ohm-meters and is different from resistance, the latter being measured in ohms. Single-point resistance logging measures the total resistance of the earth materials, while resistivity measures a specific property of the rock and the contained pore waters. The trace of a resistivity log is similar to a single-point resistance log. However, resistivity logs can be calibrated and used quantitatively.

A number of different electrode configurations are used in resistivity logging. These include the **short-normal, long-normal,** and **lateral configurations,** as shown in Figure 12.21. The three logs have similar traces (Figure 12.22). The short-normal curve indicates the resistivity of the zone close to the borehole. It is in this area that the drilling fluid may have invaded the formations. The long-normal spacing has more spacing between electrodes, and thus measures the resistivity further away from the borehole—presumably, beyond the influence of the drilling fluid. Both the short-normal and long-normal resistivity measure a greater radius of influence than the single-point resistance.

Lateral devices have very widely spaced electrodes for measuring zones that are far from the borehole. Because of the wide spacing, lateral devices will not pick out thin beds of different resistivity. For example, the 18-foot × 8-inch lateral log will be best for beds at least 40 feet thick.

12.4.5 SPONTANEOUS POTENTIAL

Along with resistance and resistivity, **spontaneous potential** is another form of electrical logging. It is a measure of the natural electrical potential that develops between the formation and the borehole fluids. It is only run on an open

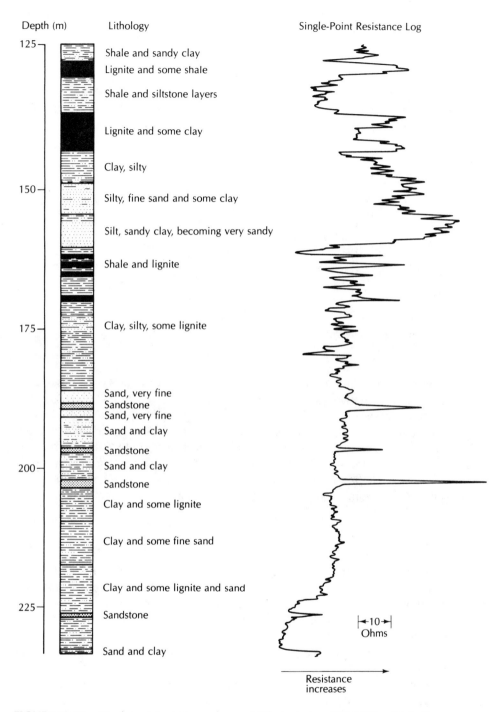

FIGURE 12.20. Single-point resistance log and lithologic log. SOURCE: W. S. Keys and L. M. McCary, in *Techniques of Water-Resources Investigations,* U.S. Geological Survey, 1971, Book 2, Chap. E1.

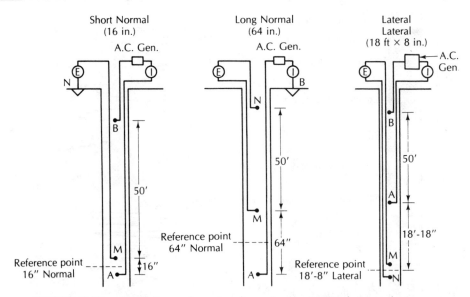

FIGURE 12.21. Electrode configuration for various resistivity logging devices.

hole filled with fluid, as are resistance and resistivity. It can be used below a casing in a partially cased well. The spontaneous-potential (SP) curve can be used for determination of bed thickness, geologic correlation, and also for delineation of permeable rocks. An SP device consists of a surface electrode and a borehole electrode with a voltmeter to measure potential.

One use of the SP curve is to distinguish shale from sandstone lithology. Shale has a positive SP response and sandstone a negative one if the salinity of the formation fluid is greater than that of the borehole fluid.

12.4.6 NUCLEAR LOGGING

Some of the most useful logging methods involve the measurement of either natural radioactivity of the rock and fluids or their attenuation of induced radiation. **Nuclear logging** can be done in either a cased or an uncased hole, and the logs are not affected by the type of drilling mud. The use of radioactive isotopes involves special safety precautions.

Radioactive decay is a process with a random component; hence, the instantaneous rate of decay fluctuates. Over a long time period, the rate of decay per time period is constant. However, as the time period decreases, the variation in the number of decay events per time period will increase. Nuclear logging records the number of disintegrations over a fixed time period, called the **time constant**. The longer the time constant, the less the likelihood that a variation in radiation intensity is due to random decay and, hence, the more likely the variation reflects different lithology.

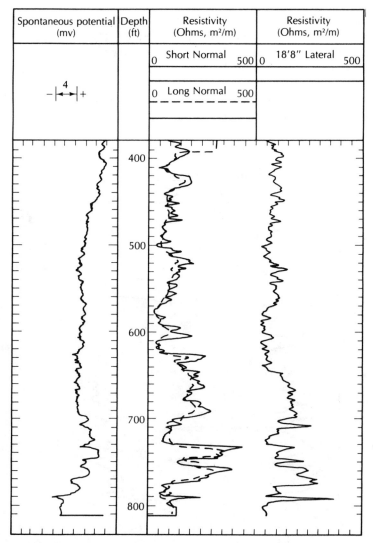

Spontaneous potential (mv)	Depth (ft)	Resistivity (Ohms, m²/m)	Resistivity (Ohms, m²/m)

FIGURE 12.22. Electric logs of a limestone well. The long-normal log is shown as a dashed line. SOURCE: W. S. Keys and L. M. MacCary, in *Techniques of Water-Resources Investigations*, U.S. Geological Survey, 1971, Book 2, Chap. E1.

Consideration must also be given to the speed at which the probe is moved up or down the hole. If the speed is too great, the probe may pass a thin bed before the time constant has elapsed. Consequently, the selection of the proper time constant and logging speed is very important and depends upon the equipment, logging technique, and lithology (17).

Nuclear logs do not have exact reproducibility, due to the statistical nature of the decay process. Repeat logging runs are necessary to determine if an observed variation represents a lithologic change or a statistical fluctuation in the decay rate. In Figure 12.23, the first two neutron-gamma logs were made going up and down the hole, respectively. The same peaks are present, but there is variability in the exact radiation count. To the right is a third log of the same hole made with a different radiation source having a longer time constant—10 seconds versus 3 seconds. The right-hand log has a poor ratio of time constant and logging speed. It does not distinguish thin beds, and the positions of the lithologic contacts are incorrect.

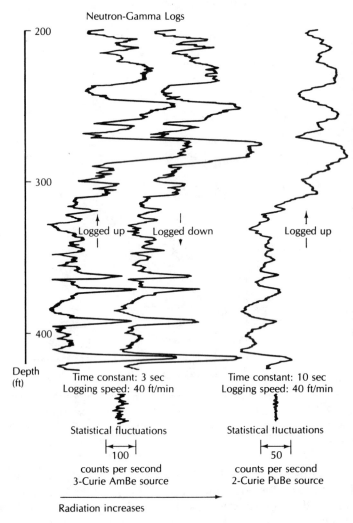

FIGURE 12.23. Statistical variation in neutron-gamma logs made of the same hole. The log on the right had a poor ratio of time constant and logging speed. SOURCE: W. S. Keys and L. M. MacCary, in *Techniques of Water-Resources Investigations*, U.S. Geological Survey, 1971, Book 2, Chap. E1.

The thickness of individual strata can be determined from nuclear logs if there is a change in lithology or porosity from one unit to the next. This is assumed to be equal to the thickness of the anomaly at one-half the maximum amplitude. This method will overestimate the thickness of thin layers. By convention, radiation increases to the right in nuclear logging. In a reversed log, it increases to the left.

There are three nuclear methods that can be used for composite identification (17). The neutron count rate increases with decreasing porosity, while the gamma-gamma count rate decreases. The natural gamma radiation increases with an increasing clay or shale content as well as with increased phosphate and K-feldspars, but has no direct relationship to porosity. A stratum with low natural gamma radiation and a low neutron count (or high gamma-gamma count) could be interpreted as a porous sandstone. A low natural gamma radiation and high neutron count could be a dense quartz sandstone or quartzite (see Figure 12.18).

Natural Gamma Radiation

This is the nuclear-logging method most commonly employed in hydrogeology. It is a measure of the natural radiation of rocks as determined by the emission of gamma activity by potassium 40, the uranium 238 decay series, and the thorium 232 decay series. These are constituent materials for some shales and clays which have high gamma activity. Certain feldspars and micas are high in ^{40}K. A natural gamma log shows increasing radiation opposite sedimentary beds that contain potassium-rich shale, or clay or phosphate rock. Thus, a shaly sandstone could be distinguished from a clean quartz sandstone. The natural gamma log can be used for lithologic determination—especially of detrital sediments—on the basis of differences in radiation intensity. No calibration of the unit is necessary in this nuclear-logging method. Another advantage is that no radiation sources need be used.

Neutron Logging

A neutron probe contains a radioactive element, such as PbBe, which is a source of neutrons, and a detector. The emitted neutrons are slowed and scattered by collisions with nuclei of hydrogen atoms. Detectors are available to measure gamma radiation produced by the neutron-hydrogen atom collision, or the number of neutrons present at different energy levels. Thus, a neutron log will be identified as a **neutron-thermal neutron**, a **neutron-epithermal neutron**, or a **neutron-gamma log** on the basis of the method of detection.

Hydrogen is present in the ground primarily in the form of water or hydrocarbons. In almost all rocks of interest to hydrogeologists, there are no hydrocarbons. Therefore, apart from hydrated minerals, water is present as moisture in the pore spaces of the rock. An increase in the amount of water will result in an increase in the number of neutrons that are captured or moderated.

As a result, saturated rocks with a high porosity will have a lower neutron count than low-porosity rocks. Above the water table, the neutron-logging equipment can be used to measure the moisture content, but not the porosity. Neutron logging can be used to determine the specific yield of unconfined aquifers (24). It also can distinguish gypsum, with a high proportion of hydrated water, from anhydrite. Both have a very low natural gamma radiation, however anhydrite has a high neutron count while gypsum has a low count.

Gamma-Gamma Radiation

In this type of logging, a source of gamma radiation, such as cobalt 60, is lowered into the borehole. Gamma photons are absorbed or scattered by all material with which the cobalt 60 comes in contact. This includes fluid, casing, and rock. The absorption is proportional to the bulk density of the earth material. **Bulk density** is defined as the weight of the rock divided by the total volume, including the porosity. Thus, gamma-gamma radiation increases with decreasing bulk density (increasing porosity). Bulk density can be determined from a calibrated gamma-gamma log. The formation porosity can be determined from the equation

$$\text{Porosity} = \frac{\text{grain density} - \text{bulk density}}{\text{grain density} - \text{fluid density}} \qquad \text{(12-14)}$$

Grain density may be determined from drill cuttings or assumed to be 2.65 grams per cubic centimeter for quartz sandstone. Fluid density is 1.000 grams per cubic centimeter for mud-free fresh water. The drilling fluid may contain additives to increase the fluid density.

 MODELS OF GROUNDWATER FLOW

12.5.1 BASIC MODEL TYPES

Models of aquifers are used to determine the actual or predicted behavior of the aquifer under different sets of conditions. They facilitate understanding of the behavior of an aquifer under historic conditions and are useful in predicting future conditions. The impact of alternative management plans can be tested via models. A **model** is a representation of a physical system, behaving in a manner similar to that of the real system.

There are three basic types of models which have been used in hydrogeology: scale models, analog models, and computer models (27). A **scale model** is made from the same materials as those of the natural system. For instance, a plastic container may be manufactured to scale and filled with sand or glass beads with a hydraulic conductivity scaled to the actual aquifer material. Dye added to the water helps the observer trace the flow. Pressure is measured in piezometers inserted through the walls. Water can be added to the model aquifer to simulate recharge and also be pumped from scale-model wells. Sand models have been used for a variety of studies (e.g., 28, 29, 30).

The flow of water through porous media is governed by equations similar to those governing the flow of electricity through a conductor. This also is true for the flow of a viscous fluid between two very closely spaced parallel plates. Models can be constructed using electrical circuits or viscous-fluid flow to simulate real or ideal aquifers. These are called **analog models**, since the model is analogous to the actual aquifer. Analog models are typically constructed to model two-dimensional flow. The models can be either horizontal or vertical. If areal flow patterns are being studied, a horizontal model is indicated. In order to study vertical flow, a cross-sectional or vertical model is the one of choice.

Electrical models (e.g., 31, 32) are made with a network of resistors scaled to represent the framework of the aquifer, with capacitors to provide for storage. The flow of electrical current in amperes through the model is representative of fluid flow. The voltage in the model corresponds to hydraulic potential, and the volume of water in storage is analogous to coulombs of electricity. Resistivity is inversely proportional to the hydraulic conductivity of the aquifer, while the capacitance network is scaled to aquifer storativity. Measurements of current and voltage made at various points in the model represent the flow of water and hydraulic head in the aquifer. Electrical analog models can be made to represent three-dimensional flow by linking a series of horizontal models together. While electrical analogs can be used to study the flow of water, they are not amenable to studying the flow in unconfined aquifers, where aquifer transmissivity decreases with pumpage as the water table declines. They are also unable to simulate mass transport, dispersion, and diffusion (27).

The **viscous-fluid model** is also known as a **Hele-Shaw model** after H. S. Hele-Shaw, who first used it. It has been applied to many problems in hydrogeology. Hele-Shaw models are especially adapted to the study of immiscible fluids with different densities; for example, as encountered in saltwater intrusion (33). Two liquids with different densities are used to simulate the fresh and salt water. The injection of wastewater into flow fields can be visually studied using Hele-Shaw models (34). Both horizontal and vertical models have been built. Because of physical constraints, it has not been possible to model three-dimensional flow with a viscous-fluid model.

All analog models, and scale models, as well, have some definite disadvantages. Not the least of these is that the models must be constructed by someone handy at carpentry, plumbing, and wiring. The time and materials cost for large models is substantial. There must be room to construct and house the models, and possibly to store them when not in use. Finally, the models are not very flexible. It is difficult to change the aquifer geometry and hydraulic characteristics built into a model.

439

Scale models and Hele-Shaw models are advantageous in that the fluid movement is visible through the use of dyes. This is particularly helpful in presentations to those not familiar with subsurface flow. Under some conditions, such as transient flow with closely spaced wells and nonlinear boundaries, the resistance-capacitance electrical analog may be more accurate than digital-computer models (27).

The most widely used groundwater models are **digital models**, consisting of a series of instructions and data values that are fed into a digital computer. One great advantage of a digital model is its flexibility. The same computer program can serve as the instructions for a large number of actual models. By contrast, separate analog models would be needed for different aquifer systems. Some computer models are extremely versatile, treating both saturated and unsaturated flow in two or three dimensions. Because of computer limitations, such programs can only deal with small basins, although aquifers can be nonhomogeneous and anisotropic (35). Computer models can also be applied to problems of diffusion and dispersion in transient multiphase flow problems. Such models have been used to study saltwater intrusion and mass transport (e.g., 37, 38, 39).

Computer models of groundwater flow are widely available in published form (40, 41, 41); thus, it is not necessary for the hydrogeologist to be able to design and write a program for a computer model of simple flow systems. However, knowledge of FORTRAN is very helpful and some familiarity with digital computers is necessary. More complex flow problems involve complicated algorithms for which there may be no published program listings. In such cases, specific training in computer modeling may be needed to develop a suitable model.

12.5.2 FINITE-DIFFERENCE METHODS

The two-dimensional, transient flow of groundwater in a confined, isotropic aquifer is governed by the partial differential equation

$$\frac{\partial}{\partial x}\left(T\frac{\partial h}{\partial x}\right) + \frac{\partial}{\partial y}\left(T\frac{\partial h}{\partial y}\right) = S\frac{\partial h}{\partial t} + Q \qquad \text{(12-15)}$$

where

T is the transmissivity

h is the head

t is the time

S is the storativity

Q is the net groundwater recharge or withdrawal

x, y are rectilinear coordinates

This equation is valid for a continuous aquifer, for which there is a value of T, S, and h everywhere. The continuous aquifer can be replaced by a network of points. If the points are a small, but finite, distance apart and the spatial changes in h are approximately linear between adjacent points, a finite-difference model can be applied. The points form a regular grid pattern (Figure 12.24A). The partial differential flow equations are replaced by a set of finite-difference equations. The space coordinates Δx and Δy have a numerical value and time is divided into discrete steps, Δt. Computer notation for the finite-difference grid is given in Figure 12.24B.

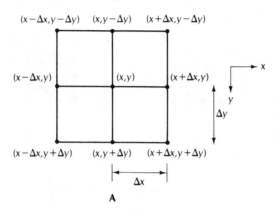

A

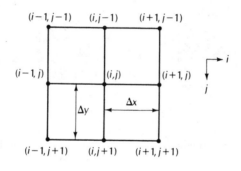

B

FIGURE 12.24. **A.** Finite-difference grid; **B.** Computer notation for finite-difference grid.

The finite-difference form of Equation (12-15) is given by (40):

$$h_{i,j}(T_{i-1,j,2} + T_{i,j,2} + T_{i,j,1} + T_{i,j-1,1} + S\Delta x^2/\Delta t) \qquad \text{(12-16)}$$
$$- T_{i-1,j,2}h_{i-1,j} - T_{i,j,2}h_{i+1,j} - T_{i,j,1}h_{i,j+1}$$
$$- T_{i,j-1,1}h_{i,j-1} = (S\Delta x^2/\Delta t)h\phi_{i,j} - Q_{i,j} + Q_n$$

where

T is the aquifer transmissivity of a volume of the aquifer close to node i, j (Figure 12.25 shows the special coordinates for each value of T)

441

$h_{i,j}$ is the head calculated at node i, j

$h\phi_{i,j}$ is the head calculated at node i, j at the end of the previous time increment, Δt

$Q_{i,j}$ is the net withdrawal rate at node i, j

Q_n is the leakage or recharge rate into the aquifer

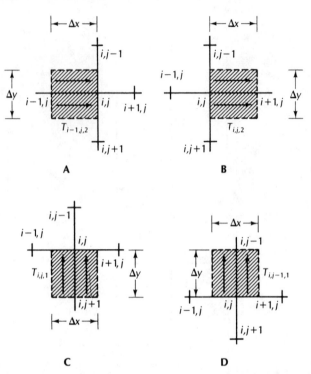

FIGURE 12.25. Vector volumes for aquifer transmissivity for Equation (12-16). SOURCE: T. C. Prickett and C. G. Lonnquist, Illinois State Water Survey Bulletin 55, 1971.

There is a difference equation of the form of Equation (12-16) for each node in the finite-difference grid. Even a small grid may have several hundred nodes, and large ones will have many thousands of equations. Thus, a large set of equations must be solved simultaneously for h at each node. A solution is obtained by solving the sets of equations for nodes along the rows or columns of the grid starting at one boundary of the model and progressing sideways or downward to another boundary. For steady flow conditions ($\Delta t=0$), the finite-difference equations may be solved to find values for h at all points in the interior of the boundaries. If the problem is transient, then the initial values of head must be known or assumed. The model then solves for the change in head with time. The accuracy of the model increases as the mesh size (Δx and Δy) decreases, and also as the time increment, Δt, becomes smaller. Of course, as these values decrease, the number of points and time

442

increments increases, which increases the cost of the model and the size of the computer needed.

There are a number of different methods of solving finite-difference equations for transient conditions. These fall into two broad categories: explicit and implicit methods. In the **explicit methods**, an unknown value of h at one node is determined from known values of h at neighboring points. Unknown values at two or more neighboring nodes are computed simultaneously by using known values at one or more nearby nodes. Explicit methods are simple and straightforward, but pose some severe restrictions on the mesh size and time increments. **Implicit methods** are more complicated, but also more versatile. They may require greater computer storage capacity, but use less running time than explicit methods. Implicit models are superior in that they permit the use of larger Δt-values. Explicit models are generally too time consuming and, hence, expensive.

12.5.3 ILLINOIS STATE WATER SURVEY MODEL FOR NONLEAKY CONFINED AQUIFERS

One extremely versatile finite-difference model has been developed by T. A. Prickett and C. G. Lonnquist (40). The Illinois State Water Survey basic aquifer simulation model is for two-dimensional non-steady–state flow in a nonhomogeneous (or homogeneous), isotropic, nonleaky confined aquifer. The aquifer boundaries may be either no-flow, constant-head, or recharge boundaries. Modifications of the basic program are available to simulate leaky confined conditions, water-table conditions, induced infiltration, groundwater evaporation, and conditions whereby a confined aquifer undergoes transformation into a water-table aquifer. The model is solved using a technique known as the **modified alternating-direction implicit method**, which is an implicit-solution technique.

Figure 12.26 gives a listing for the basic aquifer simulation program, which is coded in FORTRAN IV. The program requires 75,000 bytes of core storage and is written for an IBM 360 model 75 computer with a G- or H-level compiler. Subroutine ERRSET is an IBM 360 subroutine (ID number 0040). It may have a different name on other computer systems. The program will work with any consistent system of units. The comment cards specify head in feet, constant withdrawal rates in gallons per day, transmissivity in gallons per day per foot, time increment in days, and a storage factor in gallons per foot. The storage factor (SF1) in gallons per foot is equal to the storativity multiplied by 7.48 $\Delta X \Delta Y$:

$$SF1_{i,j} = 7.48\ S_{i,j}\Delta X \Delta Y \qquad (12\text{-}17)$$

If a consistent set of units is being used, such as head in meters, T in meters per day, Q in cubic meters per day, and Δt in days, then

$$SF1_{i,j} = S_{i,j}\Delta X \Delta Y \qquad (12\text{-}18)$$

443

```
C      ILLINOIS STATE WATER SURVEY                        0001
C      BASIC AQUIFER SIMULATION PROGRAM                   0002
C                                                         0003
C      DEFINITION OF VARIABLES                            0004
C                                                         0005
C      HO(I,J)-----HEADS AT START OF TIME                 0006
C                  INCREMENT (I,J)                        0007
C      H(I,J)------HEADS AT END OF TIME                   0008
C                  INCREMENT (FT)                         0009
C      SF1(I,J)----STORAGE FACTOR FOR                     0010
C                  ARTESIAN CONDITIONS                    0011
C                  (GAL/FT)                               0012
C      Q(I,J)------CONSTANT WITHDRAWAL                    0013
C                  RATES (GPD)                            0014
C      T(I,J,1)----AQUIFER TRANSMISSIVITY                 0015
C                  BETWEEN I,J AND I,J+1                  0016
C                  (GAL/DAY/FT)                           0017
C      T(I,J,2)----AQUIFER TRANSMISSIVITY                 0018
C                  BETWEEN I,J AND I+1,J                  0019
C                  (GAL/DAY/FT)                           0020
C      AA,BB,CC,DD-COEFFICIENTS IN WATER                  0021
C                  BALANCE EQUATIONS                      0022
C      NR---------NO. OF ROWS IN MODEL                    0023
C      NC---------NO. OF COLUMNS IN MODEL                 0024
C      NSTEPS------NO. OF TIME INCREMENTS                 0025
C      DELTA-------TIME INCREMENTS (DAYS)                 0026
C      HH,S1,QQ,TT-DEFAULT VALUES                         0027
C      I-----------MODEL COLUMN NUMBER                    0028
C      J-----------MODEL ROW NUMBER                       0029
C                                                         0030
C                                                         0031
C                                                         0032
C                                                         0033
       DIMENSION H(50,50),HO(50,50),                      0034
      1SF1(50,50),Q(50,50),T(50,50,2),                    0035
      2B(50),G(50),DL(50,50)                              0036
C                                                         0037
C      TURN OFF UNDERFLOW TRAP                            0038
C                                                         0039
       CALL ERRSET(208,256,-1,1)                          0040
C                                                         0041
C      DEFINE INPUT AND OUTPUT DEVICE NUMBERS             0042
C                                                         0043
       INTEGER OUT                                        0044
       IN=5                                               0045
       OUT=6                                              0046
C                                                         0047
C      READ PARAMETER CARD AND                            0048
C      DEFAULT VALUE CARD                                 0049
C                                                         0050
       READ(IN,10)NSTEPS,DELTA,ERROR,                     0051
      1NC,NR,TT,S1,HH,QQ                                  0052
10     FORMAT(I6,2F6.0,2I6,4F6.0)                         0053
C                                                         0054
C      FILL ARRAYS WITH DEFAULT VALUES                    0055
C                                                         0056
       DO 20 I=1,NC                                       0057
       DO 20 J=1,NR                                       0058
       T(I,J,1)=TT                                        0059
       T(I,J,2)=TT                                        0060
       SF1(I,J)=S1                                        0061
       H(I,J)=HH                                          0062
       HO(I,J)=HH                                         0063
20     Q(I,J)=QQ                                          0064
C                                                         0065
C      READ NODE CARDS                                    0066
C                                                         0067
30     READ(IN,40,END=50)I,J,T(I,J,1),                    0068
      1T(I,J,2),SF1(I,J),H(I,J),Q(I,J)                    0069
40     FORMAT(2I3,2F6.0,2F4.0,1F6.0)                      0070
       GO TO 30                                           0071
C                                                         0072
C      START OF SIMULATION                                0073
C                                                         0074
50     TIME=0                                             0075
       DO 320 ISTEP=1,NSTEPS                              0076
       TIME=TIME+DELTA                                    0077
C                                                         0078
C      PREDICT HEADS FOR NEXT                             0079
C      TIME INCREMENT                                     0080
C                                                         0081
       DO 70 I=1,NC                                       0082
       DO 70 J=1,NR                                       0083
       D=H(I,J)-HO(I,J)                                   0084
       HO(I,J)=H(I,J)                                     0085
       F=1.0                                              0086
       IF(DL(I,J).EQ.0.0)GO TO 60                         0087
       IF(ISTEP.GT.2)F=D/DL(I,J)                          0088
       IF(F.GT.5)F=5.0                                    0089
       IF(F.LT.0.0)F=0.0                                  0090
60     DL(I,J)=D                                          0091
70     H(I,J)=H(I,J)+D*F                                  0092
C                                                         0093
C      REFINE ESTIMATES OF HEADS BY IADI METHOD           0094
C                                                         0095
       ITER=0                                             0096
80     E=0.0                                              0097
       ITER=ITER+1                                        0098
C                                                         0099
C      COLUMN CALCULATIONS                                0100
C                                                         0101
       DO 190 II=1,NC                                     0102
       I=II                                               0103
       IF(MOD(ISTEP+ITER,2).EQ.1) I=NC-I+1                0104
       DO 170 J=1,NR                                      0105
C                                                         0106
C      CALCULATE B AND G ARRAYS                           0107
C                                                         0108
       BB=SF1(I,J)/DELTA                                  0109
       DD=HO(I,J)*SF1(I,J)/DELTA-Q(I,J)                   0110
       AA=0.0                                             0111
       CC=0.0                                             0112
       IF(J-1)90,100,90                                   0113
90     AA=-T(I,J-1,1)                                     0114
       BB=BB+T(I,J-1,1)                                   0115
100    IF(J-NR)110,120,110                                0116
110    CC=-T(I,J,1)                                       0117
       BB=BB+T(I,J,1)                                     0118
120    IF(I-1)130,140,130                                 0119
130    BB=BB+T(I-1,J,2)                                   0120
       DD=DD+H(I-1,J)*T(I-1,J,2)                          0121
140    IF(I-NC)150,160,150                                0122
150    BB=BB+T(I,J,2)                                     0123
       DD=DD+H(I+1,J)*T(I,J,2)                            0124
160    W=BB-AA*B(J-1)                                     0125
       B(J)=CC/W                                          0126
170    G(J)=(DD-AA*G(J-1))/W                              0127
C                                                         0128
C      RE-ESTIMATE HEADS                                  0129
C                                                         0130
       E=E+ABS(H(I,NR)-G(NR))                             0131
       H(I,NR)=G(NR)                                      0132
       N=NR-1                                             0133
180    HA=G(N)-B(N)*H(I,N+1)                              0134
       E=E+ABS(HA-H(I,N))                                 0135
       H(I,N)=HA                                          0136
       N=N-1                                              0137
       IF(N)190,190,180                                   0138
190    CONTINUE                                           0139
C                                                         0140
C      ROW CALCULATIONS                                   0141
C                                                         0142
       DO 300 JJ=1,NR                                     0143
       J=JJ                                               0144
       IF(MOD(ISTEP+ITER,2).EQ.1) J=NR-J+1                0145
       DO 280 I=1,NC                                      0146
       BB=SF1(I,J)/DELTA                                  0147
       DD=HO(I,J)*SF1(I,J)/DELTA-Q(I,J)                   0148
       AA=0.0                                             0149
       CC=0.0                                             0150
       IF(J-1)200,210,200                                 0151
200    BB=BB+T(I,J-1,1)                                   0152
       DD=DD+H(I,J-1)*T(I,J-1,1)                          0153
210    IF(J-NR)220,230,220                                0154
220    DD=DD+H(I,J+1)*T(I,J,1)                            0155
       BB=BB+T(I,J,1)                                     0156
230    IF(I-1)240,250,240                                 0157
240    BB=BB+T(I-1,J,2)                                   0158
       AA=-T(I-1,J,2)                                     0159
250    IF(I-NC)260,270,260                                0160
260    BB=BB+T(I,J,2)                                     0161
       CC=-T(I,J,2)                                       0162
270    W=BB-AA*B(I-1)                                     0163
       B(I)=CC/W                                          0164
280    G(I)=(DD-AA*G(I-1))/W                              0165
C                                                         0166
C      RE-ESTIMATE HEADS                                  0167
C                                                         0168
       E=E+ABS(H(NC,J)-G(NC))                             0169
       H(NC,J)=G(NC)                                      0170
       N=NC-1                                             0171
290    HA=G(N)-B(N)*H(N+1,J)                              0172
       E=E+ABS(H(N,J)-HA)                                 0173
       H(N,J)=HA                                          0174
       N=N-1                                              0175
       IF(N)300,300,290                                   0176
300    CONTINUE                                           0177
       IF(E.GT.ERROR) GO TO 80                            0178
C                                                         0179
C      PRINT RESULTS                                      0180
C                                                         0181
       WRITE(OUT,310)TIME,E,ITER                          0182
310    FORMAT(6H2TIME=,F6.2///,E20.7,I5)                  0183
       DELTA=DELTA*1.2                                    0184
       DO 320 J=1,NR                                      0185
320    WRITE(OUT,330)J,(H(I,J),I=1,NC)                    0186
330    FORMAT(I5,5X,10F10.4/(12X10F10.4))                 0187
C                                                         0188
       STOP                                               0189
       END                                                0190
```

FIGURE 12.26. Illinois State Water Survey basic confined aquifer simulation program listing. SOURCE: T. C. Prickett and C. G. Lonnquist, Illinois State Water Survey Bulletin 55, 1971.

444

A flowchart for the basic aquifer simulation program is given in Figure 12.27. In the program statements, 0001 through 0033 are comment cards describing the program. Starting with line 0034 through line 0072, storage space is' reserved for the data, which are read and stored. The actual simulation occurs in lines 0073 through 0178. The printout of the results is controlled by lines 0180 through 0187.

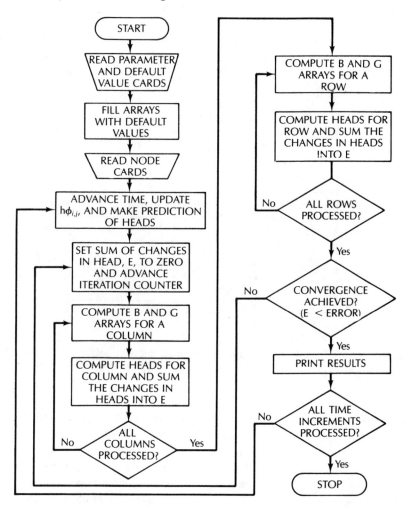

FIGURE 12.27. Flowchart for Illinois State Water Survey basic confined aquifer simulation program. SOURCE: T. C. Prickett and C. G. Lonnquist, Illinois State Water Survey Bulletin 55, 1971.

In order to prepare a model, the aquifer must be overlain by a finite-difference grid. A grid overlain on a map of an aquifer system is shown in Figure 12.28. The grid is prepared so that ΔX is equal to ΔY between every node. In order to use the Prickett-Lonnquist model as written, the grid cannot have more than 50 rows and 50 columns. The dimensions of the model are NR, the number of rows of width ΔY necessary to contain the aquifer, and NC, the number of columns of width ΔX. It is necessary to assign values for transmissivity and the storage factor for each node in the aquifer. The value of the storage factor is found from Equation (12-17) or (12-18). If the boundary of the aquifer does not coincide with the NC-by-NR matrix (for example, the aquifer of Figure 12.28), then a value of zero transmissivity is assigned to each node of the matrix that is outside of the aquifer boundary. Storage-factor values of zero should not be assigned to nodes outside the aquifer boundary, as this will cause an error in the program. An average value for the storage factor outside the aquifer should be assigned, even though this is a dummy value.

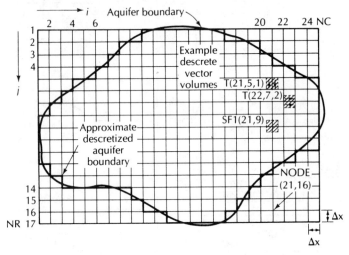

FIGURE 12.28. Finite-difference grid overlain on map of aquifer. SOURCE: T. C. Prickett and C. G. Lonnquist, Illinois State Water Survey Bulletin 55, 1971.

It is necessary to select an initial time step, DELTA. It is also necessary to define the number of time steps that will be made, NSTEPS. The given model is programmed on line 0184, so that the value of DELTA increases after each time iteration. For most transient cases, the head change is most rapid during early pumping and then slows with time. Therefore, the time steps can lengthen with increased time since the start of simulation. The total time can be found from the number of time steps, ISTEP, by multiplying the initial DELTA by an ISTEP factor:

$$\text{Time} = \text{DELTA} \times \text{ISTEP factor} \qquad \textbf{(12-19)}$$

For the given program, the ISTEP factors are listed in Table 12.2. As a result of the ISTEP factor, time values are not usually in exact days. Should exact-day values be necessary, line 0184 could be eliminated. If DELTA is in even days (or

TABLE 12.2. Values of the ISTEP function

ISTEP	ISTEP Function	ISTEP	ISTEP Function
1	1.00	26	567.37
2	2.20	27	681.84
3	3.64	28	819.21
4	5.37	29	984.05
5	7.44	30	1181.87
6	9.93	31	1419.24
7	12.92	32	1704.08
8	16.50	33	2045.90
9	20.80	34	2456.08
10	25.96	35	2948.29
11	32.15	36	3538.95
12	39.58	37	4247.74
13	48.50	38	5098.28
14	59.20	39	6118.93
15	72.03	40	7343.71
16	87.44	41	8813.45
17	105.93	42	10577.14
18	128.12	43	12693.57
19	154.74	44	15233.28
20	186.69	45	18280.93
21	225.02	46	21938.11
22	271.03	47	26326.73
23	326.23	48	31593.07
24	392.48	49	37912.68
25	471.98	50	45496.21

SOURCE: T. A. Prickett and C. G. Lonnquist, *Selected Digital Computer Techniques for Groundwater Resource Evaluation*, Illinois State Water Survey Bulletin 55, 1971.

half days, etc.), the computed values will also be in even days. However, many more iterations will result, with a corresponding increase in computer time. The convenience of results in even days would rarely be worth the extra cost of such a step.

If a test for about 100 days is to be simulated, and DELTA is 0.5 days, an ISTEP factor of about 200 is needed. From Table 12.2, an NSTEPS value of 21 has an ISTEP factor of 225.02. The actual time would be 112.51 days for the final increment. If exactly 100 days are required, an NSTEPS value of 21 with a DELTA of 0.4444 days would yield the desired result.

It is also necessary to define an error-closure value, ERROR. The error check in the program compares computed drawdowns or heads at the end of an iteration with the results of the previous iteration to see if the differ-

447

ence is less than a specified tolerance. If it is less, the model has converged to an acceptable answer; if it is more, then additional iteration(s) are performed until the error-closure value is equal to or greater than the actual error. The value of ERROR may be found from

$$\text{ERROR} = Q \times \text{DELTA}/10 \times \text{SF1} \qquad (12\text{-}20)$$

where

Q is the total net withdrawal of the model (gallons per day)

DELTA is the initial time increment (days)

SF1 is the average storage value of the model (gallons per foot)

If consistent units are used, SF1 should be computed by Equation (12-18) rather than by Equation (12-17). Equation (12-20) is a rule-of-thumb value which might need to be adjusted on a trial-and-error basis for a specific model. Smaller values of ERROR will yield solutions that are more accurate, but they will require more computer time.

Recharge boundaries can be thought of as places at which there is no drawdown. In order to simulate a recharge boundary in the model, the nodes along the finite-grid boundary should be assigned a very high storage factor, e.g., 10^{21} gallons per foot.

The model is set up with a parameter card, a default-value card, and a node-card deck. The **parameter card** contains values for NSTEPS, DELTA, and ERROR. On the **default-value card**, the size of the grid is entered with values for NR and NC, as well as the average transmissivity for all nodes, TT, the average storage factor, S1, the average initial head, HH, and the average net withdrawal rate at each node, QQ. The parameter and default-value cards are sufficient to describe an aquifer system which has an NC-by-NR rectangular shape, is homogeneous and isotropic, has the same initial head values at each node, and the same net withdrawal rate at each node.

The preceding conditions are very restrictive; thus, the model would not be very useful unless it could be modified. Modifications are made via the **node-card deck**. A separate node card is prepared for each node having properties different from those specified on the default-value card. For example, in almost every case there would be pumping from only a small number of all the nodes. Under such conditions, the value of QQ on the default-value card would be set at zero and a node card prepared for each node from which there is pumping. Each node card contains the description of the node, I and J, a value for $T_{i,j,1}$ and $T_{i,j,2}$, a storage factor for the node, SF1, and an initial head, H, as well as the pumping rate, Q. When node cards are necessary, every value must be punched onto a node card—even those that are equal to the default values. The formats for the parameter cards, the default-value card, and the node cards are shown in Figure 12.29. The job set-up consists of the program deck, parameter card, default-value card, node deck, and appropriate job-control cards.

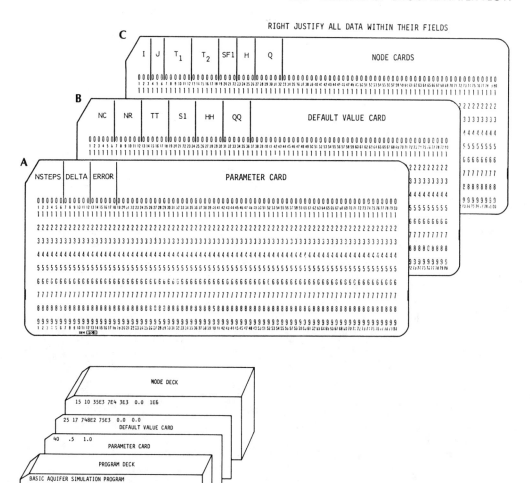

FIGURE 12.29. Input data formats for the parameter card, default-value card, and node cards for the Illinois State Water Survey basic confined aquifer simulation program. SOURCE: T. C. Prickett and C. G. Lonnquist, Illinois State Water Survey Bulletin 55, 1971.

Output from the program is printed as a matrix of drawdown values for each node at the end of each time step. For many problems, it is convenient to set the initial head at zero everywhere in the aquifer. The model will then yield drawdown values from a level datum. If the actual potentiometric surface is not level, this procedure can still be used; however it will not yield a map of the resulting potentiometric surface. Figure 12.30 shows part of the program output for a model with 7 columns and 15 rows where node 4,8 was pumped at 300,000 gallons per day.

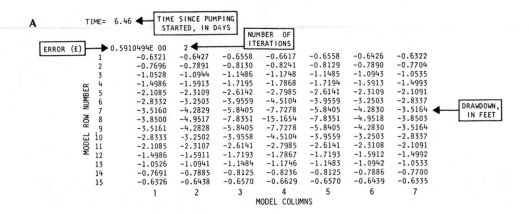

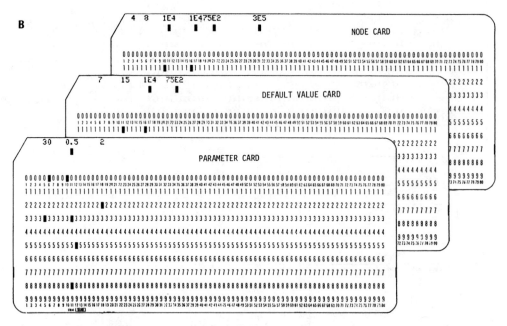

FIGURE 12.30. A. Sample of output data for a computer model; B. Input-card format in which the following parameter values are used in the computer model:

NSTEPS	= 30 days	TT	= 10,000 gal/day/ft
DELTA	= 0.5 days	S1	= 7500 gal/ft
ERROR	= 2	HH	= 0
NC	= 7	Q	= 300,000 at 4,8
NR	= 15		

SOURCE: T. C. Prickett and C. G. Lonnquist, Illinois State Water Survey Bulletin 55, 1971.

12.5.4 ILLINOIS STATE WATER SURVEY MODEL
FOR A WATER-TABLE AQUIFER

The basic Illinois State Water Survey model can easily be modified for water-table conditions. A program listing is given in Figure 12.31. The finite difference grid is overlain on the aquifer and, at each node, a hydraulic conductivity value, $PERM_{i,j}$, and the value of the elevation of the aquifer bottom, $BOT_{i,j}$, are used to compute changing aquifer transmissivity as the saturated thickness changes. The transmissivity between i, j and $i, j + 1$ is given by

$$T_{i,j,1} = PERM_{i,j,1}\sqrt{(H_{i,j} - BOT_{i,j})(H_{i,j+1} - BOT_{i,j+1})} \qquad \textbf{(12-21)}$$

The values of head, $H_{i,j}$, and $BOT_{i,j}$ are needed for each node and can be set up in two ways. Head can be assigned a zero value if the water table is level; BOT thus will have a negative value representing the distance from the water table to the aquifer bottom. As an alternative, both $H_{i,j}$ and $BOT_{i,j}$ can be set as positive values above a datum below the aquifer bottom.

For units of gallons per foot, the water-table storage factor $SF2_{i,j}$ is found from

$$SF2_{i,j} = 7.48\ S_{WT}\Delta X\Delta Y \qquad \textbf{(12-22)}$$

where S_{WT} is the specific yield at node i, j. If consistent units are used, the 7.48 factor may be dropped.

The parameter card is punched with the values for NSTEPS, DELTA, and ERROR. On the default-value card, values for NC, NR, TT, HH, QQ, S2, PP, and BOTT are entered as shown in Figure 12.32. TT is a value of the initial transmissivity for a node, HH is the initial head, PP is the average permeability of the aquifer, S2 is the average storage factor, and BOTT is the average elevation of the aquifer bottom. If the aquifer is receiving recharge, this can be simulated by a value at QQ found from

$$QQ_{i,j} = -7.48\ R_{i,j}\Delta X\Delta Y \qquad \textbf{(12-23)}$$

where

$\quad QQ_{i,j}$ is the recharge (gallons per day per square foot)
$\quad R_{i,j}$ is the recharge rate (feet per day)

If consistent units are used, the 7.48 may be dropped.

For each node at which the aquifer values are different from the default values, or at which there is pumping, a node card must be prepared as shown in Figure 12.32. The node card contains the i, j location of the node, values for T1 and T2, H, and Q, as in the basic simulation program. A storage factor for the node, SF2; permeability factors, P1 and P2, for the node; and the elevation of the bottom, BOT, are also punched onto each node card. The deck set-up is the same as for the basic program (refer back to Figure 12.29).

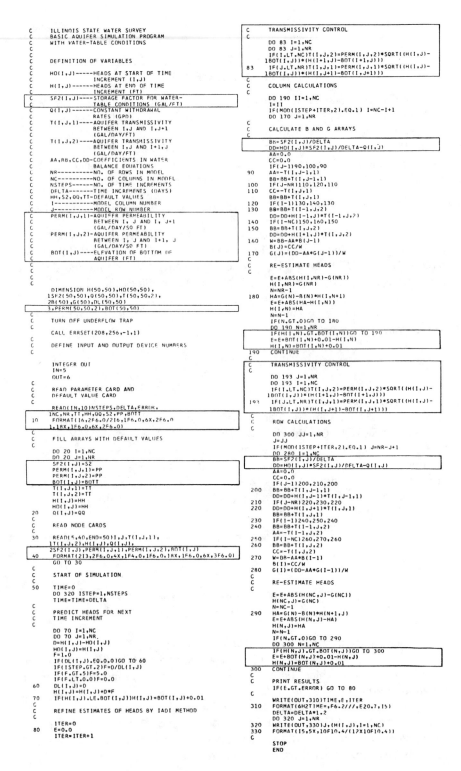

```
C     ILLINOIS STATE WATER SURVEY                  C     TRANSMISSIVITY CONTROL
C     BASIC AQUIFER SIMULATION PROGRAM             C
C     WITH WATER-TABLE CONDITIONS                        DO 83 I=1,NC
C                                                        DO 83 J=1,NR
C                                                        IF(I.LT.NC)T(I,J,2)=PERM(I,J,2)*SQRT((H(I,J)-
C     DEFINITION OF VARIABLES                      1BOT(I,J))*(H(I+1,J)-BOT(I+1,J)))
C                                                  83    IF(J.LT.NR)T(I,J,1)=PERM(I,J,1)*SQRT((H(I,J)-
C     HO(I,J)-----HEADS AT START OF TIME           1BOT(I,J))*(H(I,J+1)-BOT(I,J+1)))
C                 INCREMENT (I,J)                  C
C     H(I,J)------HEADS AT END OF TIME             C     COLUMN CALCULATIONS
C                 INCREMENT (FT)                   C
C     SF2(I,J)----STORAGE FACTOR FOR WATER-              DO 190 II=1,NC
C                 TABLE CONDITIONS (GAL/FT)              I=II
C     Q(I,J)------CONSTANT WITHDRAWAL                    IF(MOD(ISTEP+ITER,2).EQ.1) I=NC-I+1
C                 RATES (GPD)                            DO 170 J=1,NR
C     T(I,J,1)----AQUIFER TRANSMISSIVITY           C
C                 BETWEEN I,J AND I,J+1            C     CALCULATE B AND G ARRAYS
C                 (GAL/DAY/FT)                     C
C     T(I,J,2)----AQUIFER TRANSMISSIVITY                 BB=SF2(I,J)/DELTA
C                 BETWEEN I,J AND I+1,J                  DD=HO(I,J)*SF2(I,J)/DELTA-Q(I,J)
C                 (GAL/DAY/FT)                           AA=0.0
C     AA,BB,CC,DD-COEFFICIENTS IN WATER                  CC=0.0
C                 BALANCE EQUATIONS                      IF(J-1)90,100,90
C     NR---------NO. OF ROWS IN MODEL              90    AA=-T(I,J-1,1)
C     NC---------NO. OF COLUMNS IN MODEL                 BB=BB+T(I,J-1,1)
C     NSTEPS-----NO. OF TIME INCREMENTS            100   IF(J-NR)110,120,110
C     DELTA------TIME INCREMENTS (DAYS)            110   CC=-T(I,J,1)
C     HH,S2,QQ,TT-DEFAULT VALUES                         BB=BB+T(I,J,1)
C     I----------MODEL COLUMN NUMBER               120   IF(I-1)130,140,130
C     J----------MODEL ROW NUMBER                  130   BB=BB+T(I-1,J,2)
C     PERM(I,J,1)-AQUIFER PERMEABILITY                   DD=DD+H(I-1,J)*T(I-1,J,2)
C                 BETWEEN I, J AND I, J+1          140   IF(I-NC)150,160,150
C                 (GAL/DAY/SQ FT)                  150   BB=BB+T(I,J,2)
C     PERM(I,J,2)-AQUIFER PERMEABILITY                   DD=DD+H(I+1,J)*T(I,J,2)
C                 BETWEEN I, J AND I+1, J          160   W=BB-AA*B(J-1)
C                 (GAL/DAY/SQ FT)                        B(J)=CC/W
C     BOT(I,J)----ELEVATION OF BOTTOM OF           170   G(J)=(DD-AA*G(J-1))/W
C                 AQUIFER (FT)                     C
C                                                  C     RE-ESTIMATE HEADS
C                                                  C
C                                                        E=E+ABS(H(I,NR)-G(NR))
C                                                        H(I,NR)=G(NR)
      DIMENSION H(50,50),HO(50,50),                      N=NR-1
     1SF2(50,50),Q(50,50),T(50,50,2),           180   HA=G(N)-B(N)*H(I,N+1)
     2B(50),G(50),DL(50,50)                             E=E+ABS(HA-H(I,N))
     3,PERM(50,50,2),BOT(50,50)                         H(I,N)=HA
C                                                        N=N-1
C     TURN OFF UNDERFLOW TRAP                            IF(N.GT.0)GO TO 180
C                                                        DO 190 N=1,NR
      CALL ERRSET(208,256,-1,1)                          IF(H(I,N).GT.BOT(I,N))GO TO 190
C                                                        E=E+BOT(I,N)+0.01-H(I,N)
C     DEFINE INPUT AND OUTPUT DEVICE NUMBERS             H(I,N)=BOT(I,N)+0.01
C                                                  190   CONTINUE
      INTEGER OUT                                  C
      IN=5                                         C     TRANSMISSIVITY CONTROL
      OUT=6                                        C
C                                                        DO 193 J=1,NR
C     READ PARAMETER CARD AND                            DO 193 I=1,NC
C     DEFAULT VALUE CARD                                 IF(I.LT.NC)T(I,J,2)=PERM(I,J,2)*SQRT((H(I,J)-
C                                                  1BOT(I,J))*(H(I+1,J)-BOT(I+1,J)))
      READ(IN,10)NSTEPS,DELTA,ERROR,             193   IF(J.LT.NR)T(I,J,1)=PERM(I,J,1)*SQRT((H(I,J)-
     1NC,NR,TT,HH,QQ,S2,PP,BOTT                    1BOT(I,J))*(H(I,J+1)-BOT(I,J+1)))
10    FORMAT(I6,2F6.0/2I6,1F6.0,6X,2F6.0          C
     1,18X,1F6.0,6X,2F6.0)                         C     ROW CALCULATIONS
C                                                  C
C     FILL ARRAYS WITH DEFAULT VALUES                    DO 300 JJ=1,NR
C                                                        J=JJ
      DO 20 I=1,NC                                       IF(MOD(ISTEP+ITER,2).EQ.1) J=NR-J+1
      DO 20 J=1,NR                                       DO 280 I=1,NC
      SF2(I,J)=S2                                        BB=SF2(I,J)/DELTA
      PERM(I,J,1)=PP                                     DD=HO(I,J)*SF2(I,J)/DELTA-Q(I,J)
      PERM(I,J,2)=PP                                     AA=0.0
      BOT(I,J)=BOTT                                      CC=0.0
      T(I,J,1)=TT                                        IF(J-1)200,210,200
      T(I,J,2)=TT                                 200   BB=BB+T(I,J-1,1)
      H(I,J)=HH                                          DD=DD+H(I,J-1)*T(I,J-1,1)
      HO(I,J)=HH                                  210   IF(J-NR)220,230,220
20    Q(I,J)=QQ                                   220   DD=DD+H(I,J+1)*T(I,J,1)
C                                                        BB=BB+T(I,J,1)
C     READ NODE CARDS                             230   IF(I-1)240,250,240
C                                                  240   BB=BB+T(I-1,J,2)
30    READ(5,40,END=50)I,J,T(I,J,1),                    AA=-T(I-1,J,2)
     1T(I,J,2),H(I,J),Q(I,J),                     250   IF(I-NC)260,270,260
     2SF2(I,J),PERM(I,J,1),PERM(I,J,2),BOT(I,J)   260   BB=BB+T(I,J,2)
40    FORMAT(2I3,2F6.0,4X,1F4.0,1F6.0,1RX,1F6.0,6X,3F6.0)  CC=-T(I,J,2)
      GO TO 30                                    270   W=BB-AA*B(I-1)
C                                                        B(I)=CC/W
C     START OF SIMULATION                         280   G(I)=(DD-AA*G(I-1))/W
C                                                  C
50    TIME=0                                       C     RE-ESTIMATE HEADS
      DO 320 ISTEP=1,NSTEPS                        C
      TIME=TIME+DELTA                                    E=E+ABS(H(NC,J)-G(NC))
C                                                        H(NC,J)=G(NC)
C     PREDICT HEADS FOR NEXT                             N=NC-1
C     TIME INCREMENT                             290   HA=G(N)-B(N)*H(N+1,J)
C                                                        E=E+ABS(H(N,J)-HA)
      DO 70 I=1,NC                                       H(N,J)=HA
      DO 70 J=1,NR                                       N=N-1
      D=H(I,J)-HO(I,J)                                   IF(N.GT.0)GO TO 290
      HO(I,J)=H(I,J)                                     DO 300 N=1,NC
      F=1.0                                             IF(H(N,J).GT.BOT(N,J))GO TO 300
      IF(DL(I,J).EQ.0.0)GO TO 60                         E=E+BOT(N,J)+0.01-H(N,J)
      IF(ISTEP.GT.2)F=D/DL(I,J)                          H(N,J)=BOT(N,J)+0.01
      IF(F.GT.5)F=5.0                             300   CONTINUE
      IF(F.LT.0.0)F=0.0                           C
60    DL(I,J)=D                                    C     PRINT RESULTS
      H(I,J)=H(I,J)+D*F                            C
70    IF(H(I,J).LE.BOT(I,J))H(I,J)=BOT(I,J)+0.01         IF(E.GT.ERROR) GO TO 80
C                                                  C
C     REFINE ESTIMATES OF HEADS BY IADI METHOD           WRITE(OUT,310)TIME,E,ITER
C                                                  310   FORMAT(6H2TIME=,F6.2///,E20.7,I5)
      ITER=0                                             DELTA=DELTA*1.2
80    E=0.0                                              DO 320 J=1,NR
      ITER=ITER+1                                 320   WRITE(OUT,330)J,(H(I,J),I=1,NC)
                                                  330   FORMAT(I5,5X,10F10.4/(12X10F10.4))
                                                  C
                                                        STOP
                                                        END
```

FIGURE 12.31. Program listing for Illinois State Water Survey model for water-table aquifers. SOURCE: T. C. Prickett and C. G. Lonnquist, Illinois State Water Survey Bulletin 55, 1971.

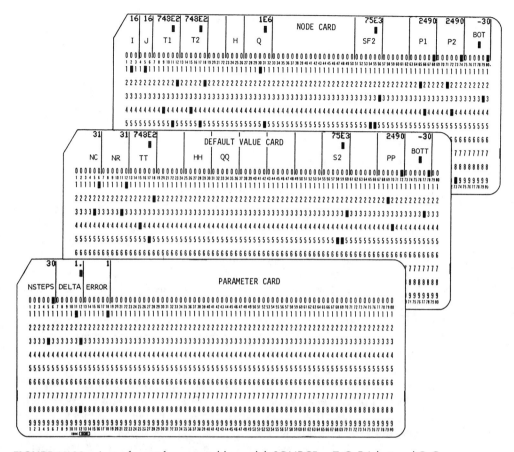

FIGURE 12.32. Input format for water-table model. SOURCE: T. C. Prickett and C. G. Lonnquist, Illinois State Water Survey Bulletin 55, 1971.

HYDROGEOLOGIC SITE EVALUATIONS 12.6

Many types of construction and similar projects require the evaluation of the hydrogeology of a site. This may be a part of a comprehensive study of the engineering geology of a proposed dam or tunnel. The siting of sanitary landfills or areas for toxic or hazardous waste disposal or storage almost always requires a hydrogeologic site evaluation. Environmental impact studies often include sections on geology and hydrogeology. Hydrogeological analysis can help ensure the integrity and safety of structures located in areas of high water table, spring discharge, or quicksand potential. Likewise, groundwater pollution generally can be avoided if sanitary landfill sites are selected and engineered on the basis of a thorough hydrogeological investigation.

While the detailed design of a hydrogeological site study will be dictated by the specific purpose, there are some elements common to all

453

studies. The first is that they are all relatively expensive. The greatest expense is in the collection of the basic data. This includes the cost of geophysical surveys, test borings, test wells, pumping tests, permeability lab tests, and so forth. It is almost always the case that data collection is more costly than the evaluation of the data and the formulation of conclusions and recommendations.

Part of the study results will be presented in map form. A base map prepared by a survey or plane-table mapping may be needed. Information can also be shown on aerial photos or standard quadrangle maps. The topography of the area must be known in order to prepare geologic cross sections and potentiometric surface maps, as well as show surface drainage. The base map should indicate the locations of test borings, outcrops, observation wells, springs, streams, and other such pertinent features.

A map of the surficial geology can be prepared using standard geologic field techniques. Attention should be given to the nature of bedding planes, fractures, and porosity of rock outcrops. A fracture-trace analysis on air photos can enhance the surficial geology map in areas of fractured bedrock. Detailed mapping of surficial geology can be facilitated if there are soil maps available. Interpretation sheets for the soil-map units indicate the parent material on which the soil formed. They also give a general range of infiltration rate for each soil type.

Surface geophysical surveys may be made as a part of the preliminary site evaluation. Seismic refraction, geoelectric soundings, and geoelectric surveys would be useful for those areas in which the unconsolidated deposits are important. The data from the geophysical surveys can guide the selection of the location of test borings and provide data to correlate between test borings. These borings, or wells, are needed to determine the geologic units present. **Borings** refer to uncased holes drilled in unconsolidated overburden, while **test wells** either have a casing or extend into bedrock. A geologist should examine the drill samples or cuttings, prepare a detailed sedimentological and petrographic description, and make a well log showing the depth and thickness of each stratum.

Slightly disturbed samples of unconsolidated deposits can be collected by driving a thin-walled steel tube (either a Shelby tube or split-spoon sampler) into the bottom of the drilled hole. In sandy soils, this must be done through hollow-stem drill pipe or auger-flights. The slightly disturbed sample can be used for laboratory testing, including permeameter tests for hydraulic conductivity and grain-size analysis. Test borings and wells can also be used for slug and bail-down tests for hydraulic conductivity.

Piezometers should be installed to determine the configuration of the potentiometric surface. These are small-diameter, cased wells with a short well screen. They may be steel or plastic pipe. A minimum of three piezometers is necessary to determine the horizontal direction of groundwater flow if the potentiometric surface is a sloping plane. If it has a more complex shape, then more piezometers are necessary. In order to determine vertical head differences, closely spaced piezometers set at different depths are needed.

A set of piezometers at different depths is called a **piezometer nest**. There should be at least one piezometer nest with separate shallow and deep piezometers for a small site.

Following their installation, piezometers should be developed by pumping, or at least by pouring water into them. If the water level is close to the surface, a plunger pushed up and down under the water level to surge the water in the piezometer will generally clear the well point and allow water to freely move into and out of the piezometer. Slug tests may also be made with piezometers. The elevation of the ground surface and the top of the casing of each piezometer should be accurately surveyed. Measurements of the depth to water of each piezometer can then be used to construct a potentiometric map. Successive water-level measurements are necessary, as the elevation and configuration of the potentiometric surface may change with time. Likewise, the amount and direction of vertical flow components may vary.

In addition to the potentiometric maps, the water table may be shown on geologic cross sections. These are prepared on the basis of the test borings and wells. Piezometers are generally installed in a test-boring hole so that both geologic and hydraulic data are collected.

Water quality sampling may be done using piezometers or test wells. There are several pumping methods, including pitcher pumps, bailing, and an air-lift device utilizing canisters of propane cooking gas (25). In areas of groundwater contamination, the location of the sampling piezometers must be such that they intercept the plume of contaminated water (26).

A project report will be prepared for the client or agency to document the field and laboratory study and to present conclusions and recommendations. The contents of a hydrogeological site report should include the following:

1. A title page describing the report, who prepared it, for whom it was prepared, and a date.

2. An introduction stating the purpose of the study, why it was made, and the general conditions under which it was made.

3. The conclusions that can be drawn about the site on the basis of the study.

4. Recommendations to the client or agency regarding the use of the site or the need for additional studies.

5. The body of the report consisting of
   (a) a review of previous work that was done on the site;
   (b) a description of the procedures and methods used in the study;
   (c) the general results of the field study and laboratory analyses;
   (d) an interpretation of the findings; and
   (e) appropriate maps and cross sections.

455

6. An appendix consisting of
   (a) acknowledgements of help given by others during the study;
   (b) a bibliography; and
   (c) technical data, computations, and other supporting evidence.

The report should be written in clear, precise, nontechnical language. If it is intended for persons not familiar with hydrogeology, it may be necessary to define terms and explain concepts in the body of the report.

References

1. LATTMAN, L. H. "Technique of Mapping Geologic Fracture Traces and Lineaments on Aerial Photographs." *Photogrammetric Engineering*, 24 (1958):568–76.

2. PARIZEK, R. R. "On the Nature and Significance of Fracture Traces and Lineaments in Carbonate and Other Terranes." In *Karst Hydrology and Water Resources*, ed. V. V. Yevjevich. Fort Collins, Colo.: Water Resources Publications, 1976, pp. 3-1–3-62.

3. LATTMAN, L. H. and R. H. MATZKE. "Geological Significance of Fracture Traces." *Photogrammetric Engineering*, 27 (1961):435–38.

4. SETZER, JOSÉ. "Hydrologic Significance of Tectonic Fractures Detectable on Air Photos." *Ground Water*, 4, no. 4 (1966):23–29.

5. SIDDIQUI, S. H. and R. R. PARIZEK. "Hydrogeologic Factors Influencing Well Yields in Folded and Faulted Carbonate Rocks in Central Pennsylvania." *Water Resources Research*, 7 (1971):1295–1312.

6. LATTMAN, L. H. and R. R. PARIZEK. "Relationship between Fracture Traces and the Occurrence of Ground Water in Carbonate Rocks." *Journal of Hydrology*, 2 (1964):73–91.

7. WOBBER, F. J. "Fracture Traces in Illinois." *Photogrammetric Engineering*, 33 (1967):499–506.

8. PARIZEK, R. R. "Lineaments and Groundwater." In *Interdisciplinary Applications and Interpretations of EREP Data within the Susquehanna River Basin*. Skylab EREP Investigation 475 Contract Report to NASA, 1976.

9. MC DONALD, H. R. and D. WANTLAND. "Geophysical Procedures in Ground Water Study." *Transactions, American Society of Civil Engineers*, 126 (1961):122–35.

10. HEIGOLD, P. C. et al. "Aquifer Transmissivity from Surficial Electrical Methods." *Ground Water*, 17, no. 4 (1979):338–45.

11. BAYS, C. A. "Prospecting for Ground Water—Geophysical Methods." *Journal of American Water Works Association*, 42 (1950):947–56.

12. ZOHDY, A. A. R., G. P. EATON, and D. R. MABEY. "Application of Surface Geophysics to Ground-Water Investigations." In *Techniques of Water-Resources Investigations,* U.S. Geological Survey, 1974, Book 2, Chap. D1.

13. MOONEY, H. M. and W. W. WETZEL. "The Potentials About a Point Electrode and Apparent Resistivity for a Two-, Three-, and Four-Layer Earth." Minneapolis: University of Minnesota Press, 1956, 145 pp., 243 plates.

14. ZOHDY, A. A. R. "Geoelectric and Seismic Refraction Investigations Near San José, California." *Ground Water,* 3, no. 3 (1965):41–48.

15. DOBRIN, M. B. *Introduction to Geophysical Prospecting,* 3rd ed. New York: McGraw-Hill Book Company, 1976, 630 pp.

16. MOTA, L. "Determination of Dips and Depths of Geological Layers by the Seismic Refraction Method." *Geophysics,* 19 (1954):242–54.

17. KEYS, W. S. and L. M. MAC CARY. "Application of Borehole Geophysics to Water-Resources Investigations." In *Techniques of Water-Resources Investigations,* U.S. Geological Survey, 1971, Book 2, Chap. E1.

18. KEYS, W. S. *Borehole Geophysics as Applied to Groundwater.* Canadian Geological Survey Economic Geology Report 26, 1967, pp. 598–614.

19. BALDWIN, A. D., JR. and J. MILLER. "Use of a Gamma Logger to Delineate Glacial and Bedrock Stratigraphy in Southwestern Ohio." *Ground Water,* 17, no. 4 (1979):385–90.

20. BROWN, D. L. "Techniques for Quality-of-Water Interpretations from Calibrated Geophysical Logs, Atlantic Coastal Areas." *Ground Water,* 9, no. 4 (1971):25–38.

21. CROSBY, J. W., III and J. V. ANDERSON. "Some Applications of Geophysical Well Logging to Basalt Hydrogeology. *Ground Water,* 9, no. 5 (1971):12–20.

22. NORRIS, S. E. "The Use of Gamma Logs in Determining the Character of Unconsolidated Sediments and Well Construction Features." *Ground Water,* 10, no. 6 (1972):14–21.

23. KEYS, W. S. and R. F. BROWN. "The Use of Temperature Logs to Trace the Movement of Injected Water." *Ground Water,* 16 (1978):32–48.

24. MEYER, W. R. *Use of a Neutron Moisture Probe to Determine the Storage Coefficient of an Unconfined Aquifer.* U.S. Geological Survey Professional Paper 450-E, 1963, pp. 174–76.

25. SMITH, A. J. "Water Sampling Made Easier with New Device." *Johnson Drillers Journal* (July-August 1970):1–2.

26. JOSEPHSON, J. "Safeguards for Groundwater." *Environmental Science and Technology,* 14, no. 1 (1980):38–41.

27. PRICKETT, T. A. "Modeling Techniques for Groundwater Evaluation." *Advances in Hydroscience,* 10 (1975):1–143.

28. KIMBLER, O. K. "Fluid Model Studies of the Storage of Freshwater in Saline Aquifers." *Water Resources Research,* 6 (1970):1522–27.

29. PETER, Y. "Model Tests for a Horizontal Well." *Ground Water,* 8, no. 5 (1970): 30–34.

457

30. JAMES. R. V. and J. RUBIN. "Accounting for Apparatus-Induced Dispersion in Analysis of Miscible Displacement Experiments." *Water Resources Research*, 8 (1972):717–21.

31. ANDERSON, T. W. *Electrical-Analog Analysis of the Hydrologic System, Tucson Basin, Southeastern Arizona.* U.S. Geological Survey Water Supply Paper 1939-C, 1972.

32. SPIEKER, A. M. *Effect of Increased Pumping of Ground Water in the Fairfield-New Baltimore Area, Ohio—A Prediction by Analog-Model Study.* U.S. Geological Survey Professional Paper 605-C, 1968.

33. COLLINS, M. A., L. W. GELHAR, and J. L. WILSON, III. "Hele-Shaw Model of Long Island Aquifer System." *Journal of Hydraulics Division, Proceedings of the American Society of Civil Engineers*, 98, HY9 (1972):1701–14.

34. PETERSON, F. L., J. A. WILLIAMS, and S. W. WHEATCRAFT. "Waste Injection into a Two-Phase Flow Field: Sand Box and Hele-Shaw Model Study." *Ground Water*, 16 (1978):410–16.

35. FREEZE, R. A. "Three-Dimensional, Transient, Saturated-Unsaturated Flow in a Ground-Water Flow System." *Water Resources Research*, 7 (1971):347–66.

36. FRIND, E. O. and G. F. PINDER. "Galerkin Solution to the Inverse Problem for Aquifer Transmissivity." *Water Resources Research*, 9 (1973):1397–1410.

37. BREDEHOEFT, J. D. and G. F. PINDER. "Mass Transport in Flowing Groundwater." *Water Resources Research*, 9 (1973):194–210.

38. PINDER, G. F. "A Galerkin Finite Element Simulation of Groundwater Contamination on Long Island." *Water Resources Research*, 9 (1973):108–20.

39. PINDER, G. F. and H. H. COOPER, JR. "A Numerical Technique for Calculating the Transient Position of the Saltwater Front." *Water Resources Research*, 6 (1970):875–82.

40. PRICKETT, T. A. and C. G. LONNQUIST. *Selected Digital Computer Techniques for Groundwater Resource Evaluation.* Illinois State Water Survey Bulletin 55, 1971, 62 pp.

41. PRICKETT, T. A. and C. G. LONNQUIST. *Aquifer Simulation Model for Use on Disk Supported Small Computer System.* Illinois State Water Survey Circular 114, 1973.

42. TRESCOTT, P. E., G. F. PINDER, and S. P. LARSON. "Finite-Difference Model for Aquifer Simulation in Two Dimensions with Results of Numerical Experiments." In *Techniques of Water-Resources Investigations*, U.S. Geological Survey, 1976, Book 7, Chap. C1.

43. REMSON, I., G. M. HORNBERGER, and F. J. MOLZ. *Numerical Methods in Subsurface Hydrology.* New York: Wiley-Interscience, 1971, 389 pp.

Appendices

1   Values of the function $W(u)$ for various values of u

2   Values of the function $F(\eta,\mu)$ for various values of η and μ

3   Values of the functions $W(u,r/B)$ and $W(u_A,r/B)$ for various values of u or u_A

4   Values of the function $H(\mu,\beta)$

5   Values of the functions $K_0(x)$ and $\exp(x)K_0(x)$

6   Values of the functions $W(u_A,\Gamma)$, and $W(u_B,\Gamma)$ for water-table aquifers

7   Table for length conversion

8   Table for area conversion

9   Table for volume conversion

10   Table for time conversion

APPENDIX 1.   Values of the function $W(u)$ for various values of u

u	$W(u)$	u	$W(u)$	u	$W(u)$	u	$W(u)$
1×10^{-10}	22.45	7×10^{-8}	15.90	4×10^{-5}	9.55	1×10^{-2}	4.04
2	21.76	8	15.76	5	9.33	2	3.35
3	21.35	9	15.65	6	9.14	3	2.96
4	21.06	1×10^{-7}	15.54	7	8.99	4	2.68
5	20.84	2	14.85	8	8.86	5	2.47
6	20.66	3	14.44	9	8.74	6	2.30
7	20.50	4	14.15	1×10^{-4}	8.63	7	2.15
8	20.37	5	13.93	2	7.94	8	2.03
9	20.25	6	13.75	3	7.53	9	1.92
1×10^{-9}	20.15	7	13.60	4	7.25	1×10^{-1}	1.823
2	19.45	8	13.46	5	7.02	2	1.223
3	19.05	9	13.34	6	6.84	3	0.906
4	18.76	1×10^{-6}	13.24	7	6.69	4	0.702
5	18.54	2	12.55	8	6.55	5	0.560
6	18.35	3	12.14	9	6.44	6	0.454
7	18.20	4	11.85	1×10^{-3}	6.33	7	0.374
8	18.07	5	11.63	2	5.64	8	0.311
9	17.95	6	11.45	3	5.23	9	0.260
1×10^{-8}	17.84	7	11.29	4	4.95	1×10^{0}	0.219
2	17.15	8	11.16	5	4.73	2	0.049
3	16.74	9	11.04	6	4.54	3	0.013
4	16.46	1×10^{-5}	10.94	7	4.39	4	0.004
5	16.23	2	10.24	8	4.26	5	0.001
6	16.05	3	9.84	9	4.14		

SOURCE: Adapted from L. K. Wenzel, *Methods for Determining Permeability of Water-Bearing Materials with Special Reference to Discharging Well Methods.* U.S. Geological Survey Water Supply Paper 887, 1942.

APPENDIX 2. Values of the function $F(\eta,\mu)$ for various values of η and μ

$\eta = Tt/r_c^2$	$\mu = 10^{-6}$	$\mu = 10^{-7}$	$\mu = 10^{-8}$	$\mu = 10^{-9}$	$\mu = 10^{-10}$
0.001	0.9994	0.9996	0.9996	0.9997	0.9997
0.002	0.9989	0.9992	0.9993	0.9994	0.9995
0.004	0.9980	0.9985	0.9987	0.9989	0.9991
0.006	0.9972	0.9978	0.9982	0.9984	0.9986
0.008	0.9964	0.9971	0.9976	0.9980	0.9982
0.01	0.9956	0.9965	0.9971	0.9975	0.9978
0.02	0.9919	0.9934	0.9944	0.9952	0.9958
0.04	0.9848	0.9875	0.9894	0.9908	0.9919
0.06	0.9782	0.9819	0.9846	0.9866	0.9881
0.08	0.9718	0.9765	0.9799	0.9824	0.9844
0.1	0.9655	0.9712	0.9753	0.9784	0.9807
0.2	0.9361	0.9459	0.9532	0.9587	0.9631
0.4	0.8828	0.8995	0.9122	0.9220	0.9298
0.6	0.8345	0.8569	0.8741	0.8875	0.8984
0.8	0.7901	0.8173	0.8383	0.8550	0.8686
1.0	0.7489	0.7801	0.8045	0.8240	0.8401
2.0	0.5800	0.6235	0.6591	0.6889	0.7139
3.0	0.4554	0.5033	0.5442	0.5792	0.6096
4.0	0.3613	0.4093	0.4517	0.4891	0.5222
5.0	0.2893	0.3351	0.3768	0.4146	0.4487
6.0	0.2337	0.2759	0.3157	0.3525	0.3865
7.0	0.1903	0.2285	0.2655	0.3007	0.3337
8.0	0.1562	0.1903	0.2243	0.2573	0.2888
9.0	0.1292	0.1594	0.1902	0.2208	0.2505
10.0	0.1078	0.1343	0.1620	0.1900	0.2178
20.0	0.02720	0.03343	0.04129	0.05071	0.06149
30.0	0.01286	0.01448	0.01667	0.01956	0.02320
40.0	0.008337	0.008898	0.009637	0.01062	0.01190
50.0	0.006209	0.006470	0.006789	0.007192	0.007709
60.0	0.004961	0.005111	0.005283	0.005487	0.005735
80.0	0.003547	0.003617	0.003691	0.003773	0.003863
100.0	0.002763	0.002803	0.002845	0.002890	0.002938
200.0	0.001313	0.001322	0.001330	0.001339	0.001348

SOURCE: After I. S. Papadopulos, J. D. Bredehoeft, and H. H. Cooper, Jr., "On the Analysis of 'Slug Test' Data," *Water Resources Research,* 9, (1973):1087–89.

NOTE: For slug tests on wells of finite diameter, $H/H_0 = F(\eta,\mu)$.

APPENDIX 3. Values of the functions $W(u, r/B)$ for various values of u

u \ r/B	0.01	0.015	0.03	0.05	0.075	0.10	0.15	0.2	0.3	0.4	0.5	0.6	0.7	0.8	0.9	1.0	1.5	2.0	2.5
0.000001	9.4425	8.6319	7.2471	6.2285	5.4228	4.8541	4.0601	3.5054	2.7449	2.2291	1.8488	1.5550	1.3210	1.1307	0.9735	0.8420	0.4276	0.2278	0.1247
0.000005	9.4413	"	"	"	"	"	"	"	"	"	"	"	"	"	"	"	"	"	"
0.00001	9.4176	8.6313	"	6.2282	"	"	"	"	"	"	"	"	"	"	"	"	"	"	"
0.00005	8.8827	8.4533	7.2450	6.2282	"	"	"	"	"	"	"	"	"	"	"	"	"	"	"
0.0001	8.3983	8.1414	7.2122	"	"	4.8530	"	"	"	"	"	"	"	"	"	"	"	"	"
0.0005	6.9750	6.9152	6.6219	6.0821	5.4062	"	"	"	"	"	"	"	"	"	"	"	"	"	"
0.001	6.3069	6.2765	6.1202	5.7965	5.3078	4.8292	4.0595	"	"	"	"	"	"	"	"	"	"	"	"
0.005	4.7212	4.7152	4.6829	4.6084	4.4713	4.2960	3.8821	3.4567	2.7428	2.2290	"	"	"	"	"	"	"	"	"
0.01	4.0356	4.0326	4.0167	3.9795	3.9091	3.8150	3.5725	3.3110	2.7104	2.2253	1.8486	"	"	"	"	"	"	"	"
0.05	2.4675	2.4670	2.4642	2.4576	2.4448	2.4271	2.3776	2.3110	1.9283	1.7075	1.4927	1.3115	1.2955	1.1210	0.9700	0.8409	"	"	"
0.1	1.8227	1.8225	1.8213	1.8184	1.8128	1.8050	1.7829	1.7527	1.6704	1.5644	1.4422	1.2955	1.1791	1.0505	0.9297	0.8190	0.4271	"	"
0.5	0.5598	0.5597	0.5596	0.5594	0.5588	0.5581	0.5561	0.5532	0.5453	0.5344	0.5206	0.5044	0.4860	0.4658	0.4440	0.4210	0.3007	0.1944	0.1174
1.0	0.2194	0.2194	0.2193	0.2193	0.2191	0.2190	0.2186	0.2179	0.2161	0.2135	0.2103	0.2065	0.2020	0.1970	0.1914	0.1855	0.1509	0.1139	0.0803
5.0	0.0011	0.0011	0.0011	0.0011	0.0011	0.0011	0.0011	0.0011	0.0011	0.0011	0.0011	0.0011	0.0011	0.0011	0.0011	0.0011	0.0010	0.0010	0.0009

SOURCE: After M. S. Hantush, "Analysis of Data from Pumping Tests in Leaky Aquifers," *Transactions, American Geophysical Union*, 37 (1956):702–14.

APPENDIX 4. Values of the function $H(\mu, \beta)$

μ \ β	0.001	0.005	0.01	0.05	0.10	0.20	0.50	1.0	2.0	5.0	10.0	20.0
0.000001	11.9842	10.5908	9.9259	8.3395	7.6497	6.9590	6.0463	5.3575	4.6721	3.7756	3.1110	2.4671
0.000005	10.8958	9.7174	9.0866	7.5284	6.8427	6.1548	5.2459	4.5617	3.8836	3.0055	2.3661	1.7633
0.00001	10.3739	9.3203	8.7142	7.1771	6.4944	5.8085	4.9024	4.2212	3.5481	2.6822	2.0590	1.4816
0.00005	9.0422	8.3171	7.8031	6.3523	5.6821	5.0045	4.1090	3.4394	2.7848	1.9622	1.3943	0.8994
0.0001	8.4258	7.8386	7.3803	5.9906	5.3297	4.6581	3.7700	3.1082	2.4658	1.6704	1.1359	0.6878
0.0005	6.9273	6.6024	6.2934	5.1223	4.4996	3.8527	2.9933	2.3601	1.7604	1.0564	0.6252	0.3089
0.001	6.2624	6.0193	5.7727	4.7290	4.1337	3.5045	2.6650	2.0506	1.4776	0.8271	0.4513	0.1976
0.005	4.6951	4.5786	4.4474	3.7415	3.2483	2.6891	1.9250	1.3767	0.8915	0.4001	0.1677	0.0493
0.01	4.0163	3.9334	3.8374	3.2752	2.8443	2.3325	1.6193	1.1122	0.6775	0.2670	0.0955	0.0221
0.05	2.4590	2.4243	2.3826	2.1007	1.8401	1.4872	0.9540	0.5812	0.2923	0.0755	0.0160	0.00164
0.1	1.8172	1.7949	1.7677	1.5768	1.3893	1.1207	0.6947	0.3970	0.1789	0.0359	0.00552	0.00034
0.5	0.5584	0.5530	0.5463	0.4969	0.4436	0.3591	0.2083	0.1006	0.0325	0.00288	0.00015	0.00002
1.0	0.2189	0.2169	0.2144	0.1961	0.1758	0.1427	0.0812	0.0365	0.00993	0.00055	0.00002	
5.0	0.00115	0.00114	0.00112	0.00104	0.00093	0.00076	0.00042	0.00017	0.00003			

SOURCE: Condensed from M. S. Hantush, "Modification of the Theory of Leaky Aquifers," *Journal of Geophysical Research*, 65 (1960):3713–25.

APPENDIX 5. Values of the functions $K_0(x)$ and $\exp(x)K_0(x)$

x	$K_0(x)$	$\exp(x)K_0(x)$	x	$K_0(x)$	$\exp(x)K_0(x)$
0.01	4.72	4.77	0.35	1.23	1.75
0.015	4.32	4.38	0.40	1.11	1.66
0.02	4.03	4.11	0.45	1.01	1.59
0.025	3.81	3.91	0.50	0.92	1.52
0.03	3.62	3.73	0.55	0.85	1.47
0.035	3.47	3.59	0.60	0.78	1.42
0.04	3.34	3.47	0.65	0.72	1.37
0.045	3.22	3.37	0.70	0.66	1.33
0.05	3.11	3.27	0.75	0.61	1.29
0.055	3.02	3.19	0.80	0.57	1.26
0.06	2.93	3.11	0.85	0.52	1.23
0.065	2.85	3.05	0.90	0.49	1.20
0.07	2.78	2.98	0.95	0.45	1.17
0.075	2.71	2.92	1.0	0.42	1.14
0.08	2.65	2.87	1.5	0.21	0.96
0.085	2.59	2.82	2.0	0.11	0.84
0.09	2.53	2.77	2.5	0.062	0.760
0.095	2.48	2.72	3.0	0.035	0.698
0.10	2.43	2.68	3.5	0.020	0.649
0.15	2.03	2.36	4.0	0.011	0.609
0.20	1.75	2.14	4.5	0.006	0.576
0.25	1.54	1.98	5.0	0.004	0.548
0.30	1.37	1.85			

SOURCE: Adapted from M. S. Hantush, "Analysis of Data From Pumping Tests in Leaky Aquifers," *Transactions, American Geophysical Union*, 37 (1956):702–14.

APPENDIX 6A. Values of the function $W(u_A, \Gamma)$ for water-table aquifers

$1/u_A$	$\Gamma = 0.001$	$\Gamma = 0.01$	$\Gamma = 0.06$	$\Gamma = 0.2$	$\Gamma = 0.6$	$\Gamma = 1.0$	$\Gamma = 2.0$	$\Gamma = 4.0$	$\Gamma = 6.0$
4.0×10^{-1}	2.48×10^{-2}	2.41×10^{-2}	2.30×10^{-2}	2.14×10^{-2}	1.88×10^{-2}	1.70×10^{-2}	1.38×10^{-2}	9.33×10^{-3}	6.39×10^{-3}
8.0×10^{-1}	1.45×10^{-1}	1.40×10^{-1}	1.31×10^{-1}	1.19×10^{-1}	9.88×10^{-2}	8.49×10^{-2}	6.03×10^{-2}	3.17×10^{-2}	1.74×10^{-2}
1.4×10^{0}	3.58×10^{-1}	3.45×10^{-1}	3.18×10^{-1}	2.79×10^{-1}	2.17×10^{-1}	1.75×10^{-1}	1.07×10^{-1}	4.45×10^{-2}	2.10×10^{-2}
2.4×10^{0}	6.62×10^{-1}	6.33×10^{-1}	5.70×10^{-1}	4.83×10^{-1}	3.43×10^{-1}	2.56×10^{-1}	1.33×10^{-1}	4.76×10^{-2}	2.14×10^{-2}
4.0×10^{0}	1.02×10^{0}	9.63×10^{-1}	8.49×10^{-1}	6.88×10^{-1}	4.38×10^{-1}	3.00×10^{-1}	1.40×10^{-1}	4.78×10^{-2}	2.15×10^{-2}
8.0×10^{0}	1.57×10^{0}	1.46×10^{0}	1.23×10^{0}	9.18×10^{-1}	4.97×10^{-1}	3.17×10^{-1}	1.41×10^{-1}		
1.4×10^{1}	2.05×10^{0}	1.88×10^{0}	1.51×10^{0}	1.03×10^{0}	5.07×10^{-1}				
2.4×10^{1}	2.52×10^{0}	2.27×10^{0}	1.73×10^{0}	1.07×10^{0}					
4.0×10^{1}	2.97×10^{0}	2.61×10^{0}	1.85×10^{0}	1.08×10^{0}					
8.0×10^{1}	3.56×10^{0}	3.00×10^{0}	1.92×10^{0}						
1.4×10^{2}	4.01×10^{0}	3.23×10^{0}	1.93×10^{0}						
2.4×10^{2}	4.42×10^{0}	3.37×10^{0}	1.94×10^{0}						
4.0×10^{2}	4.77×10^{0}	3.43×10^{0}							
8.0×10^{2}	5.16×10^{0}	3.45×10^{0}							
1.4×10^{3}	5.40×10^{0}	3.46×10^{0}							
2.4×10^{3}	5.54×10^{0}								
4.0×10^{3}	5.59×10^{0}								
8.0×10^{3}	5.62×10^{0}								
1.4×10^{4}	5.62×10^{0}	3.46×10^{0}	1.94×10^{0}	1.08×10^{0}	5.07×10^{-1}	3.17×10^{-1}	1.41×10^{-1}	4.78×10^{-2}	2.15×10^{-2}

APPENDIX 6B. Values of the function $W(u_B, \Gamma)$ for water-table aquifers

$1/u_B$	$\Gamma = 0.001$	$\Gamma = 0.01$	$\Gamma = 0.06$	$\Gamma = 0.2$	$\Gamma = 0.6$	$\Gamma = 1.0$	$\Gamma = 2.0$	$\Gamma = 4.0$	$\Gamma = 6.0$
4.0×10^{-4}	5.62×10^{0}	3.46×10^{0}	1.94×10^{0}	1.09×10^{0}	5.08×10^{-1}	3.18×10^{-1}	1.42×10^{-1}	4.79×10^{-2}	2.15×10^{-2}
8.0×10^{-4}								4.80×10^{-2}	2.16×10^{-2}
1.4×10^{-3}								4.81×10^{-2}	2.17×10^{-2}
2.4×10^{-3}								4.84×10^{-2}	2.19×10^{-2}
4.0×10^{-3}					5.08×10^{-1}	3.18×10^{-1}	1.42×10^{-1}	4.78×10^{-2}	2.21×10^{-2}
8.0×10^{-3}					5.09×10^{-1}	3.19×10^{-1}	1.43×10^{-1}	4.96×10^{-2}	2.28×10^{-2}
1.4×10^{-2}					5.10×10^{-1}	3.21×10^{-1}	1.45×10^{-1}	5.09×10^{-2}	2.39×10^{-2}
2.4×10^{-2}					5.12×10^{-1}	3.23×10^{-1}	1.47×10^{-1}	5.32×10^{-2}	2.57×10^{-2}
4.0×10^{-2}					5.16×10^{-1}	3.27×10^{-1}	1.52×10^{-1}	5.68×10^{-2}	2.86×10^{-2}
8.0×10^{-2}				1.09×10^{0}	5.24×10^{-1}	3.37×10^{-1}	1.62×10^{-1}	6.61×10^{-2}	3.62×10^{-2}
1.4×10^{-1}			1.94×10^{0}	1.10×10^{0}	5.37×10^{-1}	3.50×10^{-1}	1.78×10^{-1}	8.06×10^{-2}	4.86×10^{-2}
2.4×10^{-1}			1.95×10^{0}	1.11×10^{0}	5.57×10^{-1}	3.74×10^{-1}	2.05×10^{-1}	1.06×10^{-1}	7.14×10^{-2}
4.0×10^{-1}			1.96×10^{0}	1.13×10^{0}	5.89×10^{-1}	4.12×10^{-1}	2.48×10^{-1}	1.49×10^{-1}	1.13×10^{-1}
8.0×10^{-1}		3.46×10^{0}	1.98×10^{0}	1.18×10^{0}	6.67×10^{-1}	5.06×10^{-1}	3.57×10^{-1}	2.66×10^{-1}	2.31×10^{-1}
1.4×10^{0}	5.62×10^{0}	3.47×10^{0}	2.01×10^{0}	1.24×10^{0}	7.80×10^{-1}	6.42×10^{-1}	5.17×10^{-1}	4.45×10^{-1}	4.19×10^{-1}
2.4×10^{0}	5.63×10^{0}	3.49×10^{0}	2.06×10^{0}	1.35×10^{0}	9.54×10^{-1}	8.50×10^{-1}	7.63×10^{-1}	7.18×10^{-1}	7.03×10^{-1}
4.0×10^{0}	5.63×10^{0}	3.51×10^{0}	2.13×10^{0}	1.50×10^{0}	1.20×10^{0}	1.13×10^{0}	1.08×10^{0}	1.06×10^{0}	1.05×10^{0}
8.0×10^{0}	5.64×10^{0}	3.56×10^{0}	2.31×10^{0}	1.85×10^{0}	1.68×10^{0}	1.65×10^{0}	1.63×10^{0}	1.63×10^{0}	1.63×10^{0}
1.4×10^{1}	5.65×10^{0}	3.63×10^{0}	2.55×10^{0}	2.23×10^{0}	2.15×10^{0}	2.14×10^{0}	2.14×10^{0}	2.14×10^{0}	2.14×10^{0}
2.4×10^{1}	5.67×10^{0}	3.74×10^{0}	2.86×10^{0}	2.68×10^{0}	2.65×10^{0}	2.65×10^{0}	2.64×10^{0}	2.64×10^{0}	2.64×10^{0}
4.0×10^{1}	5.70×10^{0}	3.90×10^{0}	3.24×10^{0}	3.15×10^{0}	3.14×10^{0}	3.14×10^{0}	3.14×10^{0}	3.14×10^{0}	3.14×10^{0}
8.0×10^{1}	5.76×10^{0}	4.22×10^{0}	3.85×10^{0}	3.82×10^{0}	3.82×10^{0}	3.82×10^{0}	3.82×10^{0}	3.82×10^{0}	3.82×10^{0}
1.4×10^{2}	5.85×10^{0}	4.58×10^{0}	4.38×10^{0}	4.37×10^{0}	4.37×10^{0}	4.37×10^{0}	4.37×10^{0}	4.37×10^{0}	4.37×10^{0}
2.4×10^{2}	5.99×10^{0}	5.00×10^{0}	4.91×10^{0}	4.91×10^{0}	4.91×10^{0}	4.91×10^{0}	4.91×10^{0}	4.91×10^{0}	4.91×10^{0}
4.0×10^{2}	6.16×10^{0}	5.46×10^{0}	5.42×10^{0}	5.42×10^{0}	5.42×10^{0}	5.42×10^{0}	5.42×10^{0}	5.42×10^{0}	5.42×10^{0}
8.0×10^{2}	6.47×10^{0}	6.11×10^{0}	6.11×10^{0}	6.11×10^{0}	6.11×10^{0}	6.11×10^{0}	6.11×10^{0}	6.11×10^{0}	6.11×10^{0}
1.4×10^{3}	6.67×10^{0}	6.67×10^{0}	6.67×10^{0}	6.67×10^{0}	6.67×10^{0}	6.67×10^{0}	6.67×10^{0}	6.67×10^{0}	6.67×10^{0}
2.4×10^{3}	7.21×10^{0}	7.21×10^{0}	7.21×10^{0}	7.21×10^{0}	7.21×10^{0}	7.21×10^{0}	7.21×10^{0}	7.21×10^{0}	7.21×10^{0}
4.0×10^{3}	7.72×10^{0}	7.72×10^{0}	7.72×10^{0}	7.72×10^{0}	7.72×10^{0}	7.72×10^{0}	7.72×10^{0}	7.72×10^{0}	7.72×10^{0}
8.0×10^{3}	8.41×10^{0}	8.41×10^{0}	8.41×10^{0}	8.41×10^{0}	8.41×10^{0}	8.41×10^{0}	8.41×10^{0}	8.41×10^{0}	8.41×10^{0}
1.4×10^{4}	8.97×10^{0}	8.97×10^{0}	8.97×10^{0}	8.97×10^{0}	8.97×10^{0}	8.97×10^{0}	8.97×10^{0}	8.97×10^{0}	8.97×10^{0}
2.4×10^{4}	9.51×10^{0}	9.51×10^{0}	9.51×10^{0}	9.51×10^{0}	9.51×10^{0}	9.51×10^{0}	9.51×10^{0}	9.51×10^{0}	9.51×10^{0}
4.0×10^{4}	1.94×10^{1}	1.94×10^{1}	1.94×10^{1}	1.94×10^{1}	1.94×10^{1}	1.94×10^{1}	1.94×10^{1}	1.94×10^{1}	1.94×10^{1}

SOURCE: Adapted from S. P. Neuman, Water Resources Research, 11(1975):329–42.

465

APPENDIX 7. Table for length conversion

Unit	mm	cm	m	km	in	ft	yd	mi
1 millimeter	1	0.1	0.001	10^{-6}	0.0397	0.00328	0.00109	6.21×10^{-7}
1 centimeter	10	1	0.01	0.0001	0.3937	0.0328	0.0109	6.21×10^{-6}
1 meter	1000	100	1	0.001	39.37	3.281	1.094	6.21×10^{-4}
1 kilometer	10^6	10^5	1000	1	39,370	3281	1093.6	0.621
1 inch	25.4	2.54	0.0254	2.54×10^{-5}	1	0.0833	0.0278	1.58×10^{-5}
1 foot	304.8	30.48	0.3048	3.05×10^{-4}	12	1	0.333	1.89×10^{-4}
1 yard	914.4	91.44	0.9144	9.14×10^{-4}	36	3	1	5.68×10^{-4}
1 mile	1.61×10^6	1.01×10^5	1.61×10^3	1.6093	63,360	5280	1760	1

APPENDIX 8. Table for area conversion

Unit	cm^2	m^2	km^2	ha	in^2	ft^2	yd^2	mi^2	ac
1 sq. centimeter	1	0.0001	10^{-10}	10^{-8}	0.155	1.08×10^{-3}	1.2×10^{-4}	3.86×10^{-11}	2.47×10^{-8}
1 sq. meter	10^4	1	10^{-6}	10^{-4}	1550	10.76	1.196	3.86×10^{-7}	2.47×10^{-4}
1 sq. kilometer	10^{10}	10^6	1	100	1.55×10^9	1.076×10^7	1.196×10^6	0.3861	247.1
1 hectare	10^8	10^4	0.01	1	1.55×10^7	1.076×10^5	1.196×10^4	3.861×10^{-3}	2.471
1 sq. inch	6.452	6.45×10^{-4}	6.45×10^{-10}	6.45×10^{-8}	1	6.94×10^{-3}	7.7×10^{-4}	2.49×10^{-10}	1.574×10^{-7}
1 sq. foot	929	0.0929	9.29×10^{-8}	9.29×10^{-6}	144	1	0.111	3.587×10^{-8}	2.3×10^{-5}
1 sq. yard	8361	0.8361	8.36×10^{-7}	8.36×10^{-5}	1296	9	1	3.23×10^{-7}	2.07×10^{-4}
1 sq. mile	2.59×10^{10}	2.59×10^6	2.59	259	4.01×10^9	2.79×10^7	3.098×10^6	1	640
1 acre	4.04×10^7	4047	4.047×10^{-3}	0.4047	6.27×10^6	43,560	4840	1.562×10^{-3}	1

APPENDIX 9. Table for volume conversion

Unit	$m\ell$	liters	m^3	in^3	ft^3	gal	ac-ft	million gal
1 milliliter	1	0.001	10^{-6}	0.06102	3.53×10^{-5}	2.64×10^{-4}	8.1×10^{-10}	2.64×10^{-10}
1 liter	10^3	1	0.001	61.02	0.0353	0.264	8.1×10^{-7}	2.64×10^{-7}
1 cu. meter	10^6	1000	1	61,023	35.31	264.17	8.1×10^{-4}	2.64×10^{-4}
1 cu. inch	16.39	1.64×10^{-2}	1.64×10^{-5}	1	5.79×10^{-4}	4.33×10^{-3}	1.218×10^{-8}	4.329×10^{-9}
1 cu. foot	28,317	28.317	0.02832	1728	1	7.48	2.296×10^{-5}	7.48×10^{-6}
1 U.S. gallon	3785.4	3.785	3.78×10^{-3}	231	0.134	1	3.069×10^{-6}	10^{-6}
1 acre-foot	1.233×10^9	1.233×10^6	1233.5	75.27×10^6	43,560	3.26×10^5	1	0.3260
1 million gallons	3.785×10^9	3.785×10^6	3785	2.31×10^8	1.338×10^5	10^6	3.0684	1

APPENDIX 10. Table for time conversion

Unit	sec	min	hours	days	years
1 second	1	1.67×10^{-2}	2.77×10^{-4}	1.157×10^{-5}	3.17×10^{-8}
1 minute	60	1	1.67×10^{-2}	6.94×10^{-4}	1.90×10^{-6}
1 hour	3600	60	1	4.17×10^{-2}	1.14×10^{-4}
1 day	8.64×10^4	1440	24	1	2.74×10^{-3}
1 year	3.15×10^7	5.256×10^5	8760	365	1

Glossary

Adsorption	The attraction and adhesion of a layer of ions from an aqueous solution to the solid mineral surfaces with which it is in contact.
Advection	The process by which solutes are transported by the motion of flowing groundwater.
Aliquot	One of a number of equal-sized portions of a water sample that is being analyzed.
American Rule	A groundwater doctrine that holds that an overlying property owner has the right to use only a reasonable amount of groundwater.
Anisotropy	The condition under which one or more of the hydraulic properties of an aquifer vary according to the direction of flow.
Antecedent moisture	The soil moisture present before a particular precipitation event.
Aquifer	Rock or sediment in a formation, group of formations, or part of a formation which is saturated and sufficiently permeable to transmit economic quantities of water to wells and springs.
Aquifer, confined	An aquifer that is overlain by a confining bed. The confining bed has a significantly lower hydraulic conductivity than the aquifer.
Aquifer, unconfined	An aquifer in which there are no confining beds between the zone of saturation and the surface. There will be a water table in an unconfined aquifer. Water-table aquifer is a synonym.
Artificial recharge	The process by which water can be injected or added to an aquifer. Dug basins, drilled wells, or simply the spread of water across the land surface are all means of artificial recharge.

Barrier boundary An aquifer-system boundary represented by a rock mass that is not a source of water.

Baseflow That part of stream discharge derived from groundwater seeping into the stream.

Baseflow recession The declining rate of discharge of a stream fed only by baseflow for an extended period. Typically, a baseflow recession will be exponential.

Baseflow-recession hydrograph A hydrograph that shows a baseflow-recession curve.

Bioassay A method used to determine the toxicity of specific contaminants. A number of individuals of a sensitive species are placed in water containing varying concentrations of the contaminant for a specified period of time.

Borehole geophysics The general field of geophysics developed around the lowering of various probes into a well.

Caliper log A borehole log of the diameter of an uncased well.

Capillary forces The forces acting on soil moisture in the unsaturated zone, attributable to molecular attraction between soil particles and water.

Capillary fringe The zone immediately above the water table, where water is drawn upward by capillary attraction.

Chemical activity The molal concentration of an ion multiplied by a factor known as the activity coefficient.

Chlorosis An abnormal condition of plants in which the green parts either lose their color or turn yellow.

Cleat The vertical planes of fracture that are found in coal.

Common ion effect The decrease in the solubility of a salt dissolved in water already containing some of the ions of the salt.

Confining bed A body of material of low hydraulic conductivity that is stratigraphically adjacent to one or more aquifers. It may lie above or below the aquifer.

Contaminant Any solute that enters the hydrologic cycle through human action.

Current meter A device that is lowered into a stream in order to record the rate at which the current is moving.

Density The mass or quantity of a substance per unit volume. Units are kilograms per cubic meter or grams per cubic centimeter.

Depression storage Water from precipitation which collects in puddles at the land surface.

Diagenesis The chemical and physical changes occurring in sediments before consolidation or while in the environment of deposition.

Digital computer model A model of groundwater flow in which the aquifer is described by numerical equations with specified values for boundary conditions which are solved on a digital computer.

Direct precipitation Water that falls directly into a lake or stream without passing through any land phase of the runoff cycle.

Discharge area An area in which there are upward components of hydraulic head in the aquifer. Groundwater is flowing toward the surface in a discharge area and may escape as a spring, seep, or baseflow, or by evaporation and transpiration.

Discharge velocity An apparent velocity, calculated from Darcy's law, which represents the flow rate at which water would move through an aquifer if the aquifer were an open conduit. Also called specific discharge.

Drainage basin The land area from which surface runoff drains into a stream system.

Drainage divide A boundary line along a topographically high area which separates two adjacent drainage basins.

Drawdown A lowering of the water table of an unconfined aquifer or the potentiometric surface of a confined aquifer caused by pumping of groundwater from wells.

Duration curve A graph showing the percentage of time that the given flows of a stream will be equaled or exceeded. It is based upon a statistical study of historic streamflow records.

Dynamic equilibrium A condition in which the amount of recharge to an aquifer equals the amount of natural discharge.

Effective porosity The amount of interconnected pore space through which fluids can pass, expressed as a percent of bulk volume. Part of the total porosity will be occupied by static fluid being held to the mineral surface by surface tension, so effective porosity will be less than total porosity.

Electrical resistance model An analog model of groundwater flow based upon the flow of electricity through a circuit containing resistors and capacitors.

English Rule A groundwater doctrine that holds that property owners have the right of absolute ownership of the groundwater beneath their land.

Equilibrium constant The number defining the conditions of equilibrium for a particular reversible chemical reaction.

Equipotential line A line in a two-dimensional groundwater flow field such that the total hydraulic head is the same for all points along the line.

Equipotential surface A surface in a three-dimensional groundwater flow field such that the total hydraulic head is the same everywhere on the surface.

Equivalent weight The concentration in parts per million of a solute multiplied by the valence charge and then divided by its formula weight in grams.

Evaporation The process by which water passes from the liquid to the vapor state.

Evapotranspiration The sum of evaporation plus transpiration.

Evapotranspiration, actual The evapotranspiration that actually occurs under given climatic and soil-moisture conditions.

Evapotranspiration, potential The evapotranspiration that would occur under given climatic conditions if there were unlimited soil moisture.

Field capacity The maximum amount of water that the unsaturated zone of a soil can hold against the pull of gravity. The field capacity is dependent on the length of time the soil has been undergoing gravity drainage.

Flow net The set of intersecting equipotential lines and flowlines representing two-dimensional steady flow through porous media.

Flow, steady The flow that occurs when, at any point in the flow field, the magnitude and direction of the specific discharge are constant in time.

Flow, unsteady The flow that occurs when, at any point in the flow field, the magnitude or direction of the specific discharge changes with time. Also called transient flow or nonsteady flow.

Fluid potential The mechanical energy per unit mass of fluid at any given point in space and time.

Fracture trace The surface representation of a fracture zone. It may be a characteristic line of vegetation or linear soil-moisture pattern or a topographic sag.

Free energy A measure of the thermodynamic driving energy of a chemical reaction. Also known as Gibbs free energy or Gibbs function.

Gamma-gamma radiation log A borehole log in which a source of gamma radiation as well as a detector are lowered into the borehole. This log measures bulk density of the formation and fluids.

Groundwater The water contained in interconnected pores located below the water table in an unconfined aquifer or located in a confined aquifer.

Groundwater basin A rather vague designation pertaining to a groundwater reservoir which is more or less separate from neighboring groundwater reservoirs. A groundwater basin could be separated from adjacent basins by geologic boundaries or by hydrologic boundaries.

Groundwater, confined	The water contained in a confined aquifer. Pore-water pressure is greater than atmospheric at the top of the confined aquifer.
Groundwater flow	The movement of water through openings in sediment and rock which occurs in the zone of saturation.
Groundwater mining	The practice of withdrawing groundwater at rates in excess of the natural recharge.
Groundwater, perched	The water in an isolated, saturated zone located in the zone of aeration. It is the result of the presence of a layer of material of low hydraulic conductivity, called a perching bed. Perched groundwater will have a perched water table.
Groundwater, unconfined	The water in an aquifer where there is a water table.

Hardness	A measure of the amount of calcium, magnesium, and iron dissolved in the water.
Head, total	The sum of the elevation head, the pressure head, and the velocity head at a given point in an aquifer.
Hele-Shaw model	An analog model of groundwater flow based upon the movement of a viscous fluid between two closely spaced, parallel plates.
Heterogeneous	Pertaining to a substance having different characteristics in different locations. A synonym is nonuniform.
Homogeneous	Pertaining to a substance having identical characteristics everywhere. A synonym is uniform.
Hydraulic conductivity	A coefficient of proportionality describing the rate at which water can move through a permeable medium. The density and kinematic viscosity of the water must be considered in determining hydraulic conductivity.
Hydraulic diffusivity	A property of an aquifer or confining bed defined as the ratio of the transmissivity to the storativity.
Hydraulic gradient	The change in total head with a change in distance in a given direction. The direction is that which yields a maximum rate of decrease in head.
Hydrochemical facies	Bodies of water with separate but distinct chemical compositions contained in an aquifer.
Hydrodynamic dispersion	The process by which groundwater containing a solute is diluted with uncontaminated groundwater as it moves through an aquifer.
Hydrograph	A graph that shows some property of groundwater or surface water as a function of time.
Hydrologic equation	An expression of the law of mass conservation for purposes of water budgets. It may be stated as inflow equals outflow plus or minus changes in storage.

473

Hydrostratigraphic unit   A formation, part of a formation, or a group of formations in which there are similar hydrologic characteristics allowing for grouping into aquifers or confining layers.

Hygroscopic water   Water that clings to the surfaces of mineral particles in the zone of aeration.

Ideal gas   A gas having a volume that varies inversely with pressure at a constant temperature and that also expands by 1/273 of its volume at 0° C for each degree rise in temperature at constant pressure.

Infiltration   The flow of water downward from the land surface into and through the upper soil layers.

Infiltration capacity   The maximum rate at which infiltration can occur under specific conditions of soil moisture. For a given soil, the infiltration capacity is a function of the water content.

Injection well   A well drilled and constructed in such a manner that water can be pumped into an aquifer in order to recharge it.

Interception   The process by which precipitation is captured on the surfaces of vegetation before it reaches the land surface.

Interception loss   Rainfall that evaporates from standing vegetation.

Interflow   The lateral movement of water in the unsaturated zone during and immediately after a precipitation event. The water moving as interflow discharges directly into a stream or lake.

Intermediate zone   That part of the unsaturated zone below the root zone and above the capillary fringe.

Intrinsic permeability   Pertaining to the relative ease with which a porous medium can transmit a liquid under a hydraulic or potential gradient. It is a property of the porous medium and is independent of the nature of the liquid or the potential field.

Ion exchange   A process by which an ion in a mineral lattice is replaced by another ion which was present in an aqueous solution.

Isocon   A line drawn on a map to indicate equal concentrations of a solute in groundwater.

Isohyetal line   A line drawn on a map, all points along which receive equal amounts of precipitation.

Isotropy   The condition in which hydraulic properties of the aquifer are equal in all directions.

Kinematic viscosity   The ratio of dynamic viscosity to mass density. It is obtained by dividing dynamic viscosity by the fluid density. Units of kinematic viscosity are square meters per second.

Laminar flow　That type of flow in which the fluid particles follow paths that are smooth, straight, and parallel to the channel walls. In laminar flow, the viscosity of the fluid damps out turbulent motion. Compare with Turbulent flow.

Law of mass action　The law stating that for a reversible chemical reaction the rate of reaction is proportional to the concentrations of the reactants.

Leaky confining layer　A low-permeability layer that can transmit water at sufficient rates to furnish some recharge to a well pumping from an underlying aquifer. Also called aquitard.

Lineament　A natural linear surface feature longer than 1500 meters.

Lysimeter　A field device containing a soil column and vegetation which is used for measuring actual evapotranspiration.

Maximum contaminant level　The highest concentration of a solute permissible in a public water supply as specified in the National Interim Primary Drinking Water Standards for the United States.

Molality　A measure of chemical concentration. A one-molal solution has one mole of solute dissolved in 1000 grams of water. One mole of a compound is its formula weight in grams.

Mutual-prescription doctrine　A groundwater doctrine stating that in the event of an overdraft of a groundwater basin, the available groundwater will be apportioned among all the users in amounts proportional to their individual pumping rates.

Natural gamma radiation log　A borehole log that measures the natural gamma radiation emitted by the formation rocks. It can be used to delineate subsurface rock types.

Neutron log　A borehole log obtained by lowering a radioactive element, which is a source of neutrons, and a neutron detector into the well. The neutron log measures the amount of water present; hence, the porosity of the formation.

Nonequilibrium type curve　A plot of logarithmic paper of the well function $W(u)$ as a function of u.

Observation well　A nonpumping well used to observe the elevation of the water table or the potentiometric surface. An observation well is generally of larger diameter than a piezometer and typically is screened or slotted throughout the thickness of the aquifer.

Overland flow　The flow of water over a land surface due to direct precipitation. Overland flow generally occurs when the precipitation rate exceeds the infiltration capacity of the soil and depression storage is full. Also called Horton overland flow.

475

Permafrost Perenially frozen ground, occurring wherever the temperature remains at or below 0° C for two or more years in a row.

Permeameter A laboratory device used to measure the intrinsic permeability and hydraulic conductivity of a soil or rock sample.

Phreatic water Water in the zone of saturation.

Piezometer A nonpumping well, generally of small diameter, which is used to measure the elevation of the water table or potentiometric surface. A piezometer generally has a short well screen through which water can enter.

Piezometer nest A set of two or more piezometers set close to each other but screened to different depths.

Polar coordinates The means by which the position of a point in a two-dimensional plane is described, based upon the radial distance from the origin to the given point and the angle between a horizontal line passing through the origin and a line extending from the origin to the given point.

Pollutant Any solute or cause of change in physical properties which renders water unfit for a given use.

Pore space The volume between mineral grains in a porous medium.

Porosity The ratio of the volume of void spaces in a rock or sediment to the total volume of the rock or sediment.

Potentiometric surface A surface that represents the level to which water will rise in tightly cased wells. If the head varies significantly with depth in the aquifer, then there may be more than one potentiometric surface. The water table is a particular potentiometric surface for an unconfined aquifer.

Prior-appropriation doctrine A doctrine stating that the right to use water is separate from other property rights, and that the first person to withdraw and use the water holds the senior right. The doctrine has been applied to both ground and surface water.

Pumping cone The area around a discharging well where the hydraulic head in the aquifer has been lowered by pumping. Also called cone of depression.

Pumping test A test made by pumping a well for a period of time and observing the change in hydraulic head in the aquifer. A pumping test may be used to determine the capacity of the well and the hydraulic characteristics of the aquifer. Also called aquifer test.

Radial flow The flow of water in an aquifer toward a vertically oriented well.

Rating curve A graph of the discharge of a river at a particular point as a function of the elevation of the water surface.

Recharge area An area in which there are downward components of hydraulic head in the aquifer. Infiltration moves downward into the deeper parts of an aquifer in a recharge area.

Recharge basin A basin or pit excavated to provide a means of allowing water to soak into the ground at rates exceeding those that would occur naturally.

Recharge boundary An aquifer system boundary that adds water to the aquifer. Streams and lakes are typical recharge boundaries.

Recovery The rise in water level in a pumping well and nearby observation wells after groundwater pumpage has ceased.

Regolith The upper part of the earth's surface that has been altered by weathering processes. It includes both soil and weathered bedrock.

Resistivity log A borehole log made by lowering two current electrodes into the borehole and measuring the resistivity between two additional electrodes. It measures the electrical resistivity of the formation and contained fluids near the probe.

Reverse type curve A plot on logarithmic paper of the well function $W(u)$ as a function of $1/(u)$.

Riparian doctrine A doctrine that holds that the property owner adjacent to a surface-water body has first right to withdraw and use the water.

Root zone The zone from the land surface to the depth penetrated by plant roots. The root zone may contain part or all of the unsaturated zone, depending upon the depth of the roots and the thickness of the unsaturated zone.

Runoff The total amount of water flowing in a stream. It includes overland flow, return flow, interflow, and baseflow.

Safe yield The amount of naturally occurring groundwater which can be economically and legally withdrawn from an aquifer on a sustained basis without impairing the native groundwater quality or creating an undesirable effect such as environmental damage. It cannot exceed the increase in recharge or leakage from adjacent strata plus the reduction in discharge, which is due to the decline in head caused by pumping.

Saline-water encroachment The movement, as a result of human activity, of saline groundwater into an aquifer formerly occupied by fresh water. Passive saline-water encroachment occurs at a slow rate due to a general lowering of the freshwater potentiometric surface. Active saline-water encroachment proceeds at a more rapid rate due to the lowering of the freshwater potentiometric surface below sea level.

Sand model A scale model of an aquifer, which is built using a porous medium to demonstrate groundwater flow.

Sanitary landfill The disposal of solids and, in some instances, semisolid and liquid wastes by burying the material to shallow depths, usually in unconsolidated materials.

Saturated zone The zone in which the voids in the rock or soil are filled with water at a pressure greater than atmospheric. The water table is the top of the saturated zone in an unconfined aquifer.

Schlumberger array A particular arrangement of electrodes used to measure surface electrical resistivity.

Seepage velocity The actual rate of movement of fluid particles through porous media.

Seismic refraction A method of determining subsurface geophysical properties by measuring the length of time it takes for artificially generated seismic waves to pass through the ground.

Single-point resistance log A borehole log made by lowering a single electrode into the well with the other electrode at the ground surface. It measures the overall electrical resistivity of the formation and drilling fluid between the surface and the probe.

Slug test An aquifer test made by either pouring a small instantaneous charge of water into a well or by withdrawing a slug of water from the well. A synonym for this test, when a slug of water is removed from the well, is a bail-down test.

Soil moisture The water contained in the unsaturated zone.

Solubility product The equilibrium constant that describes a solution of a slightly soluble salt in water.

Specific capacity An expression of the productivity of a well, obtained by dividing the rate of discharge of water from the well by the drawdown of the water level in the well. Specific capacity should be described on the basis of the number of hours of pumping prior to the time the drawdown measurement is made. It will generally decrease with time as the drawdown increases.

Specific electrical conductance The ability of water to transmit an electrical current. It is related to the concentration and charge of ions present in the water.

Specific retention The ratio of the volume of water the rock or sediment will retain against the pull of gravity to the total volume of the rock or sediment.

Specific weight The weight of a substance per unit volume. The units are newtons per cubic meter.

Specific yield The ratio of the volume of water a rock or soil will yield by gravity drainage to the volume of the rock or soil. Gravity drainage may take many months to occur.

Spontaneous potential log A borehole log made by measuring the natural electrical potential which develops between the formation and the borehole fluids.

Stagnation point A place in a groundwater flow field at which the groundwater is not moving. The magnitude of vectors of hydraulic head at the point are equal but opposite in direction.

Storage, specific The amount of water released from or taken into storage per unit volume of a porous medium per unit change in head.

Storativity The volume of water an aquifer releases from or takes into storage per unit surface area of the aquifer per unit change in head. It is equal to the product of specific storage and aquifer thickness. In an unconfined aquifer, the storativity is equivalent to the specific yield. Also called storage coefficient.

Storm hydrograph A graph of the discharge of a stream over the time period when, in addition to direct precipitation, overland flow, interflow, and return flow are adding to the flow of the stream. The storm hydrograph will peak due to the addition of these flow elements.

Stream, gaining A stream or reach of a stream, the flow of which is being increased by inflow of groundwater. Also known as an effluent stream.

Stream, losing A stream or reach of a stream that is losing water by seepage into the ground. Also known as an influent stream.

Tensiometer A device used to measure the soil-moisture tension in the unsaturated zone.

Tension The condition under which pore water exists at a pressure less than atmospheric.

Throughflow The lateral movement of water in an unsaturated zone during and immediately after a precipitation event. The water from throughflow seeps out at the base of slopes and then flows across the ground surface as return flow, ultimately reaching a stream or lake.

Time of concentration The time that it takes for water to flow from the most distant part of the drainage basin to the measuring point.

Transmissivity The rate at which water of a prevailing density and viscosity is transmitted through a unit width of an aquifer or confining bed under a unit hydraulic gradient. It is a function of properties of the liquid, the porous media, and the thickness of the porous media.

Transpiration The process by which plants give off water vapor through their leaves.

Turbidity Cloudiness in water due to suspended and colloidal organic and inorganic material.

Turbulent flow That type of flow in which the fluid particles move along very irregular paths. Momentum can be exchanged between one portion of the fluid and another. Compare with Laminar flow.

Unsaturated zone The zone between the land surface and the water table. It includes the root zone, intermediate zone, and capillary fringe. The pore spaces contain water at less than atmospheric pressure, as well as air and other gases. Saturated bodies, such as perched groundwater, may exist in the unsaturated zone.

479

Vadose water Water in the zone of aeration.

Viscosity The property of a fluid describing its resistance to flow. Units of viscosity are newton-seconds per meter squared or pascal-seconds. Viscosity is also known as dynamic viscosity.

Water budget An evaluation of all the sources of supply and the corresponding discharges with respect to an aquifer or a drainage basin.

Water content The ratio of the volume of soil moisture to the total volume of the soil. This is the volumetric water content, also called volume wetness.

Water equivalent The depth of water obtained by melting a given thickness of snow.

Water table The surface in an unconfined aquifer or confining bed at which the pore water pressure is atmospheric. It can be measured by installing shallow wells extending a few feet into the zone of saturation and then measuring the water level in those wells.

Weir A device placed across a stream and used to measure the discharge by having the water flow over a specifically designed spillway.

Well, fully penetrating A well drilled to the bottom of an aquifer, constructed in such a way that it withdraws water from the entire thickness of the aquifer.

Well interference The result of two or more pumping wells, the drawdown cones of which intercept. At a given location, the total well interference is the sum of the drawdowns due to each individual well.

Well, partially penetrating A well constructed in such a way that it draws water directly from a fractional part of the total thickness of the aquifer. The fractional part may be located at the top or the bottom or anywhere in between the aquifer.

Wenner array A particular arrangement of electrodes used to measure surface electrical resistivity.

Wilting point The soil-moisture content below which plants are unable to withdraw soil moisture.

Winters Doctrine A United States doctrine holding that when Indian reservations were established, the federal government also reserved the water rights necessary to make the land productive.

Index

Acids
  defined, 315
  ionization constants, 314–17
Activity coefficient, 311
Adsorption, 327–29
  defined, 327
  Langmuir adsorption isotherm, 327–28
  phosphorous, 328–29
Advection, 338
Age dating, 335–36
Alluvial Basin groundwater region, 245–47
Alluvial valleys, hydrogeology, 197–98
Amargosa Desert, Nevada, 170
Aquifer, 92–94
  anisotropic, 162
  artificial recharge, 390–92
  confined, 93, 120–24, 130–32, 152–62, 164, 260–74
  cyclic storage, 394–97
  dynamic equilibrium, 378–80
  heterogeneous, 160–62
  isotropic, 126, 127, 152–59
  layered, 161
  management, 383–85
  mining, 394–97
  perched, 94
  unconfined, 93, 124–25, 132–39, 162–63
Arikaree Group, 248
Arsenic, 352–53
Artesian aquifer (see Aquifer, confined)
Artificial recharge, 390–92
  Alameda Co., California, 390
  Dayton, Ohio, 195
  defined, 390
  impact on safe yield, 386
  legal basis, 390

Artificial recharge (cont.)
  recharge basins, 391
  recharge wells, 391–92
  used to control salt-water encroachment, 393–94
  used to improve water quality, 328
  water spreading, 390–91
Ash Meadows, Nevada, 170
Atlantic and Gulf Coastal Plain groundwater region, 250–51
Avon Park Limestone, 174

Barium, 353
Baseflow, 5, 36
Baseflow recessions, 37–39
  defined, 37
  equation, 39
  prior to storm peak, 41
  used to compute recharge, 50–52
Basin and Range Province, 167–71
Baydon-Ghyben, W., 140
Bernoulli equation, 109
Beryllium, 353
Big Clifty Sandstone, 221
Biscayne aquifer, 173, 239, 243, 251
Boron, 353
Boussinesq equation, 124
Bright Angel Shale, 209, 210
Buried valley aquifers, 190
  case study, 190–97

Cadmium, 353–54
Calorie, 6
Cambrian-Ordovician aquifer (see Deep sandstone aquifer)
Capillarity, 79–80
Capillary fringe, 79–80

Carbonate equilibrium, 317–21
    ionization constants, 318
    in water with external pH control,
      320–21
    in water with fixed partial pressure of
      CO_2, 318–20
Carbonate rock aquifers
    case studies
      central Kentucky karst, 221
      Door Peninsula, Wisconsin, 220
      Floridan aquifer, 172–77, 341–44
      Great Basin, 167–71
      southern Indiana karst, 220
    chemical geohydrology, 341–44
    conceptual models, 220
    flow systems, 167–71, 172–77
    hydrogeology, 218–26
    water table, 223–24
    well location, 224–26
Carbon 14 dating, 335–36
Carrizo Sand, 251
Catahoula Sandstone, 251
Cedar Keys Formation, 174
Chadron Formation, 248
Chemical activities, 311–14
    defined, 311
Chemical reactions, reversible, 307–8
Cheswold aquifer, 251
Chlorinated hydrocarbons, 362–63
Chlorine, 358–59
Chlorophenoxys, 363
Chromium, 354
Chuska Mountain, New Mexico, 165
Clastic sedimentary rock aquifers, 214–18
Climatological data
    collection, 23
    sources, 24
Coal, hydrogeology, 226–27
Coastal Plain aquifers, hydrogeology, 239-44
Coconino Sandstone, 247
Colorado Plateau and Wyoming Basin
    groundwater region, 247–48
Columbia Lava Plateau groundwater region,
    247
Columbia River Basalts, hydrogeology,
    229–30
Common ion effect, 311
Condensation, 21
Cone of depression, 164
Confined aquifers (see Aquifer, confined)
Confining layer, 124, 163
Conjunctive use, 397–99
Connate water, 164
Contact spring, 165
Contaminant, 350 (see also Pollutant(s))
Continuity principle (see Law of mass
    conservation)
Copper, 355
Copper Harbor conglomerate, 186
Current meter, 52–53

Dakota Sandstone, 247, 248, 249
Dam
    flow beneath, 128–29
    flow through, 127
Darcy, Henry, 70
Darcy's experiment, 70–71
Darcy's law
    defined, 71
    force and potential, 113–17
    steady flow in confined aquifer, 131
    steady flow in unconfined aquifer, 133
    used to derive flow equations, 121,
      123
Dayton, Ohio, 190–97
Debye-Hückel equation, 313
Deep sandstone aquifer
  (Cambrian-Ordovician aquifer), 101, 387,
  396
    development, 380
    hydrogeology, 205–8
Density
    defined, 107
    fresh water, 109, 110, 111, 112, 114,
      115, 121, 122, 123, 141, 143
    saline water, 141, 143
Depression spring, 165
Depression storage, 34
Desert hydrogeology, 244
Devils Hole, Nevada, 170
Dew point, 15
Digital groundwater models, 440–53
    finite difference methods, 440–43
    Illinois State Water Survey model,
      443–53
      for nonleaky confined aquifers,
        443–51
      for water-table aquifer, 451–53
Direct precipitation, 36
Discharge area
    confined aquifer, 163
    defined, 152–53
    effect of pumping, 378–80
    unconfined aquifer, 154, 156, 159
Discharge velocity, 116
Drawdown cone (see Cone of depression)
Drawdown measurement, 298
DuCommun, Joseph, 140
Dupuit equation, 134
Dupuit flow
    assumptions, 133
    coastal aquifers, 141–43
    derivation, 132–39
Duration curves, 48–50
Dynamic equilibrium
    defined, 164
    impact of pumping, 378–80

Eau Claire Formation, 101, 208
Edwards Limestone, 248
Effective porosity
    defined, 61

Effective porosity *(cont.)*
    used in determining seepage velocity,
      116
Effluent streams, 42–44
Electrical groundwater model, 439
Entrada Sandstone, 247
Equations of groundwater flow
    confined aquifers, 120–24
    unconfined aquifers, 124–25
Equilibrium constant
    carbonate species, 318
    defined, 309
    water, 315–16
Equipotential line
    defined, 126
    flow nets, 127
    potential field for confined aquifer, 163
    potential field for unconfined aquifer,
      153, 154, 156
    refraction, 130
Equivalent weight
    conversion to molality, 307
    defined, 306
Evaporation, 14–17
Evapotranspiration
    affecting groundwater levels, 178
    affecting groundwater recharge, 379
    deserts, 244
    explained, 18–20
    groundwater discharge, 154, 200
    groundwater flow equation, 134, 136
    water-budget studies, 381–83

Faults, hydrogeology, 210–12
Fault spring, 166
Field capacity
    defined, 20
    explained, 84–86
    lysimeter, 19
Field methods
    fracture-trace analysis, 406–12
    geophysical
      borehole, 427–38
      surface, 412–27
    hydrogeologic site investigations,
      453–56
    pumping tests, 296–301
    water-sample collection, 331–35
Finite-difference methods, 440–43
Flint River Formation, 174
Floodwaves, 47–48
Floridan aquifer, 251
    chemical geohydrology, 341–44
    regional flow, 172–77
    sinkhole springs, 166
Flow nets
    explained, 127–29
    used to show groundwater flow near
      lakes, 179–82
    used to show regional groundwater
      flow, 152–63

Flow systems
    intermediate, 156, 177
    local, 156, 159, 177, 180, 181
    regional, 156, 159, 177, 180
Flow through lakes (*see* Lakes, seepage)
Fluoride, 359
Folds, hydrogeology, 212–14
Force, 106
Force potential
    Darcy's law, 114
    defined, 112–13
    hydraulic head, 125–26
Fort Union Formation, 249
Fossil water, 164
Fox Hills Sandstone, 249
Fracture traces
    analysis, 406–12
    explained, 224–26
Franconia Sandstone, 249
Fredonia aquifer, 251
Free energy, 322
Freeze, R. A., 386
Frenchman Flats, Nevada, 170
Freshwater-saline water relations
    coastal aquifers, 139–44
    oceanic islands, 144–46

Gaining streams, 42–44
Galconda Formation, 221
Galesville Sandstone, 249
Geology of groundwater occurrence
    alluvial valleys, 197–98
    coastal plains, 239–44
    deserts, 244
    glaciated terrain, 188–97
    igneous and metamorphic rocks,
      227–36
      intrusive igneous and metamorphic,
        227–29
      volcanic, 229–36
    permafrost regions, 236–39
    sedimentary rocks, 205–27
      carbonate, 218–26
      clastic, 214–17
      coal, 226–27
    tectonic valleys, 198–204
Geophysical methods, 412–38
    borehole, 427–38
      caliper, 430
      nuclear, 432–38
      resistivity, 432
      single-point resistance, 430–32
      spontaneous potential, 432–34
      temperature, 430
    surface, 412–27
      electrical resistivity, 413–17
      gravity, 426–27
      magnetics, 426–27
      seismic refraction, 417–26
Ghyben-Herzberg principle, 140–42
Girkin Limestone, 221

483

Glaciated Appalachian groundwater region, 250
Glaciated Central groundwater region, 249
Goliad Sand, 251
Grande Ronde Basalt, 230
Great Basin, 167–71, 200
Green River, Kentucky, 221, 223
Groundwater
    contamination, 365–70
    defined, 5
    lake interactions, 177–82
    models, 438–53
    occurrence, 188–244
    regions
        defined, 245
        United States, 245–51

Halomethanes, 363
Hantush, M. S., 275, 282, 286, 287
Hantush-Jacob formula, 275
Hawaiian Islands, hydrogeology, 230–36
Hawthorn Formation, 174, 176
Hele-Shaw, H. S., 439
Hele-Shaw model, 439–40
Hell Creek Formation, 249
Hernando County, Florida, 221
Herzberg, A., 140
High Plains groundwater region, 248
Homogeneity, 97–100
Horton overland flow, 34–36
Hualapai Plateau, 91, 208–11
Hubbert, M. K., 141, 153
Humidity
    absolute, 14
    relative, 14
    saturation, 14
Hydraulic conductivity, 70–76
    carbonate rocks, 218–19
    clastic sedimentary rocks, 215
    coal, 226–27
    conversion table, 73
    defined, 71–74
    glacial deposits, 188–89
    igneous and metamorphic rocks, 228–29
    rocks, 75–76
    sediments, 74–75
Hydraulic head
    defined, 109
    explained, 110–12
    force potential, 112–13
    gradient, 125–26
Hydraulic potential (see Hydraulic head)
Hydrochemical facies, 337–38
Hydrodynamic dispersion
    defined, 338–39
    equation, 339–40
    groundwater contaminants, 340–41
Hydrogen ion concentration
    definition, 314–15
    Eh-pH diagrams, 323–24

Hydrogen ion concentration (cont.)
    natural range, 326
Hydrogeologic boundaries, 294–96
Hydrogeology (see also Groundwater and Geology of groundwater occurrence)
    applied aspects, 8–9
    defined, 4
    employment opportunities, 7–9
    relation to planning, 8–9
    sources of information, 9–10
Hydrograph separation, 37–44
Hydrologic cycle, 4–6
Hydrologic equation
    defined, 7
    used to determine groundwater budgets, 381–83
    used to determine water budget of lakes, 15
Hydrology, defined, 4
Hydrophyte, 18
Hydrostratigraphic units, 100–101
Hygroscopic water, 68

Ideal gas, 308–9
    law, 308–9
    thermodynamic activity, 308
Igneous and metamorphic rock aquifers, 227–36
    case studies
        Columbia Plateau, 229–30
        Hawaiian Islands, 230–36
Illinois State Water Survey Model, 443–53
Infiltration, 81–82
    defined, 81
    Dupuit flow, 134, 136
    hydrologic cycle, 5
Influent streams, 42–44
Interflow, 34
Interstitial water, 164
Intrinsic permeability
    defined, 72
    rocks, 75–76
    sediments, 74–75
    units, 72–73
Ion activity product, 314
Ion exchange, 329–31
    cation exchangeability, 330
    cation-exchange capacity, 331
    distribution coefficient, 331
    sodium-adsorption ratio, 330
Ionic strength, 312
Iron
    groundwater, 325–26
    stability fields, 323–25
    water quality, 355
Ironton Sandstone, 249
Isohyetal lines, 25, 26, 28
    defined, 25
    use, 26, 28
Isotropy, 97–100

Jacob, C. E., 266, 275
Joint springs, 166
Jordan Sandstone, 249
Junction Creek Sandstone, 247
Juvenile water, 165

Karst
       central Kentucky, 220–22
       defined, 218
       hydrogeology, 218–26
       spring, 167
Keweenaw Peninsula, Michigan, 186–87
Kinetic energy, 108, 110
Kootenai Formation, 249

Lagarto Clay, 251
Lake
       evaporation, 16
       evaporation nomograph, 16–17
       groundwater flow, 177–82
       seepage, 177
       water balance, 15
Lake City Limestone, 174
Lake Michigan
       water budget, 36
       water quality, 350–51
Lakota Formation, 249
Laminar flow, 114
Landfills, 368–70
Land pan, 15–16
Land pan coefficient, 16
Langmuir adsorption isotherm, 327–28
Laplace equation
       defined, 124
       flow nets, 127
       steady regional groundwater flow, 152
Lapse rate, 22
Latent heat of condensation, 6, 15
Latent heat of fusion, 6
Latent heat of vaporization, 6
Law of mass action
       explained, 309–10
       expressed in activity coefficients, 313
Law of mass conservation
       stated, 120–21
       use, 129
Leachate, 368–70
Lead, 355–56
Leakage factor, 274
Leakage rate through leaky confining layer,
       124
Lissie Formation, 251
Lloyd Sand Member (Raritan Formation), 101,
       251
Long Island, New York, 91, 92, 101, 240, 242,
       391
Lonnquist, C. G., 443
Losing stream, 42–44
Lost River, Indiana, 221
Lysimeter, 19

McGuinness, C. L., 245
Madison Limestone, 222, 249
Magmatic water, 165
Magothy aquifer, 239, 251
Manganese, 356
Manning equation, 56–57
Maquoketa Shale, 205, 207, 208, 380
Marianna Limestone, 174
Mass, 106
Mathematical models
       finite-difference methods, 440–43
       Illinois State Water Survey models,
         443–53
       lake-groundwater systems, 178
       regional groundwater flow, 152
       solving groundwater flow equations,
         125
Maui tunnels, 234
Mauv Limestone, 208, 209
Meinzer, O. E., 74
Mercury, 356–57
Miami River, 195
Mining
       impact of coal mining, 227
       impact on water quality, 370
Minnelusa Formation, 249
Molality, 307
Mt. Simon Sandstone, 249
Muddy River Springs, Nevada, 168

National Atmospheric and Oceanic
    Administration, U.S. Weather Bureau, 9, 16
National Interim Primary Drinking Water
    Standards
       defined, 351–52
       inorganic nonmetals, 358–61
       metals, 352–58
       organic compounds, 362–63
Nernst equation, 322
Nevada Test Site, 170
Nickel, 357
Nitrogen, 359–60
Nubian Sandstone, 244

Oakville Sandstone, 251
Observation wells, 299–301
Ocala Limestone, 174
Ocala Uplift, Florida, 173, 174
Ogallala Formation, 248, 396
Oldsmar Formation, 174
Overland flow, 5 (see also Horton overland
    flow)
Oxidation potential, 322–26
       definition, 322
       Eh-pH diagrams, 323–24
       field measurement, 324
       natural range, 325
       Nernst equation, 322

Packing, sediment, 61–62
Pahasapa Limestone, 222, 249

Partial penetration of wells, 285–87
Pathogenic organisms, 365
Permafrost
    defined, 236
    Fairbanks, Alaska, 238–39
    hydrogeology, 237–39
    occurrence, 236–38
Permeameter, 117–20
Phosphorous, 360–61
Phreatophyte
    defined, 18
    near lakes, 178
Pierre Shale, 248
Piezometer
    defined, 110
    field use, 181–82, 454
    illustrating regional groundwater flow,
        153, 154, 160
    nests, 455
Polar coordinates, 258–59
Polk City, Florida, 177
Pollutant(s)
    defined, 350
    groundwater, 365–70
        landfills, 368–69
        list of known, 366
        mining, 370
    hydrodynamic dispersion, 340–41
    Lake Michigan, 350–51
Porosity, 60–67
    defined, 60–61
    plutonic and metamorphic rocks,
        64–65
    sedimentary rocks, 64–65
    sediments, 61–64
    volcanic rocks, 66–67
Portage Lake Lava Series, 186
Potential distribution (see Potential field)
Potential energy, 108
Potential evapotranspiration
    Blaney-Criddle method, 19
    definition, 18
    Thornthwaite method, 19
Potential field (flow field)
    explained, 125–26
    near lakes, 178–80
    regional groundwater flow, 152, 153,
        160, 161, 162
Potentiometric surface, 93
Prairie du Chien Formation, 216, 249
Precipitation
    convectional, 23
    effective depth, 25–29
    formation, 21–23
    frontal, 22
    gauge, 23
    hydrologic cycle, 5
    measurement, 23–24
    orographic, 23
Pressure, 108
Pressure head, 111

Prickett, T. A., 443
Primary porosity, 64
Principal artesian aquifer (see Floridan
    aquifer)
Project reports, 455–56
Pumping cone (see Cone of depression)
Pumping tests
    confined aquifer
        distance-drawdown method,
            268–70
        Jacob method, 266–68
        slug-test method, 270–74
        Theis method, 261–66
    design, 296–301
    leaky confined (semiconfined) aquifer
        Hantush method, 282–84
        no storage in confining layer,
            279–82
        storage in confining layer, 284–85
    steady-flow, 292–93
    water-table aquifer, 288–91

Radial groundwater flow, 258–60
Rational equation, 44–47
Raymond Basin, 390
Recharge, aquifer, 379
Recharge area
    defined, 152–53
    lake systems, 177
    regional groundwater flow, 154, 162,
        163
Regolith, 226
Return flow, 37
Reynolds number
    Darcy's law, 115
    definition, 114
Roswell Basin, New Mexico, 249
Runoff cycle, 35–37
Runoff, 5

Saddle Mountains Basalt, 230
Safe yield, 385–86
St. Genevieve Limestone, 221
St. Johns River, Florida, 176, 177
St. Louis Limestone, 221
St. Peter Sandstone, 216, 249
Saltwater encroachment (see also
    Freshwater-saline water relations)
        active, 242
        defined, 241
        Miami, Florida, 243
        passive, 241
        Pearl Harbor, Hawaii, 235–36
        prevention, 393–94
        Savannah, Georgia, 243–44
Saltwater interface, 140, 141, 143
Saltwater intrusion (see Freshwater-saline
    water relations and Saltwater
    encroachment)
San Andreas Limestone, 249
San Bernardino, California, 200–204

Sand Hills, Nebraska, 91, 248
San Jacinto Fault, 202
Santa Ana River, 200–204
Savannah, Georgia, 177
Scale models, 438–39
Secondary porosity, 64
Second law of thermodynamics, 121
Sedimentary rock aquifers, 205–27
Seepage velocity, 116
Selenium, 361
Semiconfined aquifer flow, 274–85
Septic tanks, 367
Silver, 357
Sinkhole, 166, 221
Sinkhole spring, 166
Sinuosity, 116, 117
Slope of potentiometric surface (see Hydraulic gradient)
Slug tests, 270–74
Snake River Canyon, 165, 247
Snow surveys, 24–25
Soil moisture, 83–88
Solubility product
    computation, 310
    defined, 309
South Platte River Basin, 389
Specific discharge (see Discharge velocity)
Specific retention, 68
Specific storage, 95
Specific weight, 107
Specific yield
    explained, 67–70
    relation to storativity, 96
Springs, 165–67
Stagnation points
    defined, 156
    near lakes, 178, 180, 182
    waste disposal, 158
Steady flow
    confined aquifers, 130–32
    defined, 124
    dynamic aquifer equilibrium, 378–80
    regional, 152–63
    unconfined aquifers, 132–39
Stomata, 18
Storage coefficient (see Storativity)
Streamflow measurement, 52–55
Stream gauging, 52–54
Streamline
    definition, 127
    refraction, 129–30
    regional groundwater flow, 152, 153, 154, 156
Stream order, 44–45
Storativity
    confined aquifer, 96
    defined, 95
    derivation, 121–23
    determined from pumping tests, 264–65, 267, 269, 272, 281, 284, 285, 288–90

Storativity (cont.)
    tidal effects, 147
    unconfined aquifer, 96
Stratigraphy
    impact on well location, 208–10
Sulfate and sulfide, 361
Suwannee Limestone, 174, 221
Suwannee River, Florida, 177

Tallahassee Limestone, 174
Tampa Limestone, 174
Teays River, 193
Tectonic valleys
    Great Basin, 167–71
    hydrogeology, 198–200
    San Bernardino, California, 200–204
Tensiometer, 83
Theis, C. V., 261
Theis equation, 262
Theisson method, 26–29
Thomas, H. E., 245
Throughflow, 37
Tidal effects on groundwater levels, 146–47
Time of concentration, 46
Transmissivity
    defined, 95
    determination by pumping tests, 264–65, 267, 269, 272, 279–81, 282–84, 288–90, 292
    flow equation, 123–24
Transpiration, 17–18
Trempealeau Formation, 249
Trilinear diagram, 336–37
Trinity Group, 248
Turbulent flow, 114–15

Unconsolidated aquifer hydrogeology, 187–205
Unglaciated Appalachian groundwater region, 250
Unglaciated central groundwater region, 248
Unita Basin, 248
United States Environmental Protection Agency, 9
United States Geological Survey, 9
Unsaturated flow, 88–91
Unsaturated hydraulic conductivity, 88
Unsaturated zone
    defined, 83
    flow, 88–91
    recharge, 82
    thickness, 172

Vadoze zone (see Unsaturated zone)
Viscosity, 114

Walton, W. C., 279
Wanapum Basalt, 230
Water balance, United States, 4
Water budget, aquifer, 381–83

Water content
 defined, 83
 time dependency, 85–86
 unsaturated hydraulic conductivity,
  88, 89
Water law, 387–90
 American Rule, 389
 English Rule, 388
 federal, 387–88
 mutual prescription, 390
 prior appropriation, 388
 riparian, 388
 state
  groundwater, 388–90
  surface water, 388
 water right, 387
 Winters Doctrine, 387–88
Water quality, 350–76
 criteria
  defined, 351
  dissolved oxygen, 364
  hardness, 364
  inorganic nonmetals, 358–61
  metals, 352–58
  organic compounds, 362–63
  pathogens, 365
  salinity, 363–64
  turbidity, 364
 importance, 350–52
 protection of groundwater, 392–94
Water resources management, 374–404
 artificial recharge, 390–92
 conjunctive use, 397–99
 cyclic storage, 394–96
 groundwater
  budgets, 381–83
  mining, 394–96
  quality protection, 392–94
  recharge, 390–92
 potential of aquifers, 383–85
 safe yield, 385–86
 trends, 399–400

Water spreading, 390–91
Water table
 explained, 77–79
 hydrologic cycle, 5
 recharge, 91–92
Water-table aquifer (see Aquifer, unconfined)
Water usage, United States
 per capita, 2, 3
 total, 2, 3
Weirs, 54–55
Well function, 261
Well hydraulics, 258–301
 confined aquifer, 260–74
 hydrogeologic boundaries, 294–96
 partial penetration, 285–87
 semiconfined (leaky) aquifer, 274–85
 water-table (unconfined) aquifer,
  287–93
 well interference, 293–94
Western Mountain groundwater region, 245
White River, Nevada, 168
Wilcox Group, 251
Willis Sand, 251
Wilting point
 defined, 18
 relation to evapotranspiration, 20
 relation to soil texture, 86
Wingate Sandstone, 247
Winter Haven, Florida, 177
Withlacoochee River, Florida, 176
Woodbine Sand, 248
Work, 106, 108
World water distribution, 4

Yucca Flats, Nevada, 170

Zinc, 358
Zone of aeration, 84
Zone of diffusion, 140
Zone of dispersion, 140